SPACE-PERCEPTION
AND THE PHILOSOPHY
OF SCIENCE

SPACE-PERCEPTION
AND THE PHILOSOPHY
OF SCIENCE

Patrick A. Heelan

UNIVERSITY OF CALIFORNIA PRESS

BERKELEY   LOS ANGELES   LONDON

University of California Press
Berkeley and Los Angeles, California

University of California Press, Ltd.
London, England

First Paperback Printing 1988

Library of Congress Cataloging in Publication Data

Heelan, Patrick A.
  Space-perception and the philosophy of science.

  Bibliography: p. 353
  Includes index.
1. Space perception. 2. Science—Philosophy.
I. Title.
BF469.H43   121'.3   82-4842
ISBN 0-520-05739-2   AACR2

Printed in the United States of America

The paper used in this publication meets the minimum
requirements of ANSI/NISO Z39.48-1992 (R 1997)
(*Permanence of Paper*). ∞

CONTENTS

Preface

This book is the product of three seemingly unrelated studies that are brought to a focus in the study of space perception. The first study was an inquiry into the structure of quantum mechanics pursued during the early sixties at the Palmer Laboratory, Princeton, and at the Institut Supérieur de Philosophie, Louvain,[1] which convinced me that objective empirical knowledge was always a function of context, and that context itself was both material—or embodied—and intentional, or directed by a specific prior culturally-shared interest of the subject. This study resulted in the conclusion that quantum logic, or the non-classical behavior of sentences in quantum mechanics first pointed out by Birkhoff and von Neumann,[2] was not an indication of incompleteness, but was in fact due to the contextual character of descriptive sentences in quantum mechanics: so that quantum logic—or the special logic of quantum mechanics—became, in my mind, the general logic of context-dependent discourse. Consequently, quantum mechanics was in my view a genuine scientific theory, but of a new kind, one that was explicitly "ecological," or context-dependent.[3]

The second study focused on the cognitive value of natural science: is it essential to a genuine scientific theory such as quantum mechanics that it be ontological and hermeneutical, or just that it be technically successful? Do scientific theories come to be accepted in principle because their explanatory entities become manifest in the world of perception, or merely because they are good instruments for manipulating our environment, or for other reasons? Moreover, is the empirical basis of science in perception given equally to all careful observers, or is it given only as the product of a hermeneutical enterprise, a correct reading of signs? What are the conditions of possibility and the extent of the hermeneutical aspect of explanation in natural science? The conclusion of this study is that an observation in experimental science is always contextual and hermeneutical; the scientist 'reads' an instrument like a 'text.' (Single quotes signify that the art of 'reading' the instrumentally 'written' 'texts' is similar to but not identical with the art of reading texts written in natural languages.) Scientific perception, like all perception, gains its meaning from the context of inquiry that generates the data. Thus, the perception of scientists is transformed during the course of an investigation. Similar conclusions have been stated by N. R. Hanson, S. Toulmin, T. S. Kuhn, P. Feyerabend, M. Hesse, M. Polanyi, M. Grene, C. Hooker and others. Yet except for some earlier work done by the present writer on this subject, no systematic account has been worked out of what is often spoken of as the ''theory-ladenness of observation'' in terms of the one philosophical tradition that should be most competent to tackle problems of this kind: the hermeneutical tradition stemming from the later Husserl and most strongly represented by writers such as M. Heidegger, P. Ricoeur, and H.-G. Gadamer.[4]

The third study was the study of the pictorial space of the paintings of Vincent van Gogh, which, ever since a conversation with Erwin Schrödinger about non-Euclidean geometries during the course of a seminar in Dublin many years ago, I suspected of exhibiting to our perception the structure of a Riemannian visual space—if only we could come to recognize its traits. The pursuit of this inquiry led me to Luneburg's theory of binocular vision and to an insight into the special problems relating to (1) discourse about perception and (2) the design and interpretation of experiments on perception. The context-dependent and hermeneutical character of perception came home to me again, as well as the fact, made possible by different embodiments and different intentions, that we can see both in a Euclidean (or Cartesian) way and in a hyperbolic way, but not at the same time nor in the same context. It also became clear to me that reading images is not as ''natural'' and straightforward as we generally think. On the contrary, reading images correctly involves something in the image analogous to a text that is read as well as the antecedent acquisition of the proper disposition of habit, that is, the proper subjective embodied intentionality-structure, without which we cannot truly perceive but only guess or infer. This study highlighted once again the hermeneutical character of perception and raised anew the question of the general conditions of possibility of both Euclidean and non-Euclidean forms of perception.[5]

This book attempts to focus all of these considerations on space perception. My initial intention was to apply as far as possible the most appropriate method of philosophy to each part of the analysis—phenomenological and hermeneutical methods to the intentional or subjective part of the study, and more formal and logical analytic methods to the part concerned primarily with materiality, neurophysiology, and the objects of science. As the study progressed, it gradually became clear to me that the standard methods of analytic philosophy and epistemology were simply inadequate to bridge the two parts of the study—material conditions and mental performance—and I was increasingly forced to adopt phenomenological and hermeneutical methods, as these alone enabled me to pursue the two parts of the study together without confusion. It may well be that the latter methods are more flexible, because they have not yet reached the "perfect" stage of constituting a paradigm for normal philosophy.[6] At any rate, I do not feel that the decision was arbitrary, but that it flowed naturally from the deployment of my set of interests and concerns in the problem of perception.

The theme of the book is space perception. I show that, despite the fact that we perceive a visual Cartesian world, our natural mode of unaided visual perception is hyperbolic; mediating our everyday perception of a Cartesian world is the carpentered environment that we have learned to 'read' like a 'text.' In this book, then, I use visual space perception as an exemplar for the study of the following: (1) perception (in general) as a hermeneutical enterprise, (2) the role of neurophysiology, other somatic processes of the body, and technological instruments in perception, (3) scientific observation and the way it has transformed, enriched, and taken imperial control over our ordinary modes of perception, (4) the consequences for the ontology of space— and ontology in general—of the primacy of perception within the context of the use of what I call "readable technologies"—or the possibilities of human embodiment in instruments for the purposes of perception, and (5) the general conditions of possibility of context-dependent perception.

Acknowledgments

I owe a special debt of gratitude to Chancellor Clifton Wharton of the State University of New York, to Acting President T. Alexander

Pond, and to Academic Vice President Sidney Gelber of Stony Brook, who greatly assisted the writing of this book by granting me a full year of academic leave after my term as Vice President for Liberal Studies and Dean of Arts and Sciences at Stony Brook.

I want to thank Professor David Cross of the Department of Psychology, who read and commented on many versions of the text. His kind, critical, and sustained interest over many months added immeasurably to the quality of the book, and helped to make it relevant to the interests of psychologists.

Among others who read and commented on the text in manuscript are Nini Praetorius, professor of psychology of the University of Copenhagen, whose comments on language about perception and consciousness were most helpful; Professors Robert Neville and Don Ihde of the Department of Philosophy at Stony Brook, Professor Robert Innis of the Department of Philosophy of the University of Lowell, Professor Lawrence Slobodkin of the Department of Ecology and Evolution at Stony Brook, and Professor Irma B. Jaffe of the Art Department of Fordham University. I am very grateful for the suggestions, criticisms, and comments that these readers have made. I have also profited from discussions with Dr. Nancy Gifford, and with graduate students at Stony Brook both from philosophy, among whom I should like to single out Susan Bordo, Jay Williams, and Anne-Marie Wachter, and from psychology, among whom preeminently are Arthur Houts and Hugh Foley. I want to thank Claire Crowther for doing the line drawings.

Finally, I owe a special tribute of gratitude and appreciation to my talented research assistant, Babette E. Babich, who came to understand the complicated matters discussed in this book so rapidly and so well. She was of invaluable assistance in library research and in many other ways.

I have numerous others to thank—my teachers and colleagues from whom I have learned so much at firsthand about art, psychology, physics, and philosophy, and many others from whose written works I have learned. Some of their names are included in the footnotes and references at the end of the book.

Patrick A. Heelan
Stony Brook, N.Y.
September 1981

1
Phenomenology, Hermeneutics, and Philosophy of Science

Introduction

Philosophers of science have tended to emphasize the differences between scientific and perceptual knowing, seeing them as two quite different, even contrasting, kinds of knowing. Scientific theories "explain" perceptual phenomena with respect to "imperceptibles" that "lie behind" the surface of phenomena. Such "imperceptibles" are, for example, atoms and elementary particles, postulated by theory and detected indirectly by apparatus; or they are principles seemingly paradoxical to sensible intuition, like the Principle of Uncertainty and the non-Euclidean structure of cosmological space.

Among the things that this book is concerned with are two basic structures of visual perception; first, that it is often and usually Cartesian (that is, Euclidean), and second, that it is at least episodically hyperbolic. By the latter I mean that at times and under certain circumstances the shapes of objects that we see fail to match their physical (i.e., their Euclidean) shapes, but instead match certain transforms of those shapes, namely, the appearances the objects would have in a hyperbolic non-Euclidean space. The Cartesian structure of visual perception is something so familiar and so transparently evident that we regard it as normative for ordinary observations. It is, nevertheless, as I shall claim, a product of scientific culture and an artifact of a technologically reconstructed human environment. In contrast, I shall claim that the hyperbolic structure of vision is not the product of scientific activity; nevertheless, it needs a mathematical model to make the coherence of its structure manifest.

The philosophy of science is concerned with the nature of scientific explanation, and with the problem of how a scientific image of nature relates to the basic events given in perceptual experience which are presumed, at least by empiricists, to be foundational for all knowledge and culture. This book is concerned with the odd fact that many perceptual events that are presumed to be foundational for all knowledge, are themselves permeated with elements of scientific origin; while others having equal authority as perceptual objects are nevertheless rejected as illusionary merely because, it seems, they do not obey those scientific laws against which they would stand in evidence, if the empiricist account of science were correct.

It is also usual for many philosophers of science today to assume that scientific knowing is privileged in presenting the most reliable set of paradigms for realistic knowledge; other claims to know—say, by perception—must on this assumption justify themselves on scientific grounds or not at all. Moreover, what is claimed on grounds other than the scientific, must, if true, be reducible to what science can demonstrate. This book makes a contrary claim, that perception is the only domain of valid, realistic knowing and that scientific theories become knowledge by transformations of our perceptual frameworks. To prevent misunderstanding at the start, let me state the thesis of this book emphatically: what we know is not limited to the deliverances of a unique privileged perceptual framework constituting an absolute transcultural empirical basis for all knowledge, and we can have access to a multiplicity of possible perceptual horizons, both of Euclidean and of non-Euclidean structure, grounded both in unaided perception and in the use of special technologies (''readable'' technologies) invented using scientific theories. This approach is what I call ''horizonal.'' With some important reservations as to the goal and function of philosophical analysis, the method used in this book as appropriate to this analysis is both *phenomenological* and *hermeneutical*.[1]

The empirical focus of this book is space perception. Its larger purpose, however, is philosophical: in particular it is about the philosophy of science. I present first an empirical thesis about the possible geometries, Euclidean and non-Euclidean, of visual space, a thesis that contradicts the Cartesian principles widely accepted by analytic empiricist philosophies of perception. I then attempt to develop a non-Cartesian philosophy of perception, one capable of giving an account, epistemologically and metaphysically, on the one hand, of the phenomena of diverse visual spaces and, on the other, of the fixed physical configurations that they represent and that mediate their occurrence. Such a philosophy will be both phenomenological and hermeneutical. Finally, I show that this philosophy of perception accounts for transformations of the perceptual field and of World structures mediated by scientific instruments and other "readable technologies."

Philosophies of Perception

Current philosophies of perception fall roughly into three categories: (1) empiricist/analytic philosophies, (2) naturalistic and causal philosophies, and (3) phenomenological/hermeneutic philosophies.

1. Empiricist/analytic philosophies of perception have their historical roots in Locke, Berkeley, and Hume; they are represented on the contemporary scene by traditions stemming from the work of E. Mach, J. S. Mill, B. Russell, the early L. Wittgenstein, and R. Carnap.[2] To perceive, according to these philosophies, is to be in possession of a pictorial representation that matches (or mirrors) physical reality. Such accounts are dualistic, opposing mind (or "internal consciousness") to body (or "external world") and phenomenon (or "internal" representation) to "external" physical reality. Internal pictorial representations (or phenomena) in such accounts are usually thought to be constituted out of atomic elementary sensations (sense data, impressions) to which they are reducible by analysis, or from which they are inferred and from which they receive their warrant via logical schemata alone; the sensations (sense data, impressions, etc.) belong to an absolute privileged empirical base causally connected with their external objects.

   While much contemporary empiricist/analytic philosophy has gone considerably beyond this original position, the problems, both epistemological and metaphysical, that such theories give rise to have been examined by, among others, W. V. Quine, W. Sellars, N. Goodman, H. Putnam, P. K. Feyerabend, and most recently by R. Rorty in his *Philosophy and the Mirror of Nature*.[3] Rorty concludes, on the basis of the work just mentioned, that such theories are no longer defensible, and were mistaken from the start. This is the view to which I also subscribe.

The outcome of this critique within the empiricist/analytic tradition has been the abandonment of (Cartesian) dualism, usually in favor of some form of materialistic monism, such as a mind-body identity theory or a theory of mind modeled on artificial intelligence.[4] Most analytic philosophers of perception would hold the view, held, for example, by W. Sellars and R. Rorty, that perception is a strictly physical interaction between the physical world and the perceiver. Most would concur with the current working assumption of neuroscience which holds, as W. Uttal writes, "The relation between brain activity and mental activity is direct . . . [and] best summed up in terms of the principle of *psychoneural identity* or *equivalence*, which states that the linguistic terms of psychology and neurophysiology denote exactly the same mechanisms and processes."[5]

2. Naturalistic theories of perception are concerned less with the security and authority of knowledge claims than with trying to account for the way the world looks to us. Why does the world present itself to us in the way it does, with the sorts of things and processes we perceive? Answers to this question appeal to the activity of perceivers ordering the environment, accommodating to it, and finding their way around in it. Such activity sets up and gives warrant to a perceptual repertory of things that are significant for human life. The repertory is not unchanging, nor is it independent of history and culture; neither are its more complex unities justified by logical inference alone but by pragmatic, evolutionary, and other extralogical considerations, like the aesthetic of art or play.[6] Causal theories deny that the phenomenal matches the physical, but assert that changes in the phenomenal are correlated with changes in the physical and vice versa, and that a good theory of perception has to "explain" changes in the phenomenal by changes in the physical.[7]

3. The fundamental principles of phenomenology and hermeneutics insofar as they provide a philosophy of perception more broadly useful for a philosophy of science will be discussed below. The rest of this chapter will outline and discuss briefly where necessary the approach, assumptions, and tools of the phenomenological and hermeneutical philos-

ophy I am using in this book. It aims at constituting a body of critical and consistent theory inspired by and dependent on the major works of Husserl, Heidegger, and Merleau-Ponty, but not coincident with any set of views attributable to all of these three. The general body of theory I am using does, however, express the goal of a phenomenological/hermeneutical philosophy of science in the authentic spirit of all three.

Phenomenology

A perceived object makes itself present by acting physically and causally on the physical organism of the perceiver. Human beings are objects in Worlds as well as subjects: they are physical and material to the extent that they are causal agents and patients in Worlds; they are also nonphysical and immaterial to the extent that they exercise freedom, responsibility, and rationality. I mean by "material" and "immaterial," "physical" and "nonphysical," provisionally no more than that human beings engage in the range of activities just enumerated. Phenomenology is a philosophical method that aims to bridge the dualisms of material/immaterial, mind/body, freedom/determinism by the study of the intentional forms of human activity, that is, of the intentionality-structures that underlie cognitive and deliberative acts of persons.

Cognitive acts, such as acts of perception or belief, can be studied from two standpoints: let me call them a "third-person" or "objective" standpoint, and a "first-person" or "subjective" one. The former addresses the question: what is it for a subject (human, animal, extragalactic creature, etc.) to perceive a state of the World? The methodological supposition is made that the inquirer is not also one of the perceiving subjects, and has no direct access to the kind of perceptual act that is being studied. The latter or first-person stance supposes that the inquirer is also a perceiving subject, has direct access to perceptual acts of the kind that is being studied, and can use this evidence in the inquiry. A first-person study then asks the question: what is it for one like me to perceive a state of the World?

Third-person or objective inquiry is characteristic of certain approaches within the social sciences, like Behaviorism, which model themselves on the natural sciences; it is also characteristic of much contemporary philosophy of mind. Cognitive states, in this approach, are treated as theoretical entities postulated to explain intelligent behavior, behavior being taken as any kind of gross public activity given unproblematically to the community of scientific observers. All third-person inquiry then, say, into perception, presupposes as background an unproblematically given world of possible perceptual objects and situations, manifesting itself directly in and to individual and group perceptions of the scientific community. All such third-person inquiry is inevitably influenced and controlled by the repertory of these background perceptual beliefs. The practitioners of such inquiry must know how to perform well those cognitive acts, specifically acts of perception, which reveal the common perceptual background on which all agree. If, however, serious questions are raised about the content of that background—how it is to be characterized, what is secure, what is tentative, what is provisional, and so on—or about the structure, conditions, or trustworthiness of perceptual knowledge relevant to that background, a first-person mode of inquiry into what it is to perceive is called for. One form of first-person philosophical inquiry is phenomenology.

Phenomenology, introduced by Edmund Husserl, was based, in his words, "on a return to the things themselves" as given to the subject in knowing acts.[8] Central to its concerns are (1) the apodicticity of given objects (*noemata*), (2) how the subject receives these objects in experience (*noesis*), and (3) the conditions of possibility of the *noesis-noema* structure in the perceiving subject or Ego.[9] For Husserl, *noemata* included mathematical and logical objects as well as perceptual objects: by "object," I mean something that can be named. The structure of the experience in which an object is given was, for Husserl, a set of invariants. Each invariant organizes perceptual profiles into distinct systems. These systems are ways in which we experience an object. These invariant, organizing structures are what is called "the eidetic essence," "the phenomenological essence," or simply "the essence" of the object. The essence can exhibit multiple invariants, each invariant corresponding to a different system of profiles. This usage of the term "essence" must not be confused with other more traditional uses: in particular, the phenomenological essence is not a Platonic form separated from material profiles, nor does it belong to the category of Aristotelian substances. For Husserl, the goal of philosophical analysis was to exhibit the unique rigorous foundation of all knowledge. In the earlier phase of his work, he sought the secure foundation of all knowledge in universal, absolute, and ideal essences belonging to mathematics, logic and science, but in the later phase of his work, he sought the foundation of all knowledge in the "facticity" of the Life World and in perceptual essences given de facto in that World.

The method Husserl proposed for the study of essences, in which we become critically reflective perceivers, is the *method of profile variation*. In this way we are brought through a study of the profiles to acquaintance with their invariants. These profiles will be prescriptive in the reflective attitude, and will offer a criterion by which we can judge any object of our immediate experience. In the method of profile variation, the individual's imagination, or "free fantasy," is used to evoke an adequate or representative set (or group) of profiles of the object. In reflecting upon these profiles, we come upon limits and constraints. Such limits then inform us of the invariants, or laws, among them; these invariants are that which is named in experience, they are the kinds of perceptual objects. Such invariants are then collectively "the essence" of the object under study (though essence here is not to be understood in the classical sense as simple and unique). Husserl's focus on the invariants of experience was inspired, no doubt, by the *Erlanger Programm* that placed the essence of geometry in the study of invariants under specific transformation groups. It is probably not without significance that Felix Klein, David Hilbert, and Hermann Minkowski were Husserl's colleagues in his department at Göttingen.[10] Finally, the conditions of possibility for all *noetic-noematic* structures were, for Husserl, in the transcendental Ego.

Husserl's phenomenological project underwent considerable transformation under the influence of Martin Heidegger and Maurice Merleau-Ponty. The emphasis moved from logic, mathematics, and theoretical physics—Husserl's main interests—to perceptual objects and the background horizon of the Life World (*Lebenswelt*) as the necessary condition for perception; from the *essential* forms of objects to the uncovering of *being* and *truth* in human perceptual encounters with a Life World; from *thematic phenomenology* focused on the structures of objects to *hermeneutic phenomenology* focused on signs, symbols, and language; and from the impersonal *transcendental Ego* to the freely developing *existential Ego* of individuals in community in history.[11]

Within these developments, my approach will combine the perceptual and ontological interests of Merleau-Ponty and the early Heidegger with the mathematical and logical interests of Husserl and the hermeneutical interests of the late Heidegger. While I admit to being interested in a primary way—like Merleau-Ponty—in structures of perception, and—like the early Heidegger—in the ontological character of perception, I do not hold such an inquiry to be capable of yielding, in principle, knowledge of a unique set of primordially privileged objects that could constitute an absolute epistemological foundation for all culture and cultures. My primary concern, however, is with the formation of empirical bases for both scientific and nonscientific forms of knowing. This leads me to a study of the way mathematical models, scientific theories, and technological instrumentation can influence, transform, and enrich the content of perception and thereby increase the inventory of possibilities and actualities that we take to define the order of reality.

Perception, Horizon, World, Intentionality

The perceptual object (perceptual *noema*) is never experienced as an isolated, unrelated entity: it always manifests itself within a horizon that has two components, an *outer horizon* and an *inner horizon*.[12] In any individual act of perception, the perceived object has an outer horizon, or boundary, which separates it from the background against which it appears. The background too belongs to a World but negatively: it is that part of a World which is not the object or a part of the object.[13]

A *profile* of a particular object is a particular manifestation of that object in and through perception, each profile having, of course, a foreground-background structure. Systematically associated with any profile is a manifold of different possible profiles of the object which exhibit all the various facets that the object can manifest under a certain system of variations. Only such systems of variations as possess an invariant structure or essence at their heart are significant: a profile relates to an essence, and an essence is the generative law of a system of profiles; essence and profiles mutually define each other. The *inner horizon* of an object is the set of possible profiles generated by the essence of the system. Any one of the essential structures of the object is studied by probing the variety of profiles the object can have while maintaining its identity, for example, as *this* object or as an object of *this* kind under study.

In this book, I shall take "horizon" to mean one objective domain of the World, specified by a single essence. The term "World" is capitalized to indicate that it is used in a technical sense that will gradually be made clear.

The outcome of an essential investigation results in the purification and refinement of the descriptive terms of a language. Once an essence is established in this way, its profiles become prescriptive. It is then said to be given apodictically in experience.[14]

Apodicticity is the kind of certainty that accompanies a perceptual judgment in which a perceiver recognizes an essence as manifested: in such a judgment the perceptual objective is given (1) directly as in any perceptual judgment, but (2) according to the purified descriptive norms of the language, that is, essentially; its characteristic is (3) that it can be verified in multiple acts more or less at will: by this I mean that, guided by the concept of the perceptual essence, we can manipulate the object so as to generate profiles that match those of the prescriptive set. Although an apodictic perceptual judgment is certain, with a certainty rooted both in the essential concept and in its fulfillment in the World, it is not infallible; it is, however, as securely warranted as human knowledge can be, but it carries with it the "facticity" of a World and the "constructive" character of what is found there (the given invariants).

Although the formation of the percept may be automatic, having a percept does not constitute knowledge: perceptual truth or falsity is possessed only in a perceptual judgment, and this is a free act, made within a descriptive linguistic framework, prudentially made in the presence of its justifying epistemic conditions.

The basic content of perceptual judgments is not absolute, independent of time, place, and culture. A community of knowers in its historical and cultural setting can, however, achieve a relative, partial, and temporary transparency about at least some of the essential structures of its own World. It is the task of philosophy to establish the necessary general conditions of possibility, subjective and objective, of Worlds.

The perceived object is experienced as being *given* to human experience, and it is normally accepted spontaneously without any reflective question having been raised and considered about the possible source in the subject of that recognition and acceptance. Husserl calls this attitude "the natural attitude."[15] This is the attitude that supposes that we can gaze on a World with an "innocent eye," and that what we find unexamined in this way is real and as such privileged.

In contrast to the natural attitude is the reflective-transcendental attitude—or simply what I shall call "the reflective attitude"; this is accompanied by an awareness of the role that the subject plays in knowing, through preparatory intentions that prefigure in our expectations, the horizons that "speak" to us. A perceptual object, for example, is given not atomically as an isolated experience unconnected with anything else, but as fulfilling certain enabling conditions, such as being located in time and space among the things and situations that comprise a *World*. Our World is the general background reality context that is experienced as given to our perception together with the individual objects that we perceive. Something is real for us (or simply *real*) exactly if it is experienced as given by and in accordance with that preexisting structure of actual and possible objects of our experience which is our World; among these conditions is the space and time of our perception.[16] It is what Wilfred Sellars calls a " 'manifest' image of man-in-the-world"; this is "the framework in terms of which man became aware of himself as man-in-the-world."[17] A World then fulfills the most general set of preunderstandings one has about reality. Such terms as "Life World," "lived World," "World of daily life," are all used more or less interchangeably for what I call "World," or "everyday World": there is always the implied connotation that we are talking about a World that belongs to the contemporary Western community, or some shared part of that.

A World, though singular in that it applies exclusively to a particular community at a particular place and period, is not the only World: Worlds are historical and anthropological, differentiated by peoples, times, places, and perhaps professions. A World is always intersubjective, the shared space of a historical community with a particular culture that uses a common language and a common description of reality.[18]

Material "substances," in the traditional sense, such as trees or tables, books or insects, have horizons in any culture. It is also true that horizons can be found among complexes of things or people, like ecological systems, economic institutions, the "art world"; even single cultural traditions may have their horizons. An essence then may, and usually does, involve systematic relationships between many descriptive elements; these relationships themselves contribute to the essential definitions of the related parts. We may then speak about the horizon of the book (as opposed, say, to the horizon of the spoken word), of nineteenth-century landscapes, of university life, of the Renaissance in Italy, provided, however, that in all these cases we can show that there is a single essential structure to the phenomena. I believe we can even speak of the horizon of atoms and elementary particles—but more about this below. We can have apodicticity about the structure of a horizon only in the reflective attitude, because only in this attitude can we be reflectively aware of the full range of profiles to which an essence refers.

I shall be concerned in this book principally with what I call "horizons of visual space." Since a visual space is not itself a visual object—it does not have an outer visual horizon—the term "horizon of visual space" will refer then to the spatial horizon of all horizons of visual objects, or the invariant geometrical structures exhibited by every visual profile of every visual object. These invariant geometrical structures, particularly the metric structures, constitute the essence of the particular visual space. The investigation into visual spaces will comprise two parts: a study of the possible systematic character of classes of anomalous visual perceptions, and a study of the philosophical consequences of admitting both Euclidean and non-Euclidean descriptive criteria for visual objects as perceived. The empirical part of the study will touch three different levels of phenomena: (1) what individual observers experience: Do people experience episodes of hyperbolic visual perception within the everyday World? Could hyperbolic perception become central (as the overall organizing principle) in some possible World? (2) communal anomalous structures of vision, as evidenced by common visual illusions, and (3) the "history of perception," or communal structures of vision inferred, for example, from the history of Western pictorial art.

Correlative to every horizon, there is in the perceiving subject a noetic intention called "intentionality," which is the ability to receive from a World and recognize in a World perceptual objects belonging to its objective horizons, and which gives the capacity to initiate a search for or an inquiry into such objects. Intentionalities, then, are the subjective conditions of possibility of the presence (or absence) of objective structures within human experience.[19] Intentionalities are multiple; they enable the subject to reach out to his/her World in many directions and after interacting with it appropriately, to recognize the presence (or absence) of situations belonging to any one of its multiple horizons. Intentionalities prefigure in the subject by a set of systematic anticipations related to inquiry about a World, aspects of a World that can be discovered as given to or through perceptual experience. Intentionalities then express the significant interests subjects have in their World.

The thesis which, in the language I am using states "there are no innate intentionalities," corresponds to Sellars's thesis about the "myth of the given": that there are no unlearned, unrevisable, privileged, descrip-

tive categories for sensible objects.[20] That intentionalities are multiple
and the product of learning processes raises questions of *genetic phenomenology:*[21] what kind of "deep" or "primordial" conditions, subjective and objective, must be posited to explain the possibility of new
horizons? Among these conditions, some will belong to a Body—a
technical term for bodily conditions of knowledge—and some to the
possibilities of Worlds: with regard to both, the scientific account of
bodily and worldly structures will be relevant.

Intentionality-structures, once established, are said to be *prepredicative* because they anticipate—without, however, creating—actual
predicated instances: they define a domain of real possibility, anterior
to actuality. Intentionality also functions in the interpretation of textual
or other symbolic material; it provides the *hermeneutical circles* without which such material cannot be deciphered.[22] All intentionality,
even that operative in perception, is essentially hermeneutical, since it
is concerned with making sense of our experience, whether the textual
material to be understood comprises words or scientific-technological
artifacts ('texts,' as I shall call them below), or whether the "text" is
those optical structures incident on the eye which function as perceptual
stimuli, that is, as evocative of perceptual acts.

In the reflective attitude, then, two sets (at least) of necessary
subjective and objective conditions of possibility of experience can be
differentiated: (1) those specific intentionality-structures, which specify the structure of historical subjects and historical Worlds, and
(2) "deep" or "primordial" structures that function as conditions of
possibility of all human subjectivity and all perceptual Worlds. Which
set of conditions is relevant to a particular discussion will be determined by the context in which the reference to reflection is made.

Being-in-the-World: Body

The phenomenological stance that I am taking assumes that the
human subject is, in Heidegger's words—adopted also by Merleau-
Ponty—a *being-in-the-World.*[23] The human subject is not just a piece
of irritable organic material, a "third-person process," nor just a
disembodied Cartesian spirit, but a *Body*—the term "Body" is capitalized to indicate a technical sense.[24] I take the individual human subject
to be identical at all times with a Body that he or she uses or experiences; that Body is inserted into its experienced setting, a World (for
that subject), within which it is both a noetic subject, and an object

through which physical causality flows freely without interference or pause. The human subject as Body, then, is an embodied subject connoting physicalities as well as intentionalities. A Body defines the human subject functionally in relation to a World as the ground for an interlocking set of environing horizons. Being-in-the-World implies being now related to one horizon, now to another. The Body, then, that a subject uses or experiences as his/her own may at times, as we shall see, include processes external to the organism. At all times, what is perceived is immediately and directly in contact with the perceiving Body, and its lineaments are recognized to the extent they are prefigured in that Body as a subject furnished with the appropriate intentionalities. This will be explained below.

Science and the Perceptual World

Husserl's directive to phenomenology to "return to the things themselves," was, says Merleau-Ponty, "from the start a rejection of science." Let me quote the passage in full. Merleau-Ponty writes:

> I am not the outcome or the meeting-point of numerous causal agencies which determine my bodily or psychological make-up. I cannot conceive myself as nothing but a bit of the world, a mere object of biological, psychological or sociological investigation. I cannot shut myself up within the realm of science. All my knowledge of the world, even my scientific knowledge, is gained from my own particular point of view, or from some experience of the world without which the symbols of science would be meaningless. The whole universe of science is built on the world as directly experienced, and if we want to subject science itself to rigorous scrutiny and arrive at a precise assessment of its meaning and scope, we must begin by reawakening the basic experience of the world of which science is the second-order expression. Science has not, and never will have, by its nature, the same significance *qua* form of being as the world we perceive, for the simple reason that it is a rationale or explanation of that world. I am not a "living creature" nor even a "man," nor again even a "consciousness" endowed with all the characteristics which zoology, social anatomy or inductive psychology recognize in these various products of the natural and historical process—I am the absolute source, my existence does not stem from my antecedents, from my physical and social environment; instead it moves out towards them and sustains them, for I alone bring into being for my-

self . . . the tradition which I elect to follow, or the horizon
whose distance from me would be abolished—since that distance
is not one of its properties—if I were not there to scan it with my
gaze. Scientific points of view, according to which my existence
is a moment of the world's are always naive and at the same time
dishonest, because they take for granted, without explicitly men-
tioning it, the other point of view, namely that of consciousness,
through which from the outset a world forms around me and
begins to exist for me. To return to things themselves is to return
to the world which precedes knowledge, of which knowledge
always *speaks*, and in relation to which every scientific sche-
matization is an abstract and derivative sign-language, as is geog-
raphy in relation to the countryside in which we have learnt
beforehand what a forest, a prairie or a river is.[25]

Merleau-Ponty sees science as having undertaken the project of
reducing the psychic to the physiological without remainder. This
project, he asserts, cannot be fulfilled. It is, therefore, pursued by
investigators only with naiveté, or with bad faith or dishonesty. The
reasons why this project cannot in principle be fulfilled, he says, are the
following: (1) Science presupposes the reality of its starting-point—a
World; it cannot then replace that World, which is both science's basic
epistemological foundation and that about which science ultimately
speaks. (2) The reality of perceptual horizons implies the existence of
noetic subjects who carry forward traditions of inquiry; it cannot then
substitute something else for noetic subjects. (3) Science gives no more
than a derivative, second-order, abstract, schematic mathematical
model of nature which is no more than an instrument for the control of
nature; consequently, scientific knowledge is a deficient and impov-
erished form of knowledge, alienated from the basic epistemological
foundations of all knowledge and culture, and alienated from (in
Husserl's words) "the things themselves." (4) Since scientific inquiry
cannot go beyond the reality prefigured in the existing intentionalities
of some World, it is a not a new source of knowledge.

Considering Merleau-Ponty's four reasons for "rejecting" science, I
find myself entirely in agreement with the first two, but in serious
disagreement with the last two. In my rejection of (3) and (4), I depart
fundamentally from views that unfortunately have become entrenched
in the phenomenological account of science. This account, justly critical about the cultural influence of the way science is appropriated in our

society, is however too exclusively concerned with the contemporary biases of the social and professional milieu, and fails in general to evaluate properly those invariances which science—because of its essence—does, could, or ought to exhibit in various historical social contexts.

Natural science, or familiarly just science, has a pervasive presence and influence in our culture because it, more than any other form of knowledge, seems to lay claim to the rigor, objectivity, permanence, and universality that the Greeks sought as their emancipatory goal,[26] and the search for which, Husserl claims,[27] is the special teleology of the Western community. Natural science, then, developed within the total cultural and philosophical perspective of the West, which gave it impetus and which in turn derived sustenance from its achievements. Many writers, mostly in the phenomenological tradition, have addressed themselves to the critique of that component of our culture they call "modern science," with its claims to unsurpassable rigor, its methodological abstraction from the Life World, its seeming independence of subjective human interests, and the nonhistorical character of its laws and explanations, as a rival of phenomenological rigor, a living antithesis of its principles, and a challenge to its primary concerns.

The features of modern science which make it the current antithesis to the valid concerns of a phenomenological philosophy may be summed up in three characteristics: objectivism, scientism, and technicism, which are shared by the two most influential philosophical systems that are most influenced by natural science, Cartesianism and Positivism.

Objectivism is the uncriticized assumption that objectivized knowledge, whether scientific or nonscientific, represents reality, its object, without any connotation of the human knowing subject; or more precisely, that human objectivations represent things as if they existed in themselves independently of human embodied intentionality-structures, that is, as apart from any World.[28] Principal among these objectivations of knowledge is the scientific image of nature, based upon objective processes of measurement which substitute the objective restrictions of causal interactions for the subjective discovery of meaning within Life Worlds of human persons. Reality, in the objectivist view, comes to be an objective World-Picture, to which the human spirit or Mind merely adds the cultural superstructure of a *Weltanschauung* or World-Perspective.[29] In criticism of this view, it is said that (1) the ontological dimension of individual ontic beings is systematically concealed, (2) the historicity of Worlds and of the human subjects are both lost, and (3) knowledge is conceived erroneously as a mental copy of what is antecedently out there. Objectivism's opposite is what Husserl calls ''transcendentalism'': this is the view that ''ontic meaning of the pre-given life-world and all objectivated knowledge is a subjective structure, it is the achievement of experiencing prescientific life.''[30]

The second criticism of modern science is its cultural imperialism, that is, the uncritical belief called ''scientism'' that the methodology of the positive sciences is in principle capable of answering all meaningful questions, and that philosophy is a prescientific stage in the thrust towards positive science and will wither away in a scientific culture. Scientism, then, comprises claims both about the comprehensiveness of the methodology of the positive sciences and about the superior rigor of that methodology vis-à-vis knowledge. Such claims, Rudolf Boehm concludes, are a threat to the very existence of a phenomenological philosophy.[31]

The third criticism of modern science is technicism, that is, the view that science is no more than a *techne*, albeit a very successful one for manipulating and exploiting nature. Jürgen Habermas, for instance, expresses this position: he writes, ''the cognitive interest of the empirical-analytic sciences is technical control over objectified processes.''[32] The argument is based on the use in science of functional concepts; these, it is claimed, are ways of relating ''mere entities,'' which are knowable only extrinsically as terms of relations implicitly defined by a mathematical model and which like Lockean substances, are unknowable in themselves. If the terms of science are unknowable in themselves, they can only serve the purpose of the manipulation and control of nature.

In summary, according to the phenomenological critique of modern science, it is nonhistorical, since it shares only incidentally in the intrinsic historical dimension of the human inquirer as a Being-in-the-World; it is nonhermeneutical, since it is founded on a theory of knowledge that conceives Mind as the ''Mirror of Nature'' rather than as a hermeneutical perceiver; it is nonontological, since it has no horizon of Being, but uses abstract models merely to serve man's interest in technical control. Implicit also is the critique that science is nondialectical, since progress in science comes not through the conflict of opposing intentionalities in human subjects, but exclusively by the gradual accretion of objective facts, the painful elimination of error, and the exercise of a universal and timeless logic.

To the contrary, I take the position that the essence of science, invariant across the spectrum of actual and possible historical cultural biases,[33] both general and specific, is ontological in intent and ordered toward the creation and discovery of new horizons of perception. In this respect, I follow the thrust of the early Heidegger, the Heidegger of *Being and Time*. The early Heidegger saw the central question of philosophy as that of a fundamental ontology, clarifying the Being question latent preconceptually and prepredicatively particularly in the horizons of modern science: "The question of Being is the spur of all scientific thinking,"[34] he wrote. About the question of Being, it is a priori to all ontical categories:

> The question of Being aims at ascertaining the a priori conditions not only for the possibility of the sciences which examine entities as entities of such and such a type, and, in so doing, operate with an understanding of Being, but also for the possibility of those ontologies themselves which are a priori to the ontical sciences and which provide their foundations. *Basically, all ontology, no matter how rich and firmly compacted a system of categories it has at its disposal, remains blind and perverted from its ownmost aim, if it has not first adequately clarified the meaning of Being, and conceived this clarification as its fundamental task.*[35]

The positive sciences nevertheless precede the labors of philosophy, and the labor of the philosophy of science is as it were a "recapitulation" of the scientific effort.

> Since the positive sciences neither "can" nor should wait for the ontological labors of philosophy to be done, the further course of research will not take the form of an "advance" but will be accomplished by *recapitulating* what has already been ontically discovered, and by purifying it in a way which is ontologically more transparent.[36]

Heidegger's method is thus from the start hermeneutical, that is, it is the search for the structures of meaning, explicit or latent, in the World horizons, particularly those created by science. The source of these meaningful structures is man himself, *Dasein*, whose existence is to-be-in-the-World, and who thereby posits Being.

In his later period, Heidegger's interest shifted away from ontology to meditations on the use and meaning of language. Reflecting on the sense that the Western linguistic community gave to scientific activity, he concluded that the hermeneutical "truth" about science, its general and specific bias in Western society, is that it is a "theory of the real."[37] To discover the sense of these terms, he meditates on their possible etymologies, and concludes that to our times and culture, science signifies a "compartmentalization of the real" into systems of isolated objects, and the "entrapment" of those objects—"the scientifically real"—for the purposes of control and "manipulation." He concludes that the *real* of modern science is the set of isolated manipulable objects given to us with Cartesian "certainty" and in an unreflective "factual" way. Heidegger uses the terms "real," or "scientifically real," for something in our culture which has become separated from the life of perception.

Paul Ricoeur concurs in this judgment of the way science has been historically appropriated by Western culture. He writes,

> The age of science appears to us not only as a consequence of science, but also as a new way of existing. If there has always been technology, more or less, today technology is no longer an accessory of our existence; it is the axis of it. Technology is a way of viewing the world, a means to practise it, as a universal manifestation of the available. In this sense, technology represents a new ontological regime.[38]

Among the senses of *science*, I distinguish two in particular: (1) science$_1$, when the essence of science is sought across all possible historical profiles with their general and specific biases, and its intent is seen to be basically ontological, to be the search for the real underlying structure of things;[39] and (2) science$_2$, the profile of science in Western culture as affected by the general biases of our culture and those specific to the scientific community. These latter biases turn out to be thoroughly incoherent; they include, on the one hand, the belief that science has the power of uncovering *the real uniquely* (Scientific Realism)—and not just *new horizons of the real*—and on the other the contrary belief that science does not concern itself with the real, but only with extending human power over nature (Instrumentalism). The general and specific biases concerning the profiles of science in our culture are in need of philosophical analysis and critique, particularly in any investigation aimed at discovering the essence of science.

The study of the essence of science—*science₁* above—I see as belonging to the philosophy of science proper; the study of *science₂*— the profile that science has in Western society—such as one finds, for example, in the work of Jurgen Habermas[40]—belongs to social or political philosophy.

The essence of science, as I will argue, is to specify new horizons of reality which become accessible to perception by "readable technologies," a special product of scientific theory. I shall argue that such a dominant feature of the modern World as the fact that we organize our visual space in a Cartesian/Euclidean way rather than according to (what I hold to be) the more native structure of, say, hyperbolic geometries, is a product of modern science, and the reflection in subjective intentionality of the carpentered environment we have created: but more of this below. In this way, the establishment of new scientific horizons adds to or further specifies the prepredicative intentionalities through which the real possibilities of Worlds are discovered, chosen, and then laid out for us to explore more fully. Consequently, I shall argue that if we could set aside or neutralize the sort of naiveté or bad faith Merleau-Ponty, the later Heidegger, Gadamer, and others see as all too pervasive in our culture, or better, if we could convert our colleagues to a different appraisal of science (one in keeping nevertheless with its essence), the World would not be in danger of being reduced to a set of schematic models, nor would it be diminished by science, as phenomenologists claim. Instead, to the extent that new readable technologies are constructed, its horizons would be multiplied and enriched. Worlds would be creatively renewed while the human subject would come to experience itself in new ways as it learns to live in new embodiments, as a Being-in-a-World renewed. The key move, as I shall explain, is a hermeneutic as well as a phenomenological one, one that enables us to 'read' like a 'text' those scientific and technological artifacts of material culture which give us direct perceptual access to classes of things and objects that are not otherwise possible objects of perception.

Models, Metaphors, and the "Things Themselves"

Husserl's "return to the things themselves" in the Life World was motivated by reasons similar in many ways to those expressed by Merleau-Ponty: he was aware that science was pursued with a dogmatic reductionist tendency.[41] This he saw as an attempt to replace the richness, spontaneity, and novelty of the Life World with a schematic, abstract, causal, mathematical model of it—the model of theoretical physics. He took the use of mathematical models to be symptomatic of a profound alienation from perception and reality: such models tended to serve as substitutes for what was perceived and thus for reality.[42] That there are other ways of using mathematical models, ways that (1) enrich possible Worlds, (2) increase the subtlety of our perceptual discriminations, and (3) enlarge the domains of givenness and apodicticity, will be the theme of this book. The incorporation of mathematical models into phenomenological research is a major innovation.

The "return to the things themselves" meant, for Husserl, the turning away from mathematical models as surrogates for reality and the search for eidetic essences, or simply essences in phenomena and in the praxis of structured human commerce with the Life World.[43] If, however, the essence turns out to have a complex internal structure, how is such an essence to be expressed or communicated?

Two ways of expressing the profiles of a complex essence suggest themselves: one is to list general laws or regularities that the perceptual profiles obey, the other is—if one is lucky—to find a model of the essence from which general laws for the profiles can be derived. For a complex essence, the number and variety of general laws for the profiles relevant to the expression of the essence may be too large and complicated to serve anything but a limited and private use: an *eidos* that is not perspicuous can hardly be apodictic.

The expression of complex essences, then, must be approached some other way: I suggest by way of *models*. A model capable of organizing a great deal of information can be both perspicuous and apodictic. Perspicuity follows from the essential simplicity and aesthetic elegance of a good model. But what about apodicticity?

A *model* is always a structure that is capable of being used to represent or elucidate a (or the) significant structure of something else, the *modeled*. There are various usages of the term "model" in the philosophy of science. Some make a distinction between a *theory* (ideally the axiomatic structure of a theory, with undefined primitive terms) and a *model* of this theory; any model being a set of objects—conceptual, such as geometrical points, lines, and so forth, or empirical, such as measured values of physical variables—possessing the same structure as the theory (or a substructure of the theory).[44] In this usage, that which is modeled is the theory; no theory is exhaustively represented by any one of its models, and any theory is underdetermined by any finite set of its models. This usage moves in the direction of an important aspect of theory, its semantical structure, but it does not go far enough, since it does not adequately permit an important distinction between the empirical (perceptual, or observational) semantical use of models and their theoretical semantical use in the conduct of science.[45]

I propose then to use the term "model"—or "scientific model"— in the following way: a model is a theoretically-structured set of elements (usually mathematical objects) purporting to be useful for the description or explanation of an empirical domain. The *modeled* is then some horizon of the World.

One distinguishes then between a scientific model considered as an *object* and the *semantical use* made of such a model to state truths about some World. A claim is often made (or implied) for a mathematical model that, as an object in itself, it is a true "picture" of the "real but hidden" (scientific) structure "behind" the surface of perceptual phenomena. Geometrical models in science are often presented in this way as surrogates of reality. It should be clear that no such model could possibly be apodictic (as a structure of the World), since it is ideal, since it refers to a hidden structure and is therefore only indirectly related to perception, and since the purported existence of that which it models is detected only through empirical procedures that are crass and clumsy compared with the infinite precision of geometrical objects and their analogs, the real numbers. When Husserl and Merleau-Ponty stated that such models take us away from the eidetic structures of perception, they were criticizing the view about models just expressed.

There is, however, a different semantical use for a model, that of helping to create new descriptive concepts by organizing new sets of perceptual profiles. That geometrical concepts can be used descriptively and apodictically was shown by Husserl in his essay "The Origin of Geometry." In it, Husserl grounded the geometrical model of physics *in the technical praxis* of making preferred shapes.[46] The investigation into the geometry of visual space which follows makes a similar use of hyperbolic geometry, and shows that the basis of such a representation is *a special praxis*, unaided by technical instruments. This praxis (or "embodiment," as it is called below) is the necessary but forgotten link, overlooked in the mainstream both of the philosophy of science and of analytic philosophy, between the model and the facts it represents.

The use of a quantified model of visual space should not be regarded as giving a theoretical-explanatory account of certain anomalous visual profiles—although aspects of the derivation of the model do have theoretical (that is, inferential) import—but as an aid to phenomenological description. Descriptively, it serves as a kind of quantified metaphor.[47] It suggests avenues for systematic perceptual inquiry, and provides means for investigating and expressing complex spatial relationships manifested in perception, which are less familiar than those easily intuited Euclidean relationships we use so freely and so unproblematically in descriptions of visual phenomena. It is evident that everyday terms such as "straight" or "circular," which get their sense from a different and more familiar mathematical model, are capable of apodictic use in the description of perceptual phenomena. There is no reason why other geometrical models could not be the source of new and unusual senses of these terms, available for descriptive use, should the World provide objects appropriate to their use. The following study attempts to provide a new descriptive usage based on hyperbolic Riemannian geometry for certain anomalous perceptual phenomena in which depth, distance, and shape fail to fall under descriptive Euclidean essences.

The descriptive application of ideal geometrical terms to experienced phenomena is always adjectival and, by way of the analogy of more or less, a witness to the inescapable underdetermination of theory by empirical data.[48] The abstract form—idealized and therefore unrealizable—serves as a model/metaphor capable of ordering empirical instances, by suggesting an intelligible structure that individuals will "have," but from which they will diverge at the same time nonsystematically, when ordered according to the set of cultural interests guiding the choice of the model/metaphor.[49] In the population of people six feet tall, for example, being six feet tall—no more and no less according to a precise mathematical measure—is not a property that is possessed by any actual member of the population: six feet is an abstract statistical parameter, not the height of any individual. The use of such norms connotes then comparison with a reference population ordered with reference to the abstract (statistical) norm. Individuals of the population then constitute an equivalence class, their title to membership being that they do not differ significantly—according to some agreed measure of more and less—in relation to some abstract (statistical) norm. This is one way a quantified metaphor works. Apodicticity falls on membership of the equivalence class: not on the assertion—

which would be simply untrue—that the idealized abstract form is realized in each individual empirical member of the population.

In conclusion then, if a Euclidean geometric term can be used adjectivally of experience in a descriptive and even apodictic way, so too surely can other geometrical terms. Moreover, it is not absurd to think that, despite the fact that they are drawn from an unfamiliar geometrical model, hyperbolic geometrical terms may be applicable to experience and may even achieve the status of apodicticity.

PART I:
HYPERBOLIC
VISUAL SPACE

2
Introduction to Visual Space

Introduction

Before plunging into the technical aspects of hyperbolic non-Euclidean pictorial spaces, I want to introduce some visual material to illustrate the claim that ordinary folk, artists, and in fact whole communities have been accustomed at divers times and places to see their World, or horizons of their World, as spatially organized in ways that we find hard to describe at the conscious level without using the language of illusion and distortion. The type of organization I am about to study is that which can be described and explained, with the help of graphs and figures, as the effect of reconstruing and reexperiencing the matter of visual experience in a hyperbolic non-Euclidean way rather than according to the familiar laws of Euclidean geometry (the geometry that ordinary rigid physical objects obey in practical everyday experience). What this means specifically, and what rationale we have for focusing on this geometry rather than on some other geometry—or on no geometry at all—will be discussed in the subsequent chapters. The purpose of this chapter is just to arouse the curiosity of the reader and to persuade him/her to accompany me through this study to the end despite possible moments of difficulty. This chapter then introduces Part I of the study.

The thesis proposed in Part I is about how we experience or can come to experience our World *visually*. It can be summarized this way: though we usually experience our physical environment as laid out

before us visually in an infinite Euclidean space, from time to time we
actually experience it as laid out before us in a non-Euclidean visual
space, in one belonging to the family known as "finite hyperbolic
spaces." Scenes—real scenes—construed in such visual spaces will
appear to be distorted in specific ways that will be described in summary fashion below, and with more technical detail in the following
chapters (and in the Appendix).

Evidence for this thesis falls into three main classes: (1) *everyday
phenomena*, such as the experience of urban canyons or the interiors of
churches, of the dynamic flow of shapes when driving on a highway, of
the shape of the open sea and the open sky; (2) *well-known visual
illusions*, such as the Hering, Müller-Lyer and other two-dimensional
illusions where the effect seems to be related to hyperbolic vision, and
three-dimensional illusions of the distant zone, such as the moon
illusion, and other illusions associated with rotating objects, and (3)
pictorial spaces, of ancient times, as witnessed by the architectural
refinements of Doric temples and by Roman wall paintings, and of
modern times, particularly in the paintings of Cézanne, Turner, and
van Gogh, where the pictured space has clear hyperbolic structures.

Appearances in Hyperbolic Visual Space

The distortions due to hyperbolic vision fall, broadly speaking, into a
number of categories:

 1. *Near and distant zones:* there is a qualitative distinction in
    the way objects appear (in what way and by how much they
    are distorted) when they are close to and in front of the
    viewer, and when they are far from the viewer. This results in
    a useful distinction between near and distant zones.

 2. *Near zone:* the core of the near zone is a "Newtonian
    oasis"—to use R. Arnheim's phrase—directly in front of the
    viewer, where visual shapes are clearly defined and differ
    little from their familiar physical shapes. On the periphery of
    the Newtonian oasis, depth appears to be dilated, and frontal
    surfaces appear to bulge convexly; in addition, the frontal size
    of objects appears to be smaller than life, compensating, for
    instance, for the distortion seen in close-up photographs, say,
    of a hand thrust out towards the lens of the camera. Parallel
    lines appear to diverge, as if seen in reverse perspective; this

effect can be noticed by holding a three-by-five-inch card horizontally in front of one's eyes: close by, the edges of the card receding in depth will appear to diverge and, farther away, they will appear to converge. The "refinements" engineered with such precision into the construction of the columns, stylobates, colonnades and entablatures of Doric temples in ancient Greece were introduced, according to ancient writers, for the purpose of correcting for visual illusions (see figure 2.1); the relation of these refinements to the near zone of hyperbolic visual space is an intriguing one.

3. *Distant zone:* the distant zone—extending infinitely in all directions in physical space—appears to be finite, shallow in depth, and slightly concave. Distant phenomena are experienced visually as if seen through a telephoto lens, that is, they appear to be closer, flatter, and with their surface planes turned to face the viewer. Depth appears shallow and compressed; distant regions of space seem as a consequence to be made up of increasingly shallow zones, shallower than the physical region each zone encompasses. Thus it is visually impossible, for example, for a rigid physical object such as a wire frame to rotate without the appearance of some illusion that converts the rotary motion into the appearance of a rythmic two-dimensional lateral distortion. Parallel lines, like railway tracks or the margins of a long, straight road, bend upward and come together to meet at a point in front of the viewer on the horizon and at a finite distance.

Illustrations of the telephoto effect of hyperbolic vision can be found, for example, in the paintings of Cézanne. Multiple eye levels, multiple directions of sighting, and multiple sight points characterize many of Cézanne's paintings: this is seen, for example, when his *Turning Road at Roche-Guyon* (figure 2.2a) is compared with John Rewald's photograph of the same motif (figure 2.2b). E. Loran in his classic study has derived these multiple sight points from the paintings on the assumption that all the individual motifs are represented by classical perspective in a Euclidean pictorial space;[1] however, the apparent multiplicity of sight points, and so forth, would be explained more economically by the hypothesis that the artist perceived his motif in hyperbolic space and was

Figure 2.1: Greek Doric temple.

responding to its structure as seen from one sight point in that space. The telephoto effect in that space would cause the planes of the object to be turned to face the viewer, mimicking the effects discovered by Loran's analysis. Whether one can infer from a painting how the painter or the people of his time and place saw their World is much disputed, and will be fully discussed below. I shall argue that one can at times make such a plausible inference. In the case of Cézanne, the decisive question is whether the perceived unity of Cézanne's pictorial space is geometric, or whether this unity is achieved by some other aesthetic principle.

The problem of geometric unity—even if incidental to the artist's other motivations—arises about the work of other artists such as, for example, Vincent van Gogh. The clear impression given, for example, by his *Bedroom at Arles* (figure 6.3) is of a closed and finite space, limited by the walls of the room and a shallow zone beyond, within which whatever else exists, even the infinity of physical space, must find its accommodation.

4. *Horizontal planes*: a viewer looking at an extended horizontal plane below eye level, such as the sea seen from the top of a cliff, seems to be at the center of a great bowl with its rim on the horizon. An extended horizontal plane above eye level, such as a cover of clouds, takes on the appearance of an arched but flattened vault resting on the horizon.

In many ancient works of art, the sky is represented as an arching vault, and the earth as having the shape of a saucer. Figure 2.3, for example, taken from an Egyptian sarcophagus, shows the sky as represented by the arched body of the mythological goddess Nut raised above the earth, represented by her brother Geb, by the power of Shu, the god of the atmosphere; it is plausible to conclude that the sky was experienced by the ancient Egyptians as we experience it, namely, as a vaulted structure.

5. *Size-distance*: The apparent size of very distant objects, such as the moon or clouds, depends on whether there are local cues and how these are construed. Consequently, the apparent size and the apparent distance of very distant objects may be related in a way contrary to the familiar (Euclidean)

Figure 2.2: (a) *Turning Road at Roche-Guyon*, by Paul Cézanne (Art Gallery, Smith College, Northhampton, Mass.).

(b) Photograph by John Rewald of the same motif.

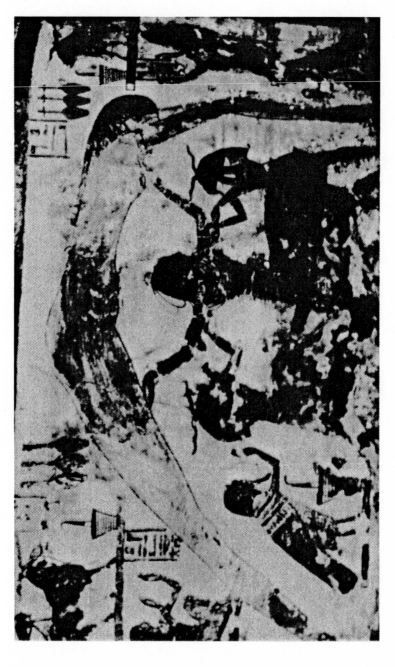

Figure 2.3: Shu, god of the atmosphere, raising his daughter Nut, representing the sky, away from her brother Geb who represents the earth, and thus creating the world. Painted coffin of Butehamon. (Museo Antichità, Turin).

size-constancy laws that characterize much, though not all, of our daily visual experience.

An illustration of this is the moon, or rather the moon illusion. On the horizon, the moon looks large and relatively close, but at its zenith, it looks small and distant; moreover, the landscape seen under the moon seems to change scale with the position of the moon.

Quasi-Stable Euclidean/Non-Euclidean Phenomena

The possible existence of two visual perceptual systems—Euclidean and hyperbolic—raises the question about their interaction within perception. Since hyperbolic visual measures are different from Euclidean physical measures, there is no logical contradiction in affirming simultaneously both "*A* and *B* are equal in size (in physical space)," and "*A* and *B* are unequal in size (in hyperbolic visual space)"; each assertion could be correct in its own proper context. But such facts cannot be perceived simultaneously. The possibility of two such diverse perceptual judgments involves something other than just an ambiguity in the relative sizes of *A* and *B*. It involves the possibility of choice between (or realization of) different spatial contexts; within each, foreground events are projected according to different geometrical laws against the background of the surrounding environment.

Where two different perceptual contexts are possible, one expects to find under suitable conditions multistable representations—now the scene is represented in a Euclidean visual space, now the same scene is represented in a hyperbolic visual space. Rudolf Arnheim seems to be alluding to this phenomenon, when he writes of entering a church:

> The interior has depth—probably somewhat less than it ought to have. Do the rows of columns look parallel and the columns equal? Yes and no. Vanishing lines converge toward the altar, but at the same time all columns are of equal size and stand on parallel tracks. All this is seen directly, not derived from some calculation.[2]

He immediately adds, "In theory this sounds paradoxical, in practice we notice no contradiction." We have all had similar experiences, whether in a church or in the urban canyons of our cities. The experience is perfectly intelligible as an illustration combining elements of Euclidean perception, such as, in Arnheim's illustration, the equality of the columns and their parallel tracks, and elements of hyperbolic perception, such as the foreshortening of depth and the convergence of the columns toward the altar. It would not be intelligible, however, to make the claim—and Arnheim in fact does not make the claim—that the perceptions are simultaneous. What we experience is a composite representation made up of two different incompatible quasi-stable facets: one sees now one, now the other but not both simultaneously, like the drawing that can be seen either as the picture of a rabbit or as the picture of a duck, but not as a picture of both simultaneously.

To complete this informal introduction to the kind of problem with which the first part of this book is concerned, I could hardly do better than introduce the reader to E. G. Ballard's fascinating study of the Jefferson Memorial Arch, called "The Gateway Arch," at St. Louis, Missouri.

The Gateway Arch

Edward G. Ballard made a study of the visual profiles of Eero Saarinen's Jefferson Memorial Arch in St. Louis (figure 2.4).[3] His purpose was to isolate the perceptual invariants related to *closeness* and *distance* in the field of vision. The Arch is 200 meters high and 200 meters wide, and it lies in a north-south plane on the west bank of the Mississippi. The Gateway to the West, as it is called, is triangular in cross-section, and is flung into the sky in one unbroken catenary of stainless steel.

To Ballard's perception, every approach to the Arch is characterized by visual ambiguity about form, solidity, orientation, and materials. He describes the first appearance of the Arch through the morning haze as a thin gray band in the sky "paper thin, no more solid than a gap in the color of the cloud."[4] From his position upriver, the Arch with its two great legs comes into view at an oblique angle, and he notices a strange phenomenon: "as soon as one leg was identified as the South leg, the Arch would 'spin around' and the appearance reversed."[5] Such perceptual ambiguity, he surmises, is an essential characteristic of the perception of distance. Accordingly, he proposes the following pair of definitions: *perceptual distance* is essentially perceptual ambiguity relative to form and orientation; *perceptual closeness* is essentially clarity of geometric form and absence of ambiguity; these he refines later.

Figure 2.4: Gateway Arch, St. Louis, Mo. Designed by Eero Saarinen.
© Art Grossman Photo.

These perceptual invariants can be analyzed by us in terms of hyperbolic space perception. The distant zone according to our calculations is shallow—consequently, we are not surprised to find that the Arch at a great distance appears to be "paper thin." In addition, the Arch should appear to occupy a zone too shallow to accommodate its shape as physically conceived, and its plane should appear to be turned so as to face the viewer more or less frontally; consequently, its two legs should appear to be roughly equidistant from the viewer wherever he/she is. Such a shape and orientation clashes with physical expectations. An unstable visual situation should be created as a consequence, and this would account for the strange phenomenon of the "spinning Arch," now showing one leg as the south leg, now the other. The "spinning Arch" is like the illusion of a rotating wire figure that appears to reverse its direction of rotation at every half turn, since both are constrained to fit into a visual zone too shallow and distorted for their physical shapes.

Returning to Ballard's account: he further discovers that what in one context is perceived as sharply defined, in another is perceived with ambiguous outlines. For example, looking at a face covered with grains of sand, one might focus on the face, and see it clearly—it would then be close; the sand, however, would be ambiguously perceived—it would be distant; or one might focus on the sand, and see it clearly—it would be close; and the face would be perceived as fuzzy and ambiguous—it would be distant. "The ambiguity or clarity, distance or closeness, of the perceived object, I want to say, is its way of appearing, and its way of appearing is a function of the object within a context which is determined or ordered by an intention."[6]

Ballard has reached the interesting refinement that *perceptual closeness* and *perceptual distance* are contextual and functions of the visual intentionality of the viewer. He writes, "closeness is contextually removed ambiguity, and distance is ambiguity elicited by a context."[7]

In the contextuality of vision, Ballard has raised a dimension of the problem of visual space which becomes explicit in hyperbolic vision. The account of hyperbolic vision given below is hermeneutical; that is, it supposes that the curvatures of hyperbolic visual space are determined contextually, by the interplay between foreground and background and the need to fulfill a particular purpose or intention by looking. As a consequence of this flexibility, a particular object or event may, within one visual context, fall within the distant zone, and within another visual context, fall within the near zone.

Ballard's description and analysis of his perceptual experiences are masterly. Nevertheless, he would have been helped in the conduct of his inquiry even *qua* phenomenological by the heuristic of a geometrical model such as the one I am proposing below. This would have suggested to him where to look for clues to other more sophisticated invariants, and it would have provided him with a set of new, complex, and interrelated descriptive forms for a more precise study of the invariants associated with visual closeness and distance. It would also have given him the means to discuss the interaction in perception between the normal Euclidean intentionality of everyday perception and the anomalous mode of hyperbolic perception, an interaction well-illustrated by the "spinning Arch," where physical anticipations clash with hyperbolic vision, producing a set of quasi-stable illusions.

One accustomed to phenomenological methods would probably object to the use of a geometrical model as distracting attention from the "things themselves," and as possibly vitiating the phenomenological study by the introduction of theoretical elements. The use of a geometrical visual model, however, need not and should not be regarded as a theoretical-explanatory study of visual phenomena, but rather as an aid to phenomenological description. It suggests avenues of inquiry, relates significant phenomena, and provides the means to describe spatial relationships more complex and less familiar than those easily intuited from daily experience, where Euclidean concepts are used freely and unproblematically. The application of hyperbolic geometrical terms to phenomena for descriptive purposes is not more problematic than the application of geometrical terms such as "straight" or "triangular" to phenomena—terms which Ballard would certainly use, and has used, for descriptive purposes. Such terms apply to phenomena only adjectivally; they designate membership in an equivalence class of empirical phenomena, and their usage may be compared with that of a quantified metaphor. Like their Euclidean counterparts, when used for descriptive purposes within a certain praxis of inquiry, they share whatever apodicticity Euclidean geometric terms can have when used for the purpose of describing empirical phenomena.

These are the phenomena I shall now proceed to study by using a particular geometrical model of visual space. I shall attempt to establish the plausibility of a certain hyperbolic model of visual space in chapter 3; chapter 4 will describe the specific structures to be looked for in hyperbolic vision, and will survey everyday evidence that supports the model; chapter 5 will study evidence from visual illusions in support of the model; and chapter 6 will study evidence in the history of art.

Visual Space: Search for a Model

Kant: Space as A Priori for Experience

For Newton and for most philosophers up to the time of Immanuel Kant, geometry was related to mechanical drawing and depended for its proofs on the feasibility of constructions with ruler and compass.[1] On the one hand, geometry was conceived to be an idealization of or abstraction from that aspect of crude, particularly visual, experience which dealt with the relationships between certain kinds of physical items such as lines and figures *as constructed* (using ruler and compass, or chains and angles, etc.) and *as movable without distortion (by translation and rotation) in space*. On the other hand, it was recognized that geometrical laws had to be normative for those standard procedures by which the physical exemplars of geometrical objects were constructed, and so geometrical laws were in some sense prior to their physical embodiments.[2]

Kant addressed himself to the resolution of this paradoxical duality:

> A new light flashed upon the mind of the first man (be he Thales or some other) who demonstrated the properties of the isosceles triangle. The true method, so he found, was not to inspect what he discerned in the figure, or in the bare concept of it, and from this, as it were, to read off its properties; but to bring out what was necessarily implied in the concepts that he himself formed a priori, and had put into the figure in the construction by which he presented it to himself. (Bxii)[3]

Kant discovered that we are constrained a priori to perceive objects of empirical intuition according to the laws of geometry: "Empirical intuition is possible only by means of the pure intuition of space and of time" (B206). Although Kant admitted that it was possible to think non-Euclidean concepts (they are noncontradictory), he held them to be without possible empirical content: "The impossibility arises not from the concept in itself, but in connection with its construction in space, that is, from the conditions of space and its determinations (A221)." That the conditions of space and its determinations were for Kant Euclidean[4] is clear from many passages: it follows from his general affirmation of Newtonian physics as well as, for example, from what he assumes about the independence of geometrical form from size (A124) and about *the* properties of *the* isosceles triangle (Bxiii). That the conditions of space are for him *necessarily* Euclidean seems to be ruled out by his admission of the conceivability of non-Euclidean forms, though it must be added that Kant's own concern was less with forms of geometry which he could not have anticipated than with the discovery that our experience is organized and structured actively by synthetic a priori principles, among which at least de facto is the Euclidean form of sensible intuition.

How have philosophers since Kant come to understand the synthetic a priori character of space? What necessity do they ascribe to Euclidean geometry? J. Vuillemin,[5] studying Kant's doctrine on space in the light of the *Erlanger Programm*[6] and the invariants of Lie transformation groups for that space, concludes that for Kant the a priori synthetic character of empirical intuition is some constant curvature geometry, one, however, that he inadvertently assumed to be Euclidean because he took geometrical form to be independent of size. P. F. Strawson[7] takes Kant to say that Euclidean laws apply with necessity (or what he prefers to call "phenomenally analytic" necessity) only to the pure visual[8] (or "phenomenal" in his words) imagination, but not to physical reality. H. Reichenbach and others[9] hold, on the contrary, that a priori Euclidean laws do not apply with necessity to the pure visual imagination, but apply de facto rather by virtue of learning, habit, convention, or the demands of objectivity. J. R. Lucas[10] insists on the normatively necessary character of Euclidean geometry in order to guarantee the *objectivity* of the notion of physically rigid bodies and rulers. G. Buchdahl, however, concludes that "Kant simply cannot be raising any claims concerning the *necessity* of a specific metric. And this is surely the case when we remember that he wants only to demonstrate the possibility of a fact—*if fact it be*—of there being a unique geometry. For in that case it would *simply be the case* that a unique metric is injected *a priori*, which is that set which is equally '*in fact'* balanced by 'what we find,' if we find it, i.e., a privileged geometry.''[11]

In summary, while all philosophy today is pursued in the light of the Kantian heritage, and philosophers accept the view that what we find in experience is prefigured in intentionality, and that the empirical object as known is actively constructed according to a priori rules, still it is evident that the undisputed content of that legacy is unclear. This is particularly so as regards the uniqueness of the geometrical rules active at the transcendental level, and their origin in the subject, or possibly in ways in which subject and World collaborate to construct space— visual, perceptual (in the broader sense), physical (of our local environment), or cosmological.

Pictorial Spaces

Pictorial vision is a special kind of vision and one that is for historical reasons privileged in our culture.[12] Though not exclusively the possession of modern culture, it is a characteristic of modern culture in the West since the Renaissance. Pictorial vision consists in the art of seeing objects as pictorial, that is, as constructed according to rational (i.e., in this case, geometrical) principles out of basic visual pictorial elements, such as points, segments of lines, patches of plane surfaces, and the color and light on these patches. These, when refined and idealized, are the objective elements of the Cartesian World-Picture; color is a secondary quality that relates the Cartesian object to the human subject. To represent a World in a pictorial way is to see objects in the World as pictorial and pictorially related: to see a World only that way is to define reality pictorially. Correlative to this ontology is a theory of knowledge: to know reality is to picture it in the Mind; Mind becomes the ''Mirror of Nature.''[13]

While holding most emphatically that the conception of *reality* as exclusively pictorial is false, both to the facts of perception and to philosophical reflection, I shall nevertheless be concerned principally with the *pictorial aspect* of perceptual objects; with the possibilities of seeing pictorial objects constructed out of basic non-Euclidean visual elements; and with the relationship of the objects of science to perception. It should be noted that my use of the term ''pictorial,'' as in ''pictorial objects'' and ''pictorial space,'' does not connote reference to image-making or works of art. As used by art historians and art critics, the term ''pictorial'' has a technical sense and signifies the space that an artist intends to create by the making of an image. I shall, of course, be concerned below with the pictorial spaces of artistic depiction, but only as part of a more general study of the pictorial aspects of perceptual objects.

How To Represent Pictorial Spaces

E. Panofsky[14] comments on how our capacities to see are constrained by the perspective system we use, that is, by our way of depicting what we see. Perhaps it is just as true to say that our ability to depict is limited by the way we see, since perspective systems are designed to construct images that, when viewed, produce in the culturally trained viewer the experience of depicted objects that match perceivable objects. What we can see, therefore, is limited by the perspective system we use only because our perspective system was designed to express a certain pictorial vision—but more about this below. To paraphrase Panofsky, the perspectival structure of a two-dimensional image is the symbol of the pictorial space of the epoch.

If the reality space of an epoch is pictorial, we may be able to determine its structure by learning to read the pictorial images of the epoch. For this purpose, historical and anthropological studies would be needed in addition to the study of the epoch's art. Then theoretical studies would have to be made of the characteristics of pictorial spaces, appropriate reasons would have to be found in the art and culture of the epoch for deciding in favor of one pictorial space rather than another, and the apodictic character of the evidence for the choice would have to be made manifest (the necessity for apodictic evidence follows from the phenomenological aims of this study).

The kinds of pictorial spaces we are dealing with are expressed in geometrical models. Each model is expressed as a geometrical transformation applied to the Cartesian/Euclidean shapes of the physical environment. Such a transformation shows how the shapes that characterize our physical environment would appear in the pictorial space of the model. A model of this kind pairs physical shapes—Euclidean by standard criteria—with their corresponding visual shapes in the pictorial space of the model. From a model arise suggestions as to where to look, what to look for, and how to order an inquiry so that the coordinated structure of the space of our pictorial perception can be understood.[15]

For the purpose of opening our eyes to what our preconceptions systematically blind us to, a graphic metaphor, like Arnheim's "pyramidal space,"[16] could serve in place of a model, but not as well. A metaphor, after all, is a kind of informal model just as a model is a kind of quantitative metaphor. I shall begin by discussing Arnheim's "pyramidal space."

Arnheim's "Pyramidal Space"

Arnheim describes the experience of visual pictorial space by means of a metaphor, "pyramidal space." "Imagine," he says, "one side of [a very large cube surrounding the observer] shrinking to the size of a point. The result will be an infinitely large pyramid. (It should be understood that I am not talking about an interior of pyramidal shape contained in the usual 'cubic' world of our reasoning, but of a world which is pyramidal itself.) Such a world is non-Euclidean."[17] Visual space, he claims, has the quality of a pyramidal space. He continues,

> There are what might be called "Newtonian oases" in perceptual space. Within a frontal plane, space is approximately Euclidean; and up to a few yards of distance from the observer, shape and size are actually seen as unchangeable. It is from these areas that our visual reasoning obtains confirmation when, at an elementary level of spatial differentiation, it conceives size, shape and speed as independent of location. But even in a more clearly pyramidal world, the relations to the framework are perceived so directly that it is all but impossible for the naive observer to "see in perspective"; because seeing in perspective means perceiving the inhomogeneous world as a distorted homogeneous one, in which the effect of depth appears as the same kind of crookedness that we observe when a twisted thing is seen in a frontal plane.[18]

Much of the substance of what Arnheim describes through the construction of a "pyramidal space," but refined and corrected as to formal structure, will be derived from the model of visual space to be introduced below.

Arnheim considers the spatial framework to be a set of (visual or imagined) standards, spread throughout space; the standards, however, are not rigid as in physical space, but systematically altered by the pyramidal structure. His rule of congruence, like the rule of congruence for physical space, is founded on local congruence between an object to be measured and the on-site, now non-Euclidean, standard. This description of a non-Euclidean reference frame, reminiscent of H. Poincaré and H. Reichenbach, as a kind of ideal reticular structure that divides space into reference boxes of unit size—*unit* by definition or formula—is, as it were, the reverse of the model I am proposing below. In this, the rule of congruence generates unit lengths anywhere in space by comparative visual estimation of the object. This model as mathematically developed will study the relationship between unit physical (Euclidean) boxes and the transforms of these boxes in visual space, or the transformation of the standard physical (Euclidean) reticular structure into its counterpart in visual space, that is, a reticular structure with points coordinated to the Cartesian network of physical space.

Pure Geometries

The work of K. Gauss, J. Bolyai, and N. Lobachevsky in the early part of the nineteenth century and of B. Riemann, H. von Helmholtz, and W. K. Clifford in the latter half[19] showed that geometries could be based on axioms other than those of Euclid. This transformed the notion of geometry by exhibiting its abstract postulational character and by removing it from any necessary connection with the structure of empirical intuition or with physical or mechanical constructions. Let us call these products of mathematical construction "pure geometries"; they do not originate as organizing forms of sensible intuition or the experience of any class of physical objects.

As for the space of physical reality, "Geometry," as Clifford remarked, "is a physical science."[20] It is a key principle about geometries that the character of a geometry is uniquely defined by the distance function, the assignment of a specific nonnegative magnitude to every pair of points in the space. If we are dealing with the space of physical objects, then the geometry of physical space will be determined by a *rule of congruence*, that is, the rule according to which distances are assigned to intervals by comparing them with a standard unit interval: this process is called the "measurement of the interval." For example, the usual rule of congruence for physical space is comparison of the object to be measured with the graduations on a rigid ruler transported in a force-free environment to the location of the object to be measured and brought to rest there (or by some process equivalent to this).

Bypassing the debates about whether or not Euclidean geometry deserves a privileged a priori status,[21] there is general agreement that Euclidean geometry alone among the "pure geometries" is prima facie related to the sensible experience of the everyday World (of middle-size things) in which we presently live. In it, lengths and distances are measured by transportable rigid rulers or their equivalent, and the standard against which rigidity is determined is a Euclidean one. Cosmological space, the very large, has its own geometry and, as for the microscopic space of quarks and elementary particles, it too may have its own peculiar geometry; but though these spatial structures are in the domain of science, nevertheless, for all but possibly a very few, neither cosmological space nor the space of quarks actually constitutes a horizon of geometrical perception. For a very few scientific investigators, however, they may indeed—according to the account of perception to be developed below—constitute genuine though very unusual perceptual horizons.[22]

Non-Euclidean Visual Space

The necessary and a priori character of our Euclidean perception and visual imagination was challenged in the last century by the experimental work of H. von Helmholtz,[23] F. Hillebrand,[24] and W. Blumenfeld.[25] It was demonstrated that, when normal observers are presented with a configuration of points of light dispersed in an otherwise dark background, they tend to construe the spatial organization of the configuration in a way not consistent with Euclidean geometry. Ernst Mach, and after him Merleau-Ponty and others, also noted the differences between the homogeneous Euclidean character of scientific space and the nonhomogeneous non-Euclidean character of visual space— what Mach called "physiological space" or the space of our sensations—but neither Mach nor Merleau-Ponty investigated the systematic structure of size, depth, and distance relationships in this space, which is best done by means of a metric model.[26]

Luneburg's Theory of Hyperbolic Visual Space

The first fully developed metric theory of visual space was worked out by Rudolf Luneburg.[27] He argued that when a stationary binocular observer, with the position of the head fixed, is presented with points of light shown against a dark background space, he will construe the scene visually as if the depth and distance of the lights from him were given as a function of binocular parallax alone.[28] He showed that it was plausible in principle to hold that the geometrical structure of visual space is a metric space of constant, probably negative, Gaussian curvature—what is called a "hyperbolic space."[29]

Free Mobility of Forms

Luneburg argued that visual space must be one of constant curvature. The reason he gave was the requirement for *free movability* or *mobility of forms*, that is, that it should be possible to construct at any position in visual space a form congruent with an arbitrary perceived form situated at some other place.[30] This condition would make possible "rigid motions" in visual space, that is, the undistorted displacement of shapes from one place to another. A "rigid motion" in visual space would not, however, be a rigid motion in physical space, nor would the opposite be true. A *physically* rigid body—that is, a body that retains its size and shape when it is moved arbitrarily in physical space, such as a cubic box—would not retain its visual shape when displaced: such a body would change shape according to definite laws when moved in the field of vision. Likewise a *visually* "rigid" body— that is, a body that retains its visual size and shape when moved arbitrarily in visual space—would not retain its physical shape under displacements in visual space. There is probably no example of a visually "rigid" body in ordinary experience. Some of the systematic ways in which shapes would be transformed will be discussed qualitatively below; the exact mathematical treatment will be found in the Appendix.

In any geometrical space, there is associated with every place (such as an extended neighborhood around a point) a definite repertory of constructible shapes, that is, of shapes that could be constructed there consonant with the geometry. Different places in general have different repertories of constructible shapes except for constant curvature spaces, where these repertories are everywhere the same. The shapes in these repertories, moreover, are characteristic of the geometry of the space, so that Euclidean repertories contain only Euclidean figures, non-Euclidean repertories contain only non-Euclidean figures. No Euclidean shape is at the same time a non-Euclidean shape of similar description in a space of equal dimensions; consequently, there is no overlap between the repertories of Euclidean three-dimensional (3D) shapes and repertories of 3D non-Euclidean shapes. Consider, for example, the surface of a globe (this surface is a 2D non-Euclidean space of constant positive curvature) and a non-Euclidean circle on this surface. The latter is a figure with its center, say, at the N (or S) pole and radii measured along the lines of longitude: it is then part of the surface of the sphere and is bounded by a parallel of latitude. One might be inclined to think of it as also a Euclidean figure, but as a Euclidean figure it is not a circle; moreover, for its existence as a Euclidean figure, it requires a 3D Euclidean space. In contrast, a Euclidean circle bounded by the same periphery as the non-Euclidean circle would lie entirely in a plane intersecting the sphere along the parallel of latitude; its center and radii would lie in this plane. The example shows that though 2D non-Euclidean figures may be mapped isomorphically into a Euclidean space, this space is of a higher order and its geometrical description is different in the higher order space. Moving to three dimensions, 3D non-Euclidean figures cannot be mapped isomorphically into 3D Euclidean space and so do not exist in this space: to exist as Euclidean figures, they would need a Euclidean space of higher dimensions in which, moreover, their geometrical descriptions would be changed. The conclusion is that there are no common elements with common geometrical descriptions in the repertories of 3D Euclidean and non-Euclidean shapes.

Since the surface of a sphere is a 2D non-Euclidean space of constant positive (Gaussian) curvature, any shape can be "moved" undistorted from place to place on the sphere: the reason for this is that among the repertory of shapes constructible at that place there is always a shape congruent with any chosen shape constructible elsewhere. On the contrary, if, instead of a sphere, one took a spheroid (a spheroid is, roughly, an egg-shaped object), then—since a spheroid does not have a uniform (Gaussian) curvature—one would find no single repertory of constructible shapes for all neighborhoods of its surface. Consequently, there is not overall free mobility of forms on a spheroidal surface; because of its rotational symmetry, however, there will be free mobility of a limited sort parallel to the lines of latitude.

The requirement of free mobility of visual forms which Luneburg considered theoretically necessary for visual space was probably motivated by the phenomenon of shape and size constancy in perception. By and large in ordinary everyday perception things in one's immediate neighborhood preserve their shapes and sizes when they are moved relative to one as an observer. Any "illusion" of distortion can be corrected by an appropriate judgment or by the directed use of the imagination if the object is a familiar one. Luneburg wanted to be able to preserve as valid for visual space some concept analogous to that of *rigid body* in physical space.

Failure of Luneburg's Model

Luneburg assumed that the structure of visual space was fixed for each individual observer, independently of hermeneutical considerations, probably by the individual's genetic heritage. This assumption implied that the parameters of visual space were independent both of the field of perceptual objects and of particular stimulus configurations. As it turned out, this was a false assumption. A. A. Blank, L. Hardy, T. Shipley, D. Williams, T. Indow, E. Inoue, K. Matsushima, and J. M. Foley,[31] experimenting with faintly illuminated points of light in a dark surrounding field, found that no fixed set of parameters was obtained for individual subjects. Subjects had been instructed to perform some standard geometrical operations from which the parameters of the assumed overall geometry were then determined.

Although some attempt was made by A. Linksz, Shipley, Williams, Hardy, Foley, and others[32] to modify the theory by introducing configuration-dependent parameters into the theory, such as the co-ordinates of the nearest and/or farthest points in the field of vision, this was not successful. The thrust of the research program moved away then from the global geometrical structure of visual space to the solution of local problems, such as how perceived size is related to physical size and so on, problems that were separate and disjoint from one another, and unrelated to the context of any overall geometric model. In the absence of an overall geometric model to coordinate experimental design and interpretation, it has become impossible to assess whether or not recent experimental findings do or do not lend support to any specific model of visual space, since published accounts generally give insufficient information to test hypotheses about the global geometric structure of visual space.

It is a pity that, with the virtual collapse of the Luneburg model and the piecemeal direction of most research on perceptual relationships, the study of the overall geometric properties of visual space has fallen into neglect. I have found that with some relaxation of the conditions that Luneburg incorporated in his model quantitative results obtained, for example, by Foley and W. C. Gogel, do fit a revised hyperbolic model.[33]

Given the partial, though inconclusive, success of the program itself,[34] and given the fact that there is much informal evidence in favor of the hyperbolic structure of visual space, one suspects that the experimental program though affected by internal problems is nevertheless fundamentally sound. These problems are, I believe, threefold. (1) The model itself lacks flexibility; Luneburg's view that for any individual the curvatures of visual space were fixed personal constants was surely too restrictive. (2) The design of the experiments assumed that visual space was unique for a given individual or caused unhermeneutically by stimuli from the environment; failing then to take into account the hermeneutical character of perception, the experiments failed to create for the subjects conditions necessary for a stable visual space. (3) Some data from the experiments express the effect of something like a multistable illusion, at one moment controlled by a Euclidean structure and at another by a hyperbolic structure; this could be another indication of the failure to consider hermeneutical effects.

Experimenters in the Luneburg research program gradually became aware of these problems, and tried to remedy at least some of them. From a model with fixed parameters for each subject, there was substituted a model with parameters dependent on some purely physical aspect of the stimulus—the nearest and/or the farthest source of optical energy. As for the possible effects of cognitive clues, there was early dispute as to whether the experiments should be done in natural surroundings (within the Life World) or in artificial surroundings where all clues were banished except for a limited number of pointlike (or rodlike) light sources against a dark background. It is also clear from Foley's summary of the achievements of the program that experimental data were not unaffected by Euclidean structures and progressively showed the effect of these as more real life clues were added.[35] What is not clear is whether the Euclidean structure begins to impose itself on a premetric or pregeometric situation or whether it gradually replaces a pre-Euclidean hyperbolic structure.

Making the assumption that a visual observer has at his disposal two kinds of spaces, a Euclidean space and a family of hyperbolic spaces, one asks: what determines the perceiver to construe a situation in one space or another? I take the answer to this question to be fundamentally hermeneutical, a response of the subject guided by an assessment, perhaps indeliberate (the term "unconscious" is sometimes used), of what makes good visual sense of the situation as a perceptual opportunity. To make good visual sense, specific sets of conditions have to be fulfilled.

For a situation to provide a Euclidean perceptual opportunity, the visual observer—independently of where he/she is physically—must be able to recognize within the visual field which visual objects are of equal physical size, and which visual distances are of equal physical measure. The measure of equality in ordinary experience is on-site coincidence with the graduations on a stationary transported rigid ruler (or some equivalent process of measurement). For a World to appear Euclidean to a visual observer, it must then be virtually populated with familiar (stationary) standards of length and distance, and be equipped with instantaneous means for communicating information about coincidences from all parts of space to the localized visual observer, wherever he/she happens to be.

For a situation to provide a hyperbolic perceptual opportunity, the visual observer must be able to use the rule of congruence which, it is claimed, is embodied in the capacity of the unaided visual system to order the sizes, depths, and distances of all objects in the unified spatial field of vision. This is done by purely visual estimation; whether the standard of length is the one intrinsic to the space or whether any fixed but arbitrary local standard can be used is something that would require empirical investigation.[36] If a local standard were used, it would presumably fulfill certain conditions relating to its significance as a standard, such as that it be an object of stable length, visually and physically.

With these considerations in mind, consider the typical experimental setup for investigating the structure of visual space: subjects are isolated from their surroundings in a dark room and presented with a distribution of pointlike light sources; they are asked to manipulate the positions of the light sources by remote control so that certain specified size, depth, or distance relationships are fulfilled according to the best of their visual judgment. The clues do not fit a Euclidean hermeneutic, for there are no on-site rigid standards of length in the field of vision.[37] However, neither are the clues apt to be recognized as appropriate for a hyperbolic geometry, since the perceptual field also lacks anything that could be interpreted as a significant local standard of length relative to which the surrounding environment could be spatially structured. There is an absence then of significant feature-specific clues in the optical field,[38] clues presumed necessary on a hermeneutical theory of perception to evoke indeliberately a determinate spatial horizon.[39] The typical experimental setup is one that is hermeneutically ambiguous or indeterminate.

Hermeneutical Visual Model

I propose, then, to modify Luneburg's research program by taking account of the hermeneutical aspect of space perception. In this hermeneutical model, I assume that the parameters are dependent both (*a*) on some overall physical structure of the configuration of stimulus information and (*b*) on the recognition or prerecognition in the visual field of cognitive spatial clues.

With respect to (*a*), I take the relevant stimulus field to be made up of three psychophysical variables, two angles (α, β) expressing orientation relative to the observer and some measure of infinitesimal visual parallax (γ).[40]

The meaning of (*b*) is that, in any perceptual situation, we operate out of that preunderstanding that makes sense of our experience, knowing that objects become manifest only when the appropriate horizon is chosen for questioning, and that the horizon is appropriate only because it makes sense of the profiles of objects that manifest their presence in experience. This search for understanding is called a "hermeneutical circle." The character and conditions of such preunderstandings will be taken up later in the book.[41]

Visual space is, as it were, the open but structured container of the visible which is a condition of possibility of perceptual objects. As the container of possible perceptual objects, it stands as a background to be differentiated from the actual objects and events that take their place in it as foreground and borrow their geometrical conditions from it. Object and space are a duality like foreground and background, and imply two kinds of cognitive visual cues or clues:[42] *primary clues* (or *space indicators*) that serve to determine the structure of the background space (e.g., by determining its curvature parameters and an appropriate scale factor), and *secondary clues* that serve merely to identify the foreground objects and events within this space. Primary clues and secondary clues relate to one another within a hermeneutical circle. Primary clues serve to complete the specification of the intentionality structure that makes a coherent space of perception possible. The laws of that space then impose themselves indeliberately and prepredicatively on whatever else is perceived, making present through the secondary clues only such forms as belong to its repertory.

It seems reasonable under most circumstances to locate the primary clues for hyperbolic vision in a group of local cognitively significant objects in the field of visual objects: these link the subject with a

World, or a horizon of a World. A field of vision, for example, may be limited to the inside of a room: it might be supposed that a fixed shape and size for the walls of the room "suggest themselves" to the observer (through a reading of the appropriate stimulus) as essential to the structure of the background space, and that other objects in the room, whether stationary like table lamps or mobile like people, "suggest themselves" as secondary or foreground objects.[43] Primary clues, in this case, the shape and size of walls, and so forth, are such that as long as they are held constant, the geometry of the spatial background is fixed in principle; such a geometry could then be tested by manipulating foreground events or objects. Since these latter are perceived to be embedded in the background space, the geometry of the background space can be inferred from the *perceived* relations between foreground objects, assuming of course that their *physical* relation is already known.

The hermeneutical model of visual space supposes that visual space can take on any one of a family of geometries depending on the hermeneutical context of foreground and background, that is, of object and containing space.[44] This variety is represented by the range of possible values that σ and τ (or κ and μ; see the Appendix) can take in the model: σ in the model is associated with the egocentric distance of a local standard of length (that is, of a ruler whose visual length could be equated with its physical length), κ is an intrinsic unit of length equal to the radius of curvature of the space, while τ and μ in the model are both associated with the variation of depth in visual space. Foley[45] has reported that experimental data argue for the existence of two such independent signals in the input stimulus, an egocentric distance signal responsible for estimates of egocentric distance and a disparity signal responsible for estimates of depth, both signals being functions of the γ-field. Two parameters of the hermeneutical model, such as σ and τ (or κ and μ), reflect these two aspects of the shaping of visual objects in visual space. In any particular visual situation, σ and τ (or κ and μ) would presumably take on some set of values that make visual sense of the contextual circumstances of vision.

With respect to the binocularity of human vision, I am adopting the viewpoint suggested by the theory of neural coding or representations, as specified, say, by William Uttal.[46] He writes of stereopsis:

> The two images are not "fused" nor "suppressed," but instead
> the incoming barrage of neural data from each eye may be thought

of as contributing to a common pool of information that leads to the emergence of a percept of a single object in depth.[47]

The neurophysiological viewpoint of Uttal is complemented by the phenomenological viewpoint of Merleau-Ponty, who writes that the unity of the object in binocular vision is "intentional": it constitutes an object in the Life World of the perceiver, to which the perceiver relates through his Body. The synthesis then is neither purely intellectual nor impersonal but comes from a sharing in the goals and intentionalities of a common World.[48]

Stability of Visual Metric

The family of metric spaces I am proposing for visual space is the two-parameter family of non-Euclidean hyperbolic metric spaces:

$$ds'^2 = \kappa^2 (\sigma^2 \, d\gamma^2 + d\alpha^2 + \cos^2\alpha d\beta^2)/\sinh^2 \sigma(\gamma + \tau)$$

where ds' is the (visually) estimated length of a small line segment in visual space. This family is identical mathematically with the spaces considered by Luneburg: the interpretation I give to the model is, however, different in the ways mentioned above. The metric as written above depends on three parameters, κ, a scale factor equal to the radius of curvature of the space, σ, and τ, but only two out of these three are independent. Assuming size and shape congruence between visual and physical space at the "true point"—the center of the "Newtonian oasis," about which more below—then the relation between κ, σ, and τ, is:

$$\kappa = \sigma \sinh (1 + \sigma\tau)$$

The psychophysical coordinates γ, α, and β are determined by *physical measurement* on the objects in the visual field (see figure 3.1 and the explanation of the mathematical model in the Appendix): each is a function of the classically measured Cartesian coordinates, x, y, and z of the point. γ is some *appropriate measure of the infinitesimal parallax* of a point relative to the visual observer, α is the azimuth and β the elevation.

The spaces are homogeneous, and this I believe is plausible to a first approximation.[49] The argument in favor of spaces of negative rather than positive curvature is surely convincing for spaces with infinite

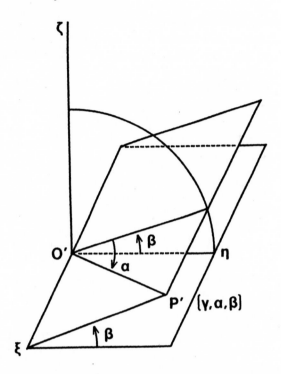

Figure 3.1: Coordinates for visual space.

physical horizons—for this argument, see the Appendix. The usual assumptions of the Luneburg program are made.[50] Considering then the family of hyperbolic visual spaces, it is clear that the number of free independent geometrical conditions that the space indicators can impose on the hyperbolic metric is two. Such conditions, for example, might be the constancy of the radius of curvature (the internal standard of length for the space), eiseikonic invariance,[51] the location of the "true point," the egocentric distance of a significant feature of the field, the visual distance of the limits of visual space (the "horizon sphere"), and so forth. If in any perceptual situation, the perceivable configurations (possible profiles of objects in the space) are subject to two independent geometrical conditions, then, that would be sufficient to determine completely the parameters of the visual space. To what extent the visual observer can establish deliberate and conscious control over the invariant geometrical conditions of visual space (and consequently of the parameters of that space) and to what extent these factors are determined indeliberately (by transcendental conditions, by past learned experience, etc.), as well as the range of values for the parameters of possible visual spaces, all remain matters to be researched empirically.

Luneburg, because of his views about the import of binocularity for depth perception, took γ to be the binocular parallax using the interpupillary distance as the base; but a consideration of the form of the equations (see the Appendix) shows that the base length for the parallax, provided it is small, does not affect the geometrical transformation outside of the very close neighborhood of the eyes, in which visual judgments are vague anyway. Thus, the model can be taken as equally monocular as binocular, since even in monocular vision there exists a measure of convergence (based on the diameter of the pupil) for a pencil of rays from the object. If then we are dealing with a single perceptual process, the greater accuracy of binocular distance estimation would be explained by the more abundant and precise input of information generated by the larger parallax angles.

It is important to note that, contrary to the views expressed in the literature, in this kind of model neither the interpupillary distance nor any other specific physical measurement of the body functions as an intrinsic measure of length for the visual space that is hyperbolic. The only true intrinsic basic measure of length in such visual spaces is the radius of curvature, κ, of the space.

Finally, although some writers conceive of the Luneburg program as a scientific-explanatory theory in which the phenomenal form of visual space is linked to a particular set of psychophysical parameters, and to structures of the retina and visual cortex, the scientific-explanatory potential of the theory is not any part of the concerns or argument of the first part of this book.[52] I am using the family of hyperbolic space transforms of physical space in a purely descriptive way: I want to see whether and to what extent they correctly describe the spatial character of a class of anomalous perceptions that are both persistent and pervasive in ordinary human experience. The development of the mathematical model and its linkage with physical space is outlined in the Appendix, where also an assessment is made of the support to be derived from experimental results obtained for the most part under the kind of specialized and artificial conditions that I have criticized above.

Hyperbolic Space: The Model

Model of Hyperbolic Visual Space

The hermeneutical Luneburg theory is a model of visual space, (1) in the form of a hyperbolic non-Euclidean metric space; (2) linked to physical space (represented by a Cartesian model) by specific transformation laws based on what are believed to be the relevant psychophysical parameters in the optical environment; (3) where it is assumed that the parameters of visual space are governed by hermeneutical considerations (what makes visual sense of the environment), such as (*a*) the need for a standard for size, depth, and distance, (*b*) the projection of this standard by a process of visual measurement across the whole of visible space, (*c*) *Gestalt*-type relationships between foreground and background, near zone and distant zone, which are functions of the purpose of perception, and so forth.

With each geometric space, there is associated a specific repertory of possible shapes which contains exactly those shapes that are constructible in principle in that particular geometric space and no other shapes. As explained above, the repertory of hyperbolic 3D space contains no Euclidean 3D element and the repertory of Euclidean 3D space contains no hyperbolic 3D element. If then it is the case that visual space is hyperbolic, a small physical cube will not appear to be a visual cube (except under conditions to be stated below), nor will its visual shape be a fixed one. Its visual shape will change with the relative location and orientation of the cube relative to the viewer.

On the assumption that visual space is hyperbolic and physical space is Euclidean, expressions for the visual shape of a small physical cube located at a point P (x, y, z) in physical space have been derived from the metric: formulae for the visual lengths of its edges, for the visual angles between its edges, for the visual shape of its faces, and for the tilt of its faces relative to visual verticals and horizontals. The mathematical work is given below in the Appendix. From the formulae for the transform of a physical cube, it is possible to derive by computation numerical models (or mappings), which yield on interpretation the typical visual shapes of many familiar engineered forms. From a study of these models, we can get a good visual sense of the way our "carpentered environment" would manifest itself to hyperbolic vision. I shall attempt to present below a qualitative summary of the principal features of these models or mappings for a variety of values for the parameters of the spaces.

The True Point and the "Newtonian Oasis"

The most important theorem about the relationship between physical and visual space is that there always exists a region surrounding some definite point directly in front of the observer in which visual and physical sizes and shapes roughly coincide. I shall call this point the "true point" for a particular space. The theorem says that the value of the hyperbolic parameter σ equals d, the distance (by physical measurement) of the true point from the observer. It is clear that if the true point is occupied, this is where the local scaling standard for size, depth, and distance is located for the space of that particular environment and intentionality-structure. Experimental studies have not yet established the extreme limits of the range of possible veridical size and shape perception; these limits would be important to establish for the hyperbolic model. H. Ono, for example, reports that veridical depth perception occurred at least up to 200 centimeters from the observer for the conditions of his experiments.[1] The region about the true point would be Arnheim's "Newtonian oasis" of visual space. In the hermeneutical model, the position of the true point relative to the observer depends on the visual purpose of the perceptual field.

Finite and Infinite Visual Spaces

The class of hyperbolic visual spaces is partitioned into finite spaces ($\tau > 0$) and infinite spaces ($\tau = 0$).

For *finite spaces*, the whole of Euclidean space is mapped on the interior of a finite hyperbolic sphere about the observer. The interior of the hyperbolic sphere is, of course, the whole of visual space: although the limit of visual possibility is at a finite distance from the viewer, visual space has no edge or periphery. The limiting sphere on which Euclidean infinity is mapped is the "horizon sphere." For finite spaces, the visual size of any object of finite physical size reduces to a point on the horizon sphere.

Parallel lines in a horizontal plane would at first appear to diverge; they would reach a maximum separation at a distance of about 2σ from the viewer, and then begin to turn inward towards each other, eventually meeting at a point on the horizon sphere at a finite distance from the viewer. All physical lines in any direction converge on a single point on the horizon sphere—a kind of antipodal point to the sight point of the observer for that viewing direction.[2]

Infinite visual spaces are the closest to Euclidean physical space; nevertheless, there are significant qualitative differences between them. An object moving at a uniform speed away from the viewer in physical space would appear to move more and more slowly in visual space: the visual distance covered being roughly as the logarithm of the physical distance. Consequently, visual depth would appear to get progressively shallower with distance from the viewer, and would finally disappear at visual infinity on the horizon sphere.

Parallel lines in a horizontal plane would appear to diverge, at first rapidly near the observer, and then more slowly until at infinity they reached their maximum separation.

Despite the fact that infinite hyperbolic spaces are closest to infinite Cartesian space, visual data suggest that visual space is usually finite.

Horizon Sphere

The *horizon sphere* is the theoretical limit of visibility: it is the spherical surface with the viewer at its center, on which Euclidean infinity is mapped. Objects on the horizon sphere are infinitely far away in physical space.

It is a common experience to perceive the sky, seen by day as the background for the sun, moon, and clouds, and by night as the setting for many heavenly bodies large and small, as a sort of limit to our visual World. It is tempting simply to identify the sky as perceived as the horizon sphere, but this identification cannot be made. (1) If visual space is finite—and there is much evidence that this is the case—then objects of finite visual size on the horizon sphere would have to be infinite in physical size and infinitely far away; now the sun, moon, and planets, clouds, and so forth are objects of finite visual size in the sky, hence the sky cannot be the horizon sphere as long as such objects are experienced as physically finite in Euclidean vision. Moreover, (2) it has often been noticed that the sky is perceived not as a sphere but as a *flattened* vaulted ceiling. Both of these phenomena will be further discussed below.

"Pyramidal Structure" of Finite Visual Space

Figures 4.1 and 4.2 display in a very schematic way the "pyramidal structure" of a finite hyperbolic visual space. It is the counterpart in our model of Arnheim's metaphor of a "pyramidal space."

Figure 4.1 is a diagram (of the visual transforms) of the vertical cross-section of a set of larger and larger nested cubes surrounding the observer. The largest—infinite—cube is transformed into the horizon sphere. P' is the point on the horizon sphere directly in front of the observer: it is the farthest one could possibly see in this direction, and represents, of course, a physical infinite distance. Smaller cubes have the shapes of truncated pyramids with sides and faces concave to the observer (broken lines). Continuations of the upper and lower horizontal lines of the cross-section (dotted in the diagram) would meet at P' on the horizon sphere.

Figure 4.2 is a diagram of the horizontal cross-section of the same set of nested cubes surrounding the observer. The contours of the transformed shapes are like those of the transformed vertical shapes, but the truncated pyramids have a wider base. Thus, the face of a large cube will appear to be wider than it is high, slightly concave to the observer, and enclosed in segments of curved contours. This gives a slight oval or spheroidal shape to the visual field.

Determination of Parameters of Model: σ

Model: Given a perceptual setting that evokes a stable visual space, how would one determine the parameters σ and τ of the appropriate hyperbolic model?

Application of the Model: The value of σ is, in principle, easy to determine: it is the distance to the point in front of the observer where the visual standard of size, depth, and distance for the whole space

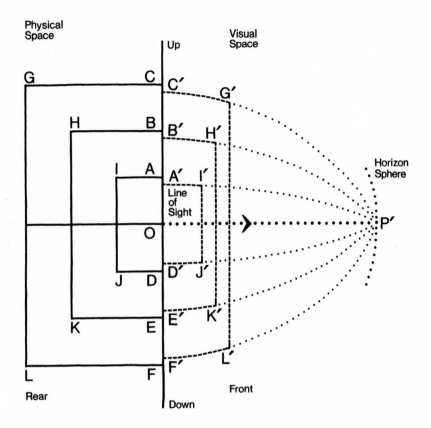

Figure 4.1: Vertical section: schematic diagram of the way vertical cross-sections of a series of larger and larger physical cubes (full lines) would be perceived in visual space (dashed lines). *O* is the observer. *A* and *A'*, etc. are corresponding points in physical and visual space. *P'* is the point on the horizon sphere directly in front of the observer. Actual values are graphed in figures 9–10 of the Appendix for a series of such hyperbolic spaces.

would be located. This is the true point where the form of a small object, for example, a cube, would be roughly congruent with its physical form.

For a small enclosed space, such as the interior of a room, the value of σ should be between $d/2$ and d, where d is the distance to the farthest visible point at eye level in the room. This estimate follows from the hermeneutical assumption that the interior of the room falls entirely

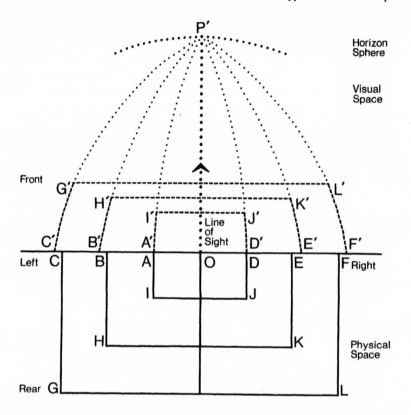

Figure 4.2: Horizontal section: schematic diagram of the way horizontal cross-sections of a series of larger and larger physical cubes (full lines) would be perceived in visual space (dashed lines). *O* is the observer. *A* and *A'*, etc. are corresponding points in physical and visual space. *P'* is the point on the horizon sphere directly in front of the observer.

within the *near zone* and the assumption—on its surface, plausible—that the true point does not lie outside the room.[3]

Near Zone and Distant Zone: τ

Model: The most significant qualitative characteristic of the model is the natural division of the visual field into a near zone and distant zone.[4]

The *near zone* is the region of the visual field which includes the true point and extends roughly a distance 2*d* in front of the observer: it would extend to right and left, and up and down, embracing a visual cone the extent of which would depend on one's choice of a measure of significant distortion. Within it, the visual dimensions would be clearly articulated and would not differ greatly from their physical dimensions.

The *distant zone* is the region of the visual field which lies outside the near zone and extends to the horizon sphere. In front of the viewer objects would be at distances greater than 2*d*; its general characteristic feature is shallowness of depth. Since our physical anticipations of shape are unrealizable in the shallow space of the distant visual zone, we may find our perception tricked by many ambiguities and illusions arising out of the tension between physical expectations and visual impossibilities.

The value of τ is related to the depth characteristics of the distant zone and—for plausible visual models—establishes the visual limits of space. For instance, when $\tau = 0$, the space is infinite and the horizon sphere is at (logarithmic) infinity; when $\tau > 0$ (it is always non-negative), the space is finite and the radial distance to the surface of the horizon sphere is finite. For finite spaces, the maximum visual depth in the distant zone is finite and a function of the value of τ.

Application of the Model: A discussion of the significance for the everyday World of the distinction between near and distant visual zones will be postponed to the end of this chapter.

Near Zone

The near zone will have certain specific characteristics in addition to those already mentioned.

Model: Although visual dimensions are clearly articulated within the near zone, their relationship to physical dimensions varies from point to point. In the domain between the viewer and the true point, size (i.e., length and breadth) visually contracts or "shrinks," while depth expands or "swells." For example, figure 4.4 shows the visual forms of two horizontal parallel lines, *AB* and *CD*, to right and left of the observer at eye level: their visual transforms, $A'P'$ and $C'P'$, diverge in the near zone and reach a maximum separation at a distance of roughly 2σ, they then begin to converge, $M'N'$ being the turning point marking the transition to the distant zone. In the distant zone, the parallel lines eventually meet at P', a point on the horizon sphere. Figure 4.4 is drawn for a finite space. For infinite spaces, however, the visual transforms of the parallel lines will diverge, rapidly at first near the observer, then more slowly to a maximum width attained on the horizon sphere. In both finite and infinite spaces, figures 4.3 and 4.4 illustrate how depth in the zone between the viewer and the true point is stretched.

Comments: The divergence of parallel lines in the near zone can be experienced by holding a small rectangular card, such as a three-by-five-inch index card, horizontally in front of the eyes. It is best to rest it on one's closed fist and to direct one's gaze at the center of the far edge of the card. By moving the card outward in depth, one can often locate the turning point where the diverging lines begin to converge: the existence of a turning point indicates, of course, a finite space. Two lines can be drawn on the card parallel to the edges and near the center line of the card, and the experience repeated. The lines and the edges of the card present when viewed from different distances a rich experience of converging, diverging, and dynamically curving shapes.

Some photographic distortions can be explained as related to phenomena of the near zone. For example, a photograph of a hand outstretched towards the camera lens appears to depict a grossly enlarged hand.[5] In real life, however, an outstretched hand does not appear to be grotesquely large; its apparent size has "shrunk": this "shrinking" is a phenomenon of the near zone. The "barrel illusion"[6] may also be explained as a phenomenon of hyperbolic space: this is the illusion that a checkerboard pattern seems to "swell out like a shield"[7] when one approaches it.

The divergence of the lines in the near zone suggests an origin for systems of reversed perspective, such as those used, for example, by early naturalistic artists of the Middle Ages. Reverse perspective could have been suggested by careful attention to visual phenomena associated with the use of small-scale models: we know that right up to the nineteenth century artists made basic drawings from such models rather than from life.[8]

Distant Zone

Model: In the first place, in both finite and infinite spaces, there would be a very significant loss of depth perception with distance. Very distant objects would appear to have little discernible depth or thickness, papered, as it were, on the inside of a large surrounding sphere. Figures 4.3 and 4.4 illustrate the foreshortening of depth in a finite space.

Comments: Given the shallowness of visual depth in the distant zone, objects with significant physical depth will appear as affected with unresolvable ambiguities of shape, size, and orientation. Since our anticipations about physical reality are Cartesian, and since, as we can confidently assume, there is a limit to our visual powers of parallax resolution, the distant zone will be characterized by perceptual ambiguity. Ballard's phenomenological study of the essential forms of *visual closeness* and *visual distance* provide remarkable confirmation of the predictions of the model.[9]

Model: In the second place, with loss of depth, all visible facets of a distant surface would seem to turn so as to face the perceiver in a more or less frontal way. Even visible surfaces that we know to be orthogonal to one another, such as the sides of a house, would appear to be turned towards the perceiver. The loss of depth and the apparent

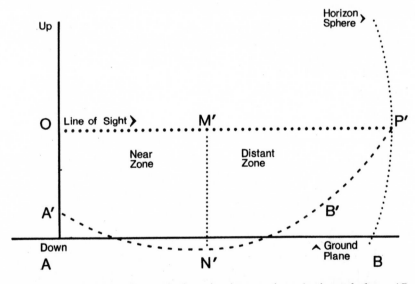

Figure 4.3: Diagram of a vertical section intersecting a horizontal plane, *AB*, below eye level: *A'B'* is the visual transform of *AB* in a finite hyperbolic model of visual space. *M'N'* marks the transition between the near and the distant zone. Contours are modeled on computed values. The mapping is not (and cannot be made) isomorphic with the hyperbolic shapes, but some size and distance relationships are preserved.

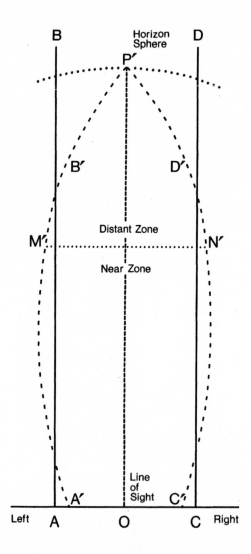

Figure 4.4: Diagram of a horizontal plane at eye level with two orthogonals, *AB* and *CD*, to the frontal plane *AOC* (*O* is the observer): *A'B'* and *C'D'* are the visual transforms of *AB* and *CD* in a finite hyperbolic model of visual space. *M'N'* marks the transition between the near and the distant zone. Contours are modeled on computed values. The mapping is not (and cannot be made) isomorphic with the hyperbolic shapes, but some size and distance relationships are preserved.

reorientation of the planes to face the viewer are characteristics of telephoto snapshots. The effect of hyperbolic vision on distant objects is very like the telephoto effect; neither, however, requires that we see parts of the rotated surface hidden from the sight point but only that the visible parts of the surface appear to have been rotated relative to the line of sight. In addition, facets of a distant surface would also show a hollow or concave face to the observer.

Comments: Distant objects are perceived with a noticeable telephoto effect; that is, they are brought closer, depth between distant objects is foreshortened, and plane facets are turned toward the viewer. In Euclidean visual space, objects so viewed would naturally appear larger in keeping with the Euclidean constancy scaling laws, but in hyperbolic space, the objects—compared with nearby objects of identical physical size—would appear smaller. This paradoxical quality of distant objects appearing to be closer and nevertheless smaller than real life is found, for example, in the paintings of van Gogh, Tintoretto, and Canaletto.[10] Aspects of the telephoto effect have already been discussed briefly in the painting *Turning Road at Roche-Guyon* of Cézanne.

Horizontal Planes

Model: Figure 4.3 represents a section of a flat physical plane below eye level, such as flat ground seen from above or the sea as seen from a boat: its visual transform is $A'P'$. The flat plane is converted into a bowllike shape with the perceiver at the center and the horizon for its edge. The visual distance of the horizon, OP', will depend on the value of τ. Figure 4.4 shows the horizontal intersection of two vertical planes parallel to the median plane. They diverge in the near zone for both finite and infinite spaces. For finite spaces (illustrated in the figure), the lines reach their maximum separation at the boundary of the near and distant zones; then they begin to bend toward each other, eventually meeting at the horizon sphere. For infinite spaces, the lines continue to diverge, but more slowly, until they reach their maximum separation at infinity on the horizon sphere.

Comments: The visual transformation described above can be illustrated by the well-known phenomenon of a mariner at sea who perceives himself as sailing at the center of a great bowl of water whose rim is the horizon. The same effect is seen when the sea is viewed from a high cliff or mountain: one gets the impression that the sea is a great wave that may break and inundate the lowland. The ground seen from the air or from the top of a mountain also gives the effect of being saucer-shaped (see figure 4.5).

For a horizontal plane above the observer, the curved shapes are inverted: the sky seems to become a flattened vault overhead with its rim resting on the horizon. For example, a horizontal layer of clouds will be perceived as a ceiling that moves down stepwise to the horizontal (see figure 4.6). The moon illusion, related to the flattened vault of the sky, will be discussed in the next chapter. Both of these phenomena support the view that visual space is usually finite.

Figures 4.3 and 4.4 can be interpreted to say something about the curious way one experiences the environment as one drives a car on a straight, flat highway. The pavement seems to dip and swell and undergo dynamic changes as it passes under and around to right and left; in front, the road far ahead seems to climb rapidly and that hill appears to retreat in contrast with the rapidly approaching foreground; the width of the road ahead decreases in the distance until the margins eventually join; closer in front, the road dips and swells to receive the moving vehicle; the road seems to unroll before the driver's gaze in one continuous swell. All of these descriptive elements are in keeping with an experience of movement in a finite hyperbolic visual space.[11]

Figures 4.3 and 4.4 can also be taken to illustrate the visual behavior of a pair of railway tracks. The railway tracks illusion will be discussed

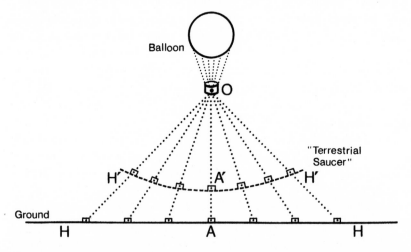

Figure 4.5: The ''terrestrial saucer'': the flat earth (*HAH*) as seen from a balloon appears to be shaped like a saucer (*H'A'H'*).

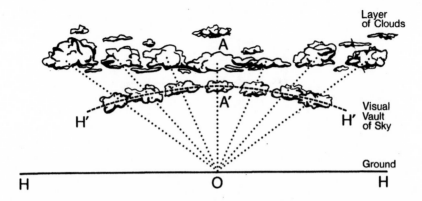

Figure 4.6: Visual vault of the heavens ($H'A'H'$) as it would appear if there were a horizontal layer of clouds (HAH) in the sky.

in the next chapter. Phenomena like these, as I have said, suggest extension in a finite hyperbolic visual space.

Frontal Planes

Model: Figure 4.7[12] tries to show how a checkered grid (sector 1 in figure), at eye level, would appear in visual space, when situated in three frontal planes: the plane of the true point (sector 2: near zone); the plane one-half as distant (sector 3: near zone), and the plane ten times more distant (sector 4: distant zone). Sector 1 is a diagram of the physical plane showing a checkerboard pattern within a circular perimeter. The center of this pattern is directly in front of the observer, and the diameter of the circular perimeter subtends in each case an angle of 90 degrees with the viewer.

Shapes in the near zone are roughly the same whether the space is finite or infinite. Only the distant zone (sector 4) is seriously affected by the value of τ, which in this case was chosen to ensure that in this sector the grid was well inside the distant zone. In an infinite space, however, the forms in sector 4 would be much more similar in shape and size to those of sector 2.

Depth distortions are not shown in the diagram: the checkered surface in sector 3, however, would be perceived as convex and in sector 4 as concave.

Sector 2 (plane of the true point): the central part of the visual pattern is congruent with the central part of the physical pattern; the circular

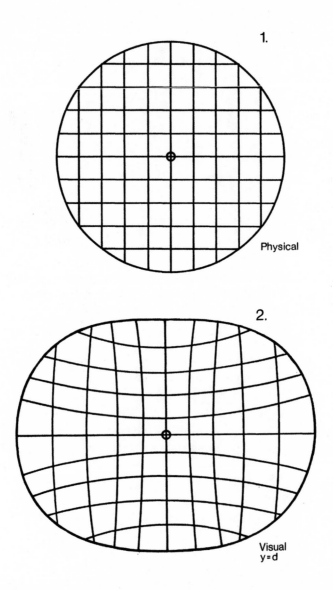

Figure 4.7: Frontal planes: composite diagram in four sectors: 1. physical frontal plane (*FP*); 2. visual transform of *FP* at distance *d* (near zone at true point); 3. visual transform of *FP* at distance *d*/2 (near zone: convex); and 4. visual transform of *FP* at distance 10*d* (distant zone: concave). *d* is the distance to the true point for the hyperbolic model. Contours are modeled on computed values for a finite hyperbolic space. The diameter of the circular

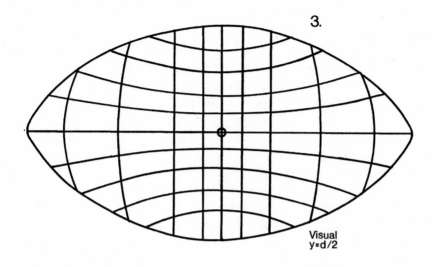

3.

Visual
y=d/2

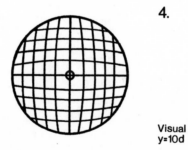

4.

Visual
y=10d

perimeter of the pattern subtends an angle of 90 degrees with the viewer. N.B. The diagram is *not* a set of images to be read visually, but a set of mappings to be interpreted with the aid of the transformation equations: the mappings are not (and cannot be made) isomorphic with the hyperbolic shapes, but some size relationships are preserved in frontal planes (or what physically are frontal planes).

outline of the physical pattern is transformed into an oval with a stretching of horizontal dimensions; the major distortion is a bending of horizontal lines diagonally upward above eye level and diagonally downward below eye level.

Sector 3 (plane at half distance of true point): the central part of the visual pattern is a checkered pattern like the physical pattern but smaller in size; the oval perimeter is more pronounced, with a flattening of the vertical dimension as well as a greater stretching of the horizontal dimension; the diagonal bending up and down of horizontal lines away from the center is also more pronounced. The surface would be perceived as convex.

Sector 4 (plane ten times as distant as the true point): the central part of the visual pattern is a checkered pattern like the physical pattern, closer to the viewer, but very much smaller in size; the perimeter is oval, but less pronounced than in sectors 2 or 3; all horizontal lines and all vertical lines are rounded and concave relative to the center of the pattern. The surface would be perceived as concave to the viewer.

Figure 4.7 shows that if a small cube is brought closer to the observer than the true point (sector 3), the side facing the observer will contract. If it is moved farther from the observer, it will first expand to a maximum size, and then diminish (sector 4), entering a funnel of space that narrows to a point on the horizon sphere. If the space were infinite, the side, however, would continue to expand to reach its maximum size on the horizon sphere.

Figure 4.7 also illustrates the way horizontal and vertical lines in a frontal plane bend in visual space. The horizontal lines sweep upward from the center diagonally to left and right of the median (sectors 2 and 3). In the near zone, vertical lines close to the median line of vision sweep upward and outward away from the median: lines to right and left farther from the median reverse in curvature, becoming concave to the perceiver (sectors 2 and 3). In the distant zone (sector 4), all lines, horizontal and vertical, become concave to the observer.[13]

"Spheroidal World of Sight"

Comments: The horizontal dilation of the visual field, like that described in the above model, is an effect well known to painters and critics, particularly since the seventeenth century: Panofsky called it the "spheroidal world of sight."[14] For example, according to J. Rothenstein, Turner exhibits in his paintings a conscious awareness of this phenomenon: "Turner's apprehension that our natural field of vision is not the rectangle of the picture frame but an oval was eventually to express itself in the use of a vortex-like composition culminating in such a work as *Snow Storm—Steam Boat Off a Harbour's Mouth*" (figure 4.8).[15] Turner's vortexlike composition is reminiscent of the funneling of space toward the horizon sphere described above as characteristic of the distant zone of a finite hyperbolic visual space.

Quasi-Stable Euclidean/Hyperbolic Phenomena

The possible existence of two spatial intentionalities—Euclidean and hyperbolic—raises the question of their interaction within visual perception. Where there is the potentiality of experiencing a scene both ways, one would expect the existence of something like multistable representations of the scene with respect to the two spaces. Rudolf Arnheim's account of this phenomenon has been quoted above in chapter 2. However, some aspects of the experience described by Arnheim are surprising.

Supposing that a particular perceptual situation, such as the interior of the church described by Arnheim, contained sufficient clues to support perceptual construals in the two spaces, then one might suppose that the interior would be perceived at one time as Euclidean and at another as hyperbolic. The two appearances could possibly present themselves one after the other in rapid succession. The enlightened viewer might even be able to get control of the two *Gestalts* and learn to switch from one to the other at will. We are familiar with such multistable visual phenomena, such as the drawing that can be the picture of a rabbit or of a duck, but not of both simultaneously. What is puzzling, however, about this phenomenon is not precisely that it has a certain likeness to other multistable perceptions but that it is so unlike them; the two spatial construals, though mutually incompatible, do not seem to be perceived in succession. Instead, even in rapid succession, one seems somehow to "glide" or "flow" into the other without the conscious sense of an abrupt transition. For this reason, I have chosen to use the term "quasi-stable" rather than "multistable." A possible explanation for the continuous character of this transition will be suggested at the end of chapter 10.

What happens to the apodicticity of the phenomena when its multistable or quasi-stable character is realized? As Don Ihde remarks, the *significance* of apodicticity changes: neither form can any longer claim

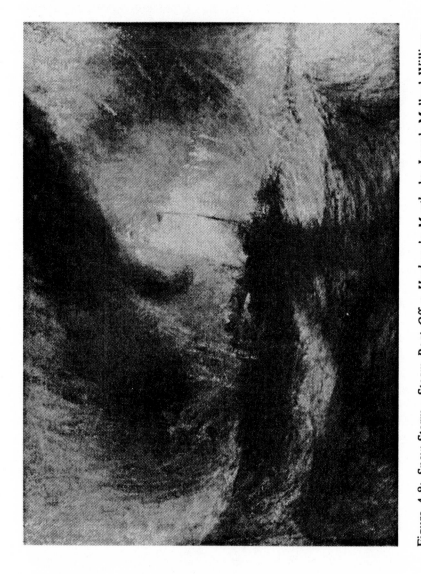

Figure 4.8: *Snow-Storm—Steam Boat Off a Harbour's Mouth*, by Joseph Mallord William Turner (The Tate Gallery, London).

absoluteness or exhaustiveness for the possibilities of perception.[16] Each form of the phenomenon can be separately apodictic, that is, each can manifest itself immediately and directly according to shared descriptive norms, and the perceiver can always return to verify anew the description once given. Each form then is apodictic, but neither absolute nor exhaustive; thus, taken together, both are more adequate to the possibilities of perception than either one alone.

Near Zone and Distant Zone in the Everyday World

Potentially the most significant qualitative characteristic of the model of a finite hyperbolic visual space is the natural division of the visual field into near and distant zone. This division is useful for a phenomenology of the everyday World because it gives a plausible account (1) of the flexibility of vision from the point of view of an individual perceiver and (2) of certain communal structures of the everyday World.

1. Let us suppose that there are circumstances in everyday experience when the Euclidean rule of congruence is suspended, and the perceiver comes to perceive objects and situations—foreground—as occurring within a hyperbolically structured environment—background. Whether such circumstances exist should be discoverable by an attentive analysis of the viewer's own perceptual experience. What is foreground and what background depends from moment to moment on interest and attention, and with each change the parameters of visual space may also change. As one's focus of visual attention changes, certain phenomena are noticed: sometimes the background scale of objects remains the same, but at other times the scale changes as if our vision of background objects were equipped with a zoom lens; one surmises that such differences are due to changes in the intrinsic length scale (κ) of the space. Or consider the following visual experiment: one focuses on a book in a bookcase on the far side of the room, then one switches one's attention to the tip of one's finger held six inches from one's nose. As one's interest and attention shifts, a transformation of the contours of background objects will be noticed indicative of changing parameters of visual space. Vision is always highly contextual, and in each context the foreground-background relation is different; the parameters of visual space seem to depend on this relation.

2. All everyday communal Worlds have some general structures borrowed from the common experience that individuals have of hyper-

bolic visual space. The phenomenology of the spatial arrangement of everyday communal Worlds begins by recognizing that such a World is divided naturally into two zones: A. Schutz calls them "the zone within actual or potential reach," and "the zone not within actual nor potential reach."[17] Being within actual reach includes one or more of the following: being here and now present, being ready-to-hand to use, being visible in a clear and unambiguous way, being within earshot, and above all, being something that shows resistance to touch or to being manipulated. The zone of potential reach is divided, for example, into the zone of restorable reach—the zone that was once attained and could be restored to reach; and the zone of attainable reach—the zone never before attained but which could in principle be reached with ready-to-hand equipment. Within the zone of actual or potential reach is what Mead calls the "manipulative zone"[18]—the zone of what could be touched as well as perceived: this, for him, is the core of what we experience as real. It is clear that all objects in the near visual zone also belong to the zone of actual or potential reach: they manifest themselves in the near space of the viewer with clear and unambiguous forms, ready-to-hand for use. Being in the near visual zone then is a sufficient condition for belonging to the zone of actual or potential reach of a subject. It is not, however, any longer necessary.

In a nontechnological naturalistic World where all action is immediate and direct and instruments are few and simple, the zone of actual or potential reach coincides, more or less, with the local visual space of members of the community; this is a kind of generalized near zone constituted by whatever is actually in or could potentially fall into the near visual zone of a typical member of the local community living in that community. With the development of technology, particularly that of communications, it has become possible to manipulate the distant visual zone from within local space and even to make distant objects manifest themselves in local space without being transported physically there. The re-presentation of these distant objects in local space will, of course, follow the metric of local space. Clearly the notions of *closeness* and *distance* are capable of being transformed by technology. The distinction between near and distant visual zones does not any longer then coincide with the distinction between the two zones of our everyday manipulative World. *Visual closeness* and *distance* are primitive kinds of closeness and distance, adapted to primitive cultures, to a simple way of life, or to a simple and direct orientation in life toward the objects given visually in experience. Visual closeness and visual distance are analogs of a primordial differentiation within the everyday World between the zone that is in actual or potential reach and the zone that is not within actual or potential reach.[19] Whether the value of this distinction could be maintained in a technologically sophisticated World will be discussed at the end of this book.

The character of various communal visual spaces has been studied in many ways: in the cognitive images that people form of the cities in which they live, what Lynch[20] calls "imageability," in which the central city is usually imagined as expanded and more precisely mapped than the more outlying areas; in various noninstrumental navigation systems of tribal peoples at sea, or forest dwellers, or tribes living in deserts, mountains, and plains. Much of this work on behavior and perception in strange environments has been summarized by Helen Ross.[21] It is among work of this kind that evidence would be sought for hyperbolic structures in the organization of communal visual spaces.

5

Evidence from Perceptual Illusions

Informal Evidence

The informal and qualitative evidence for the phenomenon of hyperbolic perception comes from everyday experiences, optical illusions, and the history of art. The preceding chapter has summarized much of the everyday experience in favor of visual space. The present chapter will treat the evidence from perceptual illusions, and evidence from the history of art will be taken up in chapter 6.

Perceptual Illusions

The psychological literature on visual illusions is immense. There are no comprehensive theories about the nature and genesis of visual illusions as such; current research is guided by a variety of theoretical orientations, and the literature is mostly concerned with the exploration of illusions that, like the Müller-Lyer and the horizontal-vertical illusions, are taken as paradigmatic for the study of such phenomena. I shall not attempt to review this extensive literature. My aim in this chapter is merely to show the relevance of the hypothesis of hyperbolic visual space to the problem of visual illusions and to show in what way hyperbolic vision may actually be a factor in (some versions of) the paradigmatic cases mentioned above.

Among perceptual illusions, R. L. Gregory[1] distinguishes (1) those due to malfunctions of neurophysiological mechanisms and (2) those due to the use of inappropriate cognitive strategies; it is with his latter category that I am principally concerned. Gregory assumes that perception is always an active, though not generally deliberate, process of construal of data according to "norms" for the constitution of a percept. Such "norms," he says, comprise an antecedent "logic" for the percept, and they are "unconscious" in the sense that the particular "norms" used in an individual case are not deliberately chosen: the "norms" are, however, in general not innate but the outcome of past learning processes. Gregory states that perception, like science, follows the Popperian way of "conjectures" and "refutations," where perception presents percepts as "hypotheses" to be verified or refuted by experience. (By a perceptual "hypothesis" Gregory means both the underlying synthetic norms and the percept produced by those norms in an individual case.) There is a difference, however, between perception and science; in perception the percept as a perceptual "hypothesis" is usually the product of an "unconscious," that is, indeliberate process and is tested in the domain of immediate and direct presentations; in science, hypothesis formation and testing is always the outcome of deliberate, intelligent, and consciously planned strategies, and involves the design and use of instrumental technologies—but more about this in chapter 11 below.

Where there are illusions in geometrical relationships, Gregory and E. Gombrich would trace them generally to the indeliberate use of an *inappropriate constancy scaling law*.[2] Constancy scaling laws are "perceptual hypotheses" about the relationship between apparent distance and apparent size, and they assume that the apparent size of mobile objects in the field of vision is a definite function of apparent distance. The notion *apparent size* is not a clear one; attempts to clarify it generally refer it to (or attempt to "explain it by") relationships between physical size, angular size, and size of image projected, say, on the retina or on some other surface. For example, on the Euclidean hypothesis normally made, the scaling law is said to be that size and distance should be linearly related, so that, for example, if you double the estimated distance, the estimated size should be doubled. There is some psychological evidence, for example, with aftereffect images, that this kind of scaling law does sometimes operate in perception.

If, however, apparent size and apparent distance sometimes relate to hyperbolic visual spaces, then we should inquire into the form that the appropriate size-distance scaling laws should take in these spaces. In finite hyperbolic visual spaces, there are two such laws: one is for constant physical size given by equation (20*b*) in the Appendix, and this is illustrated in (*a*) and (*a'*) of figure 5.1; the other is for constant angular size given by equation (20*a*) in the Appendix, and this is illustrated in (*b*) and (*b'*) of figure 5.1.

The former applies to cases where the physical size of the perceived object is kept constant. In Euclidean space, the appropriate relationship is one of size constancy for all distances. Something of the character of the relationship in a finite hyperbolic visual space can be inferred from (*a*) and (*a'*) of figure 5.1 below. Graphed lengths and distances in figure 5.1 are drawn proportional to their physical values in (*a*) and to

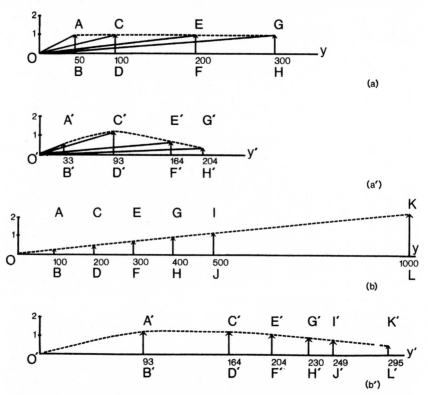

Figure 5.1: Size-constancy relationships in Euclidean and hyperbolic spaces. (*a*) Constant object size—physical space; (*a'*) corresponding size and distance in hyperbolic visual space (values derived from table 12 of the Appendix); (*b*) constant angular size—physical space; (*b'*) corresponding size and distance in hyperbolic visual space (values taken from table 12 of the Appendix). Note the distortion of the line of sight $O'A'C' \ldots K'$; this is due to the mapping that is not (and cannot be) isomorphic with the visual configuration.

their estimated visual values in (a'). Inside the near zone ($y < 100$), visual size diminishes close to the viewer. Outside the near zone ($y > 100$), however, size is inversely related to distance in a complex way: the visual size of an object diminishes steadily with increased distance and eventually disappears on the horizon sphere. These relationships are characteristic of finite hyperbolic visual spaces; for an infinite hyperbolic visual space, on the contrary, the visual size of an object of fixed physical size increases steadily with distance from the viewer and approaches a finite limit on the horizon sphere.

In the latter case, the angular size subtended at the viewer by the object is constant. In a Euclidean space, size would grow linearly with distance as shown in (b) above. In a finite hyperbolic visual space, however, the relationship of size to distance is illustrated in (b') above. Note that (in a finite hyperbolic visual space) at sufficiently large distances from the viewer, the visual size of an object may appear to be smaller than a nearby object of equal angular size. This unexpected outcome is relevant to the interpretation of some illusions, notably the Müller-Lyer Illusion, as will be shown below. Note also that in the mapping chosen to illustrate this effect, some visual relationships are not what they appear to be from the figure. For example, $O'A' \ldots K'$ is a straight line of sight in visual space even though represented in the figure by a curved line: it is not possible to make a map of a non-Euclidean plane that is isomorphic with the Euclidean plane or any part of it.

I shall now discuss a class of well-known visual illusions that may depend for their origin and persistency on the hyperbolic character of visual spaces.

Two-Dimensional Illusions

I shall discuss three classes of perceptual illusions. The first includes many illusions created by two-dimensional linear figures, such as the Hering Illusion, the Müller-Lyer Illusion, the Poggendorff Illusion, the Ponzo Illusion, the Zöllner Illusion, the convergence of railway tracks, and phenomenal regression to the real object. If hyperbolic shapes are sometimes manifested through visual illusions, then there are two quite different ways in which this could occur depending on how the hyperbolic shape is associated with the figure. (1) The figure may function as a physical object which is simply reinterpreted visually as a hyperbolic object, or (2) the figure may function as an image representing an illusionary visual object—not therefore itself—located in an illusionary hyperbolic visual space. In the first case, the perceptual space is hyperbolic; in the latter, it is the illusionary space that is hyperbolic. It may not always be easy to differentiate these two cases from each other. Whenever nongeometrical descriptive features are recognized among the objects that present themselves to the viewer, the figure is clearly serving as an image: roads, railways, buildings, or elephants, for example, are clearly not constituted of lines on paper. But, when no clear descriptive nongeometrical features are presented to the viewer, and when the visual objects consist only of geometrical forms, then the case is more problematic, even when these forms suggest solid bulk and the closure of three-dimensional surfaces.

Hering Illusion: The Hering Illusion (figure 5.2)[3] illustrates the kind of ambiguity that lies at the determination of the origin of illusionary phenomena. The figure comprises a pencil of lines radiating from a central point or vertex and two horizontal lines that intersect the radial lines symmetrically above and below the vertex. The illusion consists in the visual bowing of the horizontal lines. Considered as an object to be viewed, all of the figure's lines, radial as well as horizontal, will lie in one continuous two-dimensional surface that is the visual transform of the physical plane of the paper; on this surface, the visual shape of the figure will depend on whether it is found in the near visual zone or in the distant zone of the viewer. If the figure falls in the distant zone, the shape of the horizontals will be bowed in the way the figure appears in the normal Hering Illusion (see sector 4 of figure 4.7 of the preceding chapter). This is one possible accounting of the illusion using hyperbolic visual space.

What is seen, however, is not always the figure as object. The figure can come to serve as an image; then what is seen is a depicted illusionary object. For example, it is possible to see the vertex of the figure as representing the vertex of a cone (seen from a point on the axis

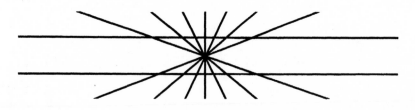

Figure 5.2: Hering Illusion.

of the cone either inside of the cone or outside) from which originates a pencil of diverging lines (generators of the cone). The visual object in this case is not just the diagram on the paper but a visual interpretation of the figure as image; it is an illusionary object displayed in some illusionary visual space. The Hering figure can be read in a variety of different purely geometrical ways. In most of these readings, the horizontals will continue to be experienced as curved lines, but with different curvatures. This fact may be taken provisionally to indicate that the illusionary geometrical spaces are non-Euclidean. However, further research is needed to determine the full set of profiles of these imaginary geometrical spaces and to find out whether any are compatible with a Euclidean interpretation.

Don Ihde has studied the phenomenon of multistable illusionary visual forms particularly when descriptive (nongeometrical) attributes are brought into play.[4] He has shown that a variety of different descriptive perceptual forms can be perceived in an image, each form being a function of the choice—deliberate or indeliberate—about what is interpreted to constitute foreground and background. He has shown that if the viewer is prepared in advance by a suitable interpretative narrative, he can often come to see in stable fashion quite different three-dimensional illusionary situations imaged by the same two-dimensional diagram. The interpretative narrative sets up the context for visual interpretation and strongly influences not only what kinds of things are seen in the image but also their geometrical shapes. For example, in the Hering Illusion, the radiating lines can be interpreted as representing the walls of a long straight tunnel down which one is looking; if the gaze is then focused strongly at the far end of the tunnel (represented by the vertex in the diagram), which is virtually at infinity in the represented physical space, then the horizontal lines become visually straight—which is, of course, congruent with the physical shape of that which is being represented. What is particularly interesting about this kind of experiment is that it shows how the interpretation we give to the figure—the intentionalities we use—influence the spatial structure of the illusionary pictorial objects we see.

The Müller-Lyer Illusion: The Müller-Lyer Illusion[5] has many forms: one of the forms is illustrated below in figure 5.3, where the two horizontal lines are drawn equal in measured length. The illusion consists in the impression that the line with the arrowhead serifs (upper in figure 5.3) is shorter than the one with the reverse serifs (lower in figure 5.3).

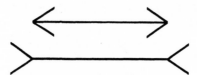

Figure 5.3: Müller-Lyer Illusion.

The first question that suggests itself is whether there is any feature of the visual transform of the Müller-Lyer figure (as a physical object) that could account for the illusion. The horizontal lines will transform with little significant change, likewise the arrowhead serifs and the reverse serifs. It seems likely, then, that *if the illusion were connected with visual space*, it would have to belong to the way this figure functions as an image capable of being visually interpreted in a pictorial way, and so the structure of an illusionary pictorial space. For the same reasons, one would be inclined to draw a similar conclusion about the vertical/horizontal illusion.

On careful inspection of the pictorial object seen when viewing the Müller-Lyer figure, it will be noticed that, in the version presented in figure 5.3, the upper and shorter line appears (or usually appears) to be at a greater distance from the viewer in illusionary space than the lower and longer line. Since the optical projections of both lines in the plane of the paper are equal (as well as their retinal projections), the physical configuration in the horizontal plane of the illusionary object must be something like that represented in figure 5.4. Ordinarily, as the figure shows, with such a configuration (*a*), the constancy scaling laws should lead to the judgment that the more distant object is the larger. The fact that it is judged visually to be smaller shows that the visual object is not being construed according to Euclidean constancy scaling laws (*a′*): it is not a Euclidean object.

Somewhat similar to the Müller-Lyer Illusion is the moon illusion, which will be discussed in detail below. There are, however, two differences. In the moon illusion one sees now one configuration (e.g., the horizon moon) and then the other configuration (the zenith moon) *in possibly different spaces*; in the Müller-Lyer Illusion both horizontal lines are seen together in the same space. The second difference is that while what one sees in the Müller-Lyer Illusion is, as I have argued above, not the lines themselves but something that the lines represent, namely, a visual object in some illusionary non-Euclidean pictorial

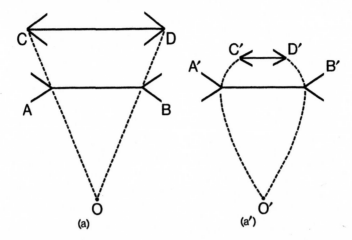

Figure 5.4: Given the observed distance relation (top line farther away than the bottom line) the relative sizes of the two lines of the Müller-Lyer Illusion should appear as in (*a*) if judged in Euclidean space, or they could appear as in (*a'*) in hyperbolic visual space. Note the distortion of the lines of sight $O'A'C'$ and $O'B'D'$ in (*a'*): this is due to the mapping, which is not (and cannot be made) isomorphic.

space, what one sees in the moon illusion is the moon itself in two different configurations.

Clearly then, with respect to the illusionary pictorial space of the Müller-Lyer Illusion, the arrowhead serifs (in relation to the reverse serifs) must be cues that, *in the form presented in figure 5.3*, indicate relative length deficiency and/or relative distance, while the reverse serifs are cues that indicate relative length excess and/or relative closeness. If the form in figure 5.3 is rotated through 90 degrees, the same relation with respect to relative size is noticed while relative distance may appear the same or appear reversed. In this vertical form, it has often been noted that the arrowhead serifs and the reverse serifs in the figure can easily be seen as perspectival cues suggesting depth and, through depth, size. If, however, figure 5.3 is rotated through 180 degrees, the visually shorter line of the two (with the arrowhead serifs) is now seen to be closer than the visually longer line (with the reverse serifs). In other forms of the Müller-Lyer Illusion where the lines are combined and each is an extension of the other, the serifs are connected with the illusion of relative size in the usual way but neither part of the line seems to be closer to the viewer than the other. To compare these illusionary shapes with the repertory of hyperbolic shapes, more extensive determinations would have to be made of the type already undertaken by Gregory.[6]

The fact that in this case the illusion affects not the visually transformed shape but what the figure is interpreted to represent is significant and suggests that the strength of the illusion, even perhaps its presence, is affected by human culture. Cultural anthropologists like M. H. Segall, D. T. Campbell, and M. J. Herskovits[7] have found that cultures that do not employ linear "carpentered" forms of the sort that are pervasive in Western urban environments but, say, circular forms, like the Zulus, hardly experience the Müller-Lyer Illusion, suggesting that the significance of the figure with its "perspectival" connotations may be mostly cultural and therefore learned.[8]

The Poggendorff Illusion: In the Poggendorff Illusion (figure 5.5),[9] the interrupted diagonal line appears to be offset by the vertical strip that separates the two parts. Unlike the Müller-Lyer Illusion, the visually transformed shape could possibly account for the visual effect. The clue we follow is that the vertical strip appears to protrude into the foreground space.

Figure 5.6*a* is a drawing of the Poggendorff figure, with *PQ* and *RS* representing the two segments of the diagonal (*PQ* and *RS* belong to the same straight line that passes through *O* at the center of the figure). From the physical coordinates of the figure, with the aid of the equations in the Appendix, the visual shape of the transformed figure in hyperbolic space can be calculated. The Poggendorff Illusion can be understood as a result of such a transformation of the figure, *provided the plane of the figure lies in the near foreground relative to the true point.* For example, let us say that the true point is 100 centimeters

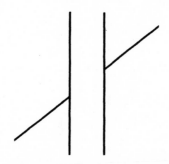

Figure 5.5: Poggendorff Illusion.

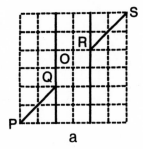
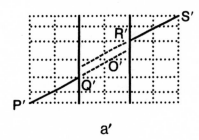

a   a'

Figure 5.6: (*a*) Two-dimensional map of diagram of Poggendorff Illusion in a Euclidean plane; (*a'*) possible two-dimensional map of the same diagram in a hyperbolic visual plane (based on table 4 of the Appendix).

from the viewer ($\sigma = 100$), and let the plane of the paper be 50 centimeters from the viewer. (The visual space, as has been said, is such that the true point lies behind the plane of the paper.) Let the physical coordinates (in cms) of the points be, say, $P = (-20, -20)$, $Q = (-10, -10)$, $R = (10, 10)$ and $S = (20, 20)$. Since the value of τ is virtually immaterial to the transformation, we can set it equal to 0. Figure 5.6a' then represents (on a flat Euclidean map) the visual transform of figure 5.6a. The coordinates of the transformed points, taken from table 4 of the Appendix, are the following: $P' = (-21, -17)$, $Q' = (-8.6, -7.3)$, $R' = (8.6, 7.3)$, and $S' = (21, 17)$. The physical straight line *PQROS* is converted into an S-shaped figure, *P'Q'O'R'S'*, in visual space, and it is apparent from the new figure that *P'Q'* and *R'S'* are vertically offset in the manner associated with the Poggendorff Illusion. If this account is correct, the magnitude and perceived direction of the effect is conditioned by the visual interpretation given to it by the perceiver, namely, that it is a hyperbolic figure that lies in the near foreground relative to the true point of that space. This interpretation needs to be further tested by varying the width of the vertical strip and by comparing the perceived direction and amount of the illusion with the values computed from the equations.

The Ponzo Illusion:[10] In the classical Ponzo Illusion—figure 5.7*a*— the upper of the two horizontal lines appears to be longer than the lower, while in the modified version—figure 5.7*b*—where the lines are now vertical, the illusion disappears and the two (now vertical) lines appear to be equal.

According to the prevailing view among psychologists, the Ponzo Illusion is due to the structure of the illusionary object represented by

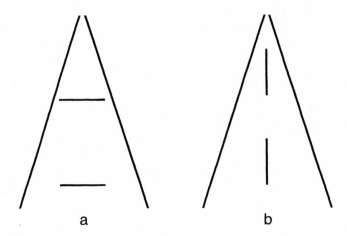

Figure 5.7: Ponzo Illusion: (*a*) the classical illusion, (*b*) the modified illusion.

the figure, the converging lines being read as an image drawn in perspective, say, of something in depth like a road down which one is looking. The absence of the illusion in the modified version has been difficult to explain, however, since if it were true that the figure is read as a drawing in perspective, the illusion should be present also in the modified version, *provided, of course, that the illusionary pictorial space is Euclidean.* Careful inspection of the visual object does, however, support the reading that what one is seeing is an illusionary object in depth, but it also suggests that the illusionary space has hyperbolic characteristics. The absence of the illusion in the modified version can be explained by the fact that in a hyperbolic visual space, depth is foreshortened with distance from the viewer (consult figure 4.3). Consequently, two equal visual depths—representing, of course, two unequal physical depths—could be represented plausibly to hyperbolic vision by two equal optical projections in the plane of the paper, provided that the upper vertical line were seen as the projection of a line in the distant visual zone. The two versions of the Ponzo Illusion, then, are consistent with an interpretation of the diagram as representing situations in an illusionary finite hyperbolic visual space.

The Zöllner Illusion:[11] The Zöllner Illusion (figure 5.8) produces the unsettling feeling that what you see does and does not lie flat on the surface of the paper. The long lines seem (1) to diverge and converge alternately, while (2) maintaining a constant separation. Such properties are inconsistent with a purely Euclidean field. They would be

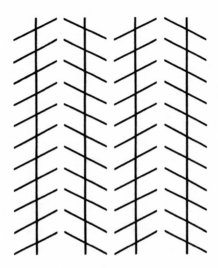

Figure 5.8: Zöllner Illusion.

consistent, however, with two multistable readings of the diagram: (1) a hyperbolic reading, in which the visual lines are no longer coplanar, and (2) a Euclidean reading, where the lines lie flat on the surface of the paper.

Railway tracks—like the sides of a long flat road—are perceived to converge to a point on the horizon. Convergence is experienced despite our knowledge that the tracks remain parallel all the way. Various explanations of this phenomenon have been given: some speak of "attending to the retinal image" as if this were in fact what we see when we see the railway tracks converging.[12] Phrases like "attending to the retinal image" are at best metaphors and at worst confusions of language: it is not *what* one perceives when one is perceiving, and if, for explanatory reasons, one posits that it is, the history of philosophy shows how this position leads to insolvable problems regarding both the perceiving subject and the real World that is perceived. It is, of course, evident that perception involves retinal stimulation in an essential way, but how this is related to the act of perceiving requires a philosophical analysis like the one to be offered in Part II of this work.

One alternative explanation that has been offered for the experience of visual convergence is the suggestion that what one is seeing is not in fact the railway tracks but an image of the railway tracks projected on the concave inside surface of a large spherical screen (in Euclidean space) surrounding the observer. The image projected on such a screen would indeed converge; but it would be two-dimensional, while we experience the railway tracks themselves as extending before us in full three-dimensional depth and converging, nevertheless, on the horizon. This proposal then does not work.

The solution follows from noticing that what we in fact see is the railway tracks themselves. However, the objects as perceived are not the railway tracks as Euclidean objects; what is perceived has all the characteristics of railway tracks construed in a finite hyperbolic visual space. Figures 4.3 and 4.4 show that in this kind of space the visual transform of the railway tracks would extend in three dimensions to a finite surrounding horizon—the horizon sphere—and that the tracks would curve upward in the distant zone and come together at a point at a finite distance from the viewer. This is actually what we experience, and constitutes what we call "the railway track illusion." In a sense to be discussed below, it is not an illusion at all, but the presentation of the railway tracks as a perceptual object within a finite hyperbolic visual space.

Phenomenal regression to the real object is the illusion that makes us see a flat object, like a circular plate, as having a shape somewhere between that of the projected image (elliptical in this case) and the real shape (circular in this case). According to Thouless,[13] who investigated this phenomenon, the perceived shape is a compromise between the shape of the projected image (he referred to the retinal image) and the true physical shape. Thouless attributed the phenomenon to an unconscious compromise between sensation structured by the retinal image and the concept of the real shape vying for expression. Such an account, however, is unsatisfactory because it rests on a metaphor and the abuse of categories. There is a simple solution in terms of visual space: what we see is a tilted oval object that is (or appears on the surface to be) congruent with the three-dimensional visual shape of a circular plate as construed in hyperbolic visual space.

Illusions of Rotation and Optical Images

The second class of illusions is associated with rotating three-dimensional objects. Consider, for example, two wire squares, facing each other and linked rigidly by a connecting wire; when these are rotated about a vertical axis, the two forms are seen to glide flatly over each other, or alternatively they may be seen to oscillate about a vertical axis, reversing direction of rotation at every half-turn.[14] Such phenomena can be understood if we assume that the objects are perceived in a shallow zone of hyperbolic space at such a distance from the perceiver that the depth available for the object to move in is insufficient to accommodate a complete rotation with a plausible sense of shape/size constancy for the object. The perception of a rigid rotation would consequently be ruled out by the structure of visual space itself, and the shallowness of the space available would permit no more than the following two options: (1) the perception of a flat nonrigid object of fixed height and pulsating width, or (2) the perception of a rigid figure performing a shallow oscillatory motion about a vertical axis. These are just the alternatives that are observed. I have discussed earlier Ballard's description of the Saarinen Gateway Arch at St. Louis. There, the shallowness of available space would also account for the reversals in orientation of the Arch perceived when the Arch is viewed obliquely from a distance.

Yet another class of phenomena that exhibits anomalous visual structures of a hyperbolic kind is that of images seen with the aid of optical apparatus, such as curved mirrors, lenses, telescopes, and microscopes. Vasco Ronchi[15] has made a study of the widespread illusions associated with the use of such equipment. The existence of such illusions, he says, was well known before the seventeenth century, but the viewer's interpretative role in using optical systems was gradually forgotten in the wake of Kepler's work on optics and Galileo's success in turning the telescope into a scientific instrument. The existence of telescopic illusions seemed to trouble Galileo less than it did many of his contemporaries.[16]

Ronchi shows that neither ray nor wave optics account for what is *seen* (Ronchi's ''effigy'') when one uses an optical system. The ''effigy'' is usually not congruent with the optical image; it seems rather to be the outcome of a hermeneutical process in which the viewer interprets the optical stimulus in the context of other surrounding clues. For example, the visual position of the image seen through a magnifying glass usually does not agree with the position of that image as construed with the aid of geometrical optics; the distance of the image from the viewer seems to be fairly constant, and equal to the physical distance of the object from the viewer no matter what the distance is between the lens and the viewer.

Such phenomena could all be explained in terms of transformations between hyperbolic visual spaces. I surmise, for example, that the case of the magnifying glass can be treated in some way like the following: let space$_1$ be the space in which the direct object is seen, and space$_2$ the space in which the imaged object is seen, each object being at the true point of its respective space, then since each would appear to be located at the same distance from the viewer, the σ's of the two spaces must be equal. The experienced magnification would then be a function of the two scale factors κ_1 and κ_2. Such a solution as that just proposed would of course have to be tested experimentally.

Three-Dimensional Illusions

The third class of illusions deals with three-dimensional shapes. I have already mentioned some of these in the preceding section: the flattened vault of the sky by day or night, the moon illusion, the illusion of the Gateway Arch of St. Louis, the dynamic flow of visual forms one perceives when one drives on a well-engineered highway, the experience of a mariner at sea being at the center of a great bowl of water, and the "refinements" of ancient Greek architecture.

It has been noted from antiquity that the vault of the sky looks to be at a finite distance and to have the shape not of a hemisphere but of a flattened dome closest to us directly overhead.[17] Something of this experience is expressed in the ancient Egyptian painting, reproduced in figure 2.3, which illustrates the mythological origins of the sky. These phenomena have an explanation in terms of features common to finite hyperbolic visual spaces. It will be noted that, since the perceived size of any object of finite physical size would vanish on the horizon sphere, no *visual* object of finite size would be experienced as actually located on the horizon sphere in visual space unless in physical space it were infinitely large (subtending a finite angular size but infinitely far away). We assume that such objects would be as problematic in visual space as they would be in physical space. Consequently, as long as the heavens are perceived to contain visual objects of finite size, such as the moon, clouds, planets, even stars (since they too appear to the naked eye to be of finite size), the heavens should be perceived as an arched ceiling dipping in all directions to meet the horizontal plane at the farthest imaginable distance, which is at the horizon sphere. The shape of such a ceiling is as figure 4.6 shows, a flattened vault of finite dimensions.

The Moon Illusion

The moon illusion[18] also has a natural interpretation within hyperbolic visual space. The moon's disk subtends a constant solid angle with the eye whether it is on the horizon or at its zenith; nevertheless, it looks larger when it is on the horizon than when it is at its zenith. This of itself is not surprising, since the moon is so far away that it is beyond the threshold of parallax resolution by our unaided senses. The apparent size and distance of the moon must then be functions of other visual clues. What is surprising, however, is that the horizon moon appears to be closer to us than the zenith moon, since this is inconsistent with the familiar (Euclidean) scaling laws.

The contradictory elements can be reconciled, however, with the family of finite hyperbolic visual spaces. Let Δ be the visual size of a very distant spherical object like the moon, and $\bar{\omega}$ the "diameter" of visual space, that is, twice the maximum visible distance in the space (twice the distance to the horizon sphere).

Let the subscript "1" (in Δ_1, ω_1, κ_1, μ_1, σ_1, τ_1, etc.—see Appendix) refer to the hyperbolic visual space, S_1, of the horizon moon, and subscript "2" (for the same parameters) refer to the hyperbolic visual space, S_2, of the zenith moon, then the conditions of the moon illusion are:

a. $\Delta_1 > \Delta_2$ (the apparent size of the horizon moon is greater than the apparent size of the zenith moon), and

b. $\omega_1 < \omega_2$ (the apparent diameter of visual space of the horizon moon is less than the apparent diameter of visual space of the zenith moon—assuming that the distance estimates of the moon correlate with the apparent diameters of their visual spaces).

These conditions are fulfilled only if the following inequalities hold:

$$\kappa_1/\mu_1 > \kappa_2/\mu_2 \tag{1}$$

$$\kappa_2 > \kappa_1 \tag{2}$$

$$\mu_2 > \mu_1 \tag{3}$$

(assuming μ_1 and μ_2 are small quantities).

Now (2) and (3) together (with equations of the Appendix), imply

$$\sigma_2 > \sigma_1 \tag{4}$$

The first inequality, (1), states that distant physical objects will appear larger when the moon is on the horizon than when the moon is at its zenith (this follows from equations 4–6 of the Appendix). This is indeed what we experience; distant trees and houses appear to change their size, appearing smaller when the moon is high than when it is close to the horizon. This is an effect of changing scale, and it can be seen even in photographs, where the scale of the depicted objects appears to change with the position of the moon in the photograph. Inequality (2) states just this, that the intrinsic visual scale factor for the estimation of the size of objects should be greater when the moon is at its zenith than when it is on the horizon.

The third inequality (3) states that depth and distance differentiation should be more pronounced when the moon is high than when it is close to the horizon. It is easy to think of this effect as implied by the earlier one, since we generally associate greater distance with "smaller apparent size" and smaller size with "greater apparent distance." This prevalent association is difficult to explain if, as is usually supposed, we are accustomed to represent the World according to the laws of Euclidean geometry. As long as its shape and motion are represented in a Euclidean space, a test object, such as a unit physical cube, should not appear to get smaller with distance from us, it should appear to be always the same size. This would not be the case, however, if the test object were represented in a hyperbolic visual space; apparent size and apparent distance would be correlated in a definite way depending on which of the family of hyperbolic visual spaces was in control of vision and on whether the object was found in the near or distant zone of that space. It is not the case that smaller apparent size would under all circumstances be correlated with greater apparent distance—smaller size (in an infinite hyperbolic space) could go along with smaller distance (in that space)—but whenever some one *finite* member of that family was in control of the visual field, then there would be a negative correlation between apparent size and apparent distance in the distant zone relative to the viewer (see figure 5.1, *a* and *a'*). Apparent size and apparent distance would then be associated in the way we are accustomed in real life to associate them visually. It is significant then to note that objects appear to be both smaller and more distant when the moon is at its zenith than when it is close to the horizon. These phenomena, though not independent when taken within one space, are otherwise independent of one another, and the two inequalities, (1) and (2) above (from which [3] follows), are needed to explain the fact and possibility of those features of the moon illusion which are so puzzling.

Inequality (4) states that the true point of S_1, the space of the horizon moon, is closer than the true point of S_2, the space of the zenith moon. There is a plausible explanation for this: for the space of the horizon moon, intervening objects probably act as space markers, bringing the true point closer to the observer and also diminishing the "diameter" of visual space; for the space of the zenith moon, the absence of intervening objects probably has the effect of moving the true point farther from the observer and also of expanding the "diameter" of visual space. There are other cases where distant objects are viewed without intervening space markers, for example, where a town is seen from the air, or a shoreline from a boat; if the explanation is correct, these should look too small and too distant, like the moon at its zenith. These phenomena are also confirmed.

The phenomenal size-distance relation that is characteristic of the moon illusion is evidently not one which is compatible with a Euclidean space; neither is it, however, compatible with an infinite hyperbolic visual space, for the visual size of a test object in such a space would remain nearly the same even if the object were moved to infinity. The only visual spaces of constant curvature in which the moon illusion can be successfully explained are those that are finite and hyperbolic.

The Radiant Sun

Confirmatory evidence that when we look at the sky we construe what we see in a finite hyperbolic space is found in the phenomenon of the sun's visible radiance. This is seen usually near sunrise or sunset, when the sun appears momentarily at the edge of a cloud as a dazzling ball of light from which rays emerge darting in all directions across the sky. In fact, rays coming from the sun originate 93 million miles away, a virtual infinity relative to the parallaxes that we can resolve, and by the time these rays reach the earth, they are in sheaves of virtually parallel rays. One would expect then to see the sun's rays as forming virtually infinite corridors of light terminating at the sun's disk. Instead, one sees what appears to be a disklike source of dazzling light situated at most a few miles away just behind the clouds, radiating streamers of light which cross the vault of the heavens. This is the "radiant sun." The image of the radiant sun, or *Gloria*, usually representing the glory of God, was depicted in eighteenth-century baroque art as a circular disk from which golden rays emanated like the spokes of a wheel. The visual phenomenon, however, includes rays that pass overhead, curve over the zenith, and are imagined to bend down behind the viewer's back to meet at an imagined point on the horizon opposite the rising or setting sun. The appearance as so described and depicted is what one would expect were the sun perceived as a distant object in a finite hyperbolic visual space with rays bending to follow trajectories of the kind illustrated in figures 4.3 and 4.4. Illusions of the kind just described could also account for some of the visual problems aircraft pilots experience, for example, in judging the distance of the touchdown point when coming in to land.

Other Confirmatory Evidence

Many authors—ancient and modern—have held that the refinements that were engineered with such precision into the construction of the entablature, stylobate, columns, and colonnades of many ancient Greek temples [19] were made for the purpose of compensating for visual distortions of straight lines. Others,[21] however, have challenged this view by pointing out that the refinements draw attention to themselves; far from concealing deviations from straight vertical and horizontal lines, they assert their presence. Moreover, it is argued that, if designed for the purpose of correcting optical illusions, no correction would serve to remove distortion for all viewers. In response, however, we can say that it would not be out of keeping with the mathematical Greek mind to design its buildings so as to be seen as a harmonious geometrical ensemble from one (mathematically) privileged position, itself geometrically determined in relation to the architectural ensemble. The fact that the columns were corrected by tapering and by an inward tilt instead of by a reverse refinement, and that the entablature and stylobate were corrected by raising the center slightly instead of lowering it, indicates that the position from which each building is to be viewed is close by, in the near zone. On the one hand, it would be interesting to calculate the geometrical position of this viewing point from the data at hand about the magnitude of the refinements.[22] On the other hand, such a calculation has interest only if the supposition is correct that the magnitude of the refinements is related to the actual size of the buildings and not, say, to the size of a small scale model of the building[21]. We know too little about the methods of architectural design and the intentions of the architects of the Greek Doric temples to be able to ascertain whether the supposition is correct.

Finally, some of the drawings of Maurits Escher illustrate and support the thesis that there is a class of visual illusions based on hyperbolic visual spaces. Some of Escher's drawings, for example, *Ascending and Descending*, and his version of the "impossible triangle,"[22] picture "impossible objects." These are objects which are experienced as perfectly coherent visually, but turn out nevertheless to be incoherent as 3D physical objects because their structure would violate the laws of physical space.[23] These strange visual objects can be construed as hyperbolic visual transforms of possible 3D physical objects viewed in such a way that all or part of the physical object falls into the distant zone of zero (or near-zero) visual depth.

6

Evidence from the History of Art

Pictures and Pictured Objects

Evidence for hyperbolic perception from the history and practice of art depends on being able to establish the following theses:[1] (1) that some artists and the epochs to which they belonged regarded reality in a *pictorial way*; (2) that they conceived the goal of image making to be to generate a congruence between the way the *pictured object* was constructed from its elementary pictorial parts and the way the *perceived object* was constructed of its parts; and (3) that of those who painted in a pictorial way, at least some attempted to construct an art, or to modify traditional ways of constructing images with refinements, for the purpose of creating a "pictorial" vision of objects constructed of elementary pictorial parts that belong not to a Euclidean space but to a hyperbolic space.

What is given to perception as real may be given through any one of the senses: pictorial reality, however, is given to perception only through vision. What is given to vision in any epoch is a function of what is regarded as visually significant at that epoch and therefore of the expectations and anticipations of human perceivers. This topic will be discussed more fully below. In much of antiquity and right down to the end of the Middle Ages, the natural universe was read as a moral, religious, or metaphysical text; it was a divine sermon, or it illustrated a religious mythology, or it incorporated in a metaphysical model the astronomy and science of the time. These readings were illustrated in works of art and technology, and in artifacts of all kinds, as well as in stories, literary texts, and inscriptions. The effort to decipher the surviving relics of these and other premodern cultures such as those in

tribal Africa has led to the brilliant work of anthropologists such as
Marcel Griaule, classicists such as Hertha von Dechend, historians of
science such as Giorgio de Santillana, and the musicologist Ernest
McClain.[2] From their work they derived the view that ancient and
recently surviving artistic and mythological representations "speak" of
their Worlds in realistic terms, even in scientific—or, at least, model-
and number-related—terms, though generally not in a pictorial way.

In various epochs in Western culture, notably in ancient Greece and
Rome and again in the Renaissance, artists tried to represent a localized
part of their World *in a pictorial way*.[3] A *pictorial object* is one that has
a pictorial essence, that is, it presents itself to visual perception as
having been constructed according to rational (i.e., geometrical) principles out of basic pictorial or visible elements: those elementary
visibles were, for example, for the Renaissance, points, segments of
Euclidean straight lines, patches of Euclidean plane surfaces bounded
by segments of straight lines, and the color and light that fell on these
patches. To represent a World in a pictorial way is to perceive objects
in the World as pictorial and pictorially related. It was only *as a
consequence* of this mode of visual perception, as Snyder has convinc-
ingly shown, that the Renaissance conceived the goal of constructing
scientifically an image on a two-dimensional surface that, when
viewed, would produce in the viewer a "quasi-hallucinatory" perception of a three-dimensional object-as-pictured that was guaranteed (in
some sense) to match the object as it would be perceived in relation to
the elementary visibles into which this mode of vision deconstructed its
object.[4]

In art-historical and art-critical writing, the term "two-dimensional
image" means a configuration of marks on a flat surface crafted to
fulfill one of the following conditions: (1) there is no illusionary
pictorial space (e.g., geometrical designs); (2) the illusionary pictorial
space is so shallow as to be virtually two-dimensional (e.g., many
primitive or primitivizing art styles); or (3) the illusionary pictorial
space has no more depth than the surface of the canvas itself (e.g.,
schematic flowers in a decorative border). "Three-dimensionality," in
this context, means a configuration of marks on a flat surface crafted to
produce an illusion of depth, or more precisely, an illusion of a
three-dimensional pictorial space. What is missed by this terminology
is the material identity between the marks on the surface *as marks*, and
those marks *as images*. All marks are two-dimensional in character and
so, by definition, all *images* are two-dimensional in character. Some

two-dimensional images when viewed produce an illusion of three-dimensional pictorial objects, but the distinction between two-dimensional and three-dimensional effects can not be expressed by the terms "two-dimensional" and "three-dimensional images," since the images in both cases are flat. The term "image," in a strict sense, means an image *of something*, either in the real World, such as a horse, or in an imaginary realm, such as a unicorn. I may paint a picture of a real room, or of an imagined room: in either case, the picture is an image of something, and in both cases the image is flat. A configuration of marks crafted so that there is no illusionary space does not yield an image: abstract designs, for example, are self-referential in this respect, they just stand for themselves. Only by extending the word "image" to an ideal or metaphorical realm can one speak of abstract or other self-referential marks as images.

I have no problem with the common understanding of shallow illusionary space as virtually three-dimensional, but a word must be said about those images that depict objects in a flat, schematized way. The space of such images may seem to be like that of self-referential designs, but there is an important difference. They are really images of things in some pictorial space, and their "spacelessness" and the "spacelessness" of their environment is not absolute but relative, as long as they retain their appearance of referring to a class of pictorial objects. Furthermore, it may be pointed out that, wherever there is an illusionary pictorial space, the image, by which I mean the set of marks themselves, is never an object in that space, it is always located in *real* space. Thus, in this chapter, I shall use "two-dimensional image" to mean a configuration of real marks on a flat surface which refer to pictorial objects in an illusionary space, whether of two or three dimensions.

Artificial Perspective

Following the Renaissance goal of constructing pictorial images, Alberti (1435)[5] developed, and his contemporaries quickly adopted, the system of linear perspective called "mathematical" or "artificial" perspective. A system of perspective is a way of crafting an image capable of representing the geometrical forms of individual objects as well as the forms of the empty regions between them as articulated parts of one systematic whole, namely, space.[6] I shall refer to mathematical or artificial perspective occasionally as "classical perspective," meaning by this the way Renaissance artists organized space predominantly according to the rules of mathematical perspective.[7]

Albertian perspective was based on two sets of principles: on the medieval optics of *perspectiva* and on Aristotelian faculty psychology. According to the latter, the "visual pyramid," that is, the set of visual rays emanating from the object (or alternatively, as some thought, emanating from the eye and going to the object) produces with the help of the *sensus communis* and the agent intellect a mental object—the object-as-perceived: this representation is, moreover, according to Aristotle, infallible provided the medium is not disturbed and the sense organs are in a healthy condition.[8]

It is important to note that, by Alberti's time, objects as perceived were already pictorial objects in an infinite Euclidean space; that is, composed of elementary surface areas rationally (geometrically) related to one another.[9] Thus, the problem set by artificial perspective was to invent some standard procedure that would enable the artist to construct an image that was guaranteed by its mode of scientific construction to be seen (by one, we would add, who shared the art and goals of pictorial vision of this kind) as depicting the object as it was perceived in pictorial vision and also, of course, as it was believed to be. The technique of regarding the canvas as a "window" through which the viewer looked at the depicted object was Alberti's genial invention that solved the problem of perspective; the canvas was to act like a window by sending to the eye rays reflected from it, just like those rays that were transmitted through a real window. The artist, by adopting a fixed spatial relation to the picture frame, made the image coincide in geometrical outline with the intersection of the "visual pyramid" with the plane of the canvas (see figure 6.1). The viewer, by adopting the same spatial relation to the picture frame, then with the help of the *sensus communis* reconstructed the depicted object out of its elementary visible parts.

Snyder finds confirmation of this account in the sixteenth-century adaptation of the medieval *camera obscura* to the task of producing, in the best technical and scientific manner, the kind of image that painters were attempting to construct by following the rules of mathematical perspective. In the seventeenth and eighteenth centuries elaborate systems of lenses and mirrors were added to the camera to make it a more useful instrument for painters. Snyder claims that pictorial vision was

Figure 6.1: Use of perspective frame, from "*Unterweisung der Messung*" by Albrecht Dürer (Print Collection, New York Public Library).

not a consequence of the development of the camera but, on the contrary, that the camera, like mathematical perspective, was developed to serve a pictorial vision that already defined the World to be of a certain kind and to assist painters to express this. The logical consequence of this mode of pictorial seeing was the *trompe l'oeil*—the painting of an image capable of producing in the viewer something, in Gombrich's words, "akin to visual hallucination."

Perspective as "Symbolic Form"

The perspective system used by an artist says a great deal about the kind of extended reality that the artist and his epoch perceived: "perspective," says Panofsky,[10] "is 'symbolic form.' " A perspective system always implies the notion that reality is *pictorial*, and sets the goal of image making to be congruence between pictorial objects as perceived and the same as pictured; all alike constructed out of a repertory of basic pictorial forms. For Renaissance perspective, the basic visible forms were Euclidean and the image that mediated between them was an optical projection on a planar surface in Euclidean space. It is of course not necessary for a pictorial mode of vision that the basic pictorial elements and the pictorial space itself be Euclidean.

Panofsky makes the point that people can become so conditioned by the perspective of their time that they fail to see perceptual possibilities that contradict it. He reports that Kepler did not notice that straight lines were sometimes perceived as curved until he discovered that we come to know reality not through the planar image of classical perspective but through a retinal image that is curved. The discovery that the retina was curved led Kepler to notice that straight lines were sometimes seen as curved. Panofsky, speaking of the curvature of the retina, says it is a prepsychological fact that sets up a fundamental discrepancy between reality and what one perceives reality to be like. Whether Panofsky is here defending the view that the projection on the retina is in some sense a subsidiary object standing for the real object on which the "real I" gazes from the secret recesses of the Mind is not clear: if he is, this nevertheless is an old and oft-refuted view that, alas!, is still around, despite the classic refutations given by Descartes, Diderot, and Condillac.[11] It is possible that Panofsky meant that the curvature of the retinal image suggests in some metaphorical sense the nonclassical rounding and modulation of forms in vision (seen, for example, in the motifs pictured by Turner, Cézanne and van Gogh[12]), or perhaps that, if the curvature of the physical retina places constraints on vision, these constraints will manifest themselves in the shapes of visual objects.

"Fishbone" Perspective and Curved Space

Following Ernst Mach, Panofsky asserted that the space represented by classical perspective—isotropic, homogeneous, and infinite—could not be the space of visual perception, or "physiological space," as Mach called it,[13] for the latter is neither isotropic, nor homogeneous, nor infinite. Up and down directions, for example, in physiological space are not sensed as having equal significance, hence physiological space is not isotropic; all places are not identical relative to what they can contain, hence it is not homogeneous; and physiological space is not infinite. In this highly qualitative space, we may presume that there exist in certain contexts of perception some stable relationships between sizes, depths, and distances, and, therefore, a model geometry.

Panofsky takes the principal evidence in favor of a curved pictorial space to be the use, in ancient Greece and Rome, of the so-called fishbone system[14] of perspective, based on Euclid's *Optics*. In this system, the size of an image is drawn to be proportional to the angle subtended by the object at the eye of the viewer, instead of—as in Albertian perspective—proportional to the object's projection on a flat picture plane perpendicular to the line of sight. Orthogonals in the image then converge, not on one point, as in Albertian perspective, but on a series of points distributed along a vertical axis, making a fishbone pattern. Such a perspective system used, for example, in the wall paintings of Pompeii and Herculaneum, is symbolic, Panofsky says, of the naturally curved forms of visual objects; he writes of the visual space of the ancients:

> A straight line is seen as curved and a curved line as straight; columns in order *not* to appear curved must have their *entasis* (for the most part, relatively slight in this classical period, as is well known); epistyle and stylobate, in order to *avoid* the impression of being curved, are to be curved by the builder; and the celebrated curvatures, at least of the Doric temples, give evidence of the carrying out in practice of such knowledge.[15]

The ancients, he concludes, were accustomed to make such observations, and the system of perspective they used, he says, symbolized the pictorial structure of objects as they were then perceived and pictured. These observations, he holds, can be verified by us in our own experience, though perhaps no longer with the same ease. Such a view about the spatial World of the ancient Greeks before Plato is supported by F. Cornford; "in these earliest cosmologies," he writes, "the universe of being was finite and spherical, with no endless stretch of emptiness beyond. Space had the form of that which filled space—the form of a sphere with center and circumference."[16] Infinite Euclidean space, as he says, was the invention of the atomists and not part of the common sense of those times.

From Image-Structure to Space-Structure: Debate

That artists at certain epochs attempted to represent pictorial space as curved, has not been uniformly accepted by critics. Gombrich, for example, in his masterly work, *Art and Illusion*, sets forth the problematic of spatial illusion in such a way as to raise doubts about the possibility of making inferences with regard to the kind of perception an artist had from the way he painted images.[17]

Assuming a realistic pictorial mode of visual perception, how does the painted surface mediate the perception of the depicted object? It is often assumed that, all things being equal, if the stimulus produced by the image surface is *in all relevant aspects* identical with that produced by the object itself, then the image when viewed is capable of producing in the viewer a realistic pictorial object—that is, is capable of being the occasion for an experience of "something akin to visual hallucination, as Gombrich calls it[18]—one that pictures some object as it is perceived to be.

There are several problems connected with this position. (1) It is not usually known what the relevant specific stimulus is; (2) whether known or not, the mere reception of the relevant specific stimulus is not sufficient for perception since, in addition, (3) the viewer needs to possess subjectively the embodied intentionality-structure necessary to perform the expected hermeneutic of experience; this, in turn, involves (4) the sharing of a common World. Given all these conditions, and furthermore given the interest and attention of the viewer, then we may expect that perception of the depicted object will occur.

Except in very artificial cases, it is not usually known what the relevant specific stimulus is for seeing a particular object in a particular way. Gombrich,[19] for example, has pointed out that the two-dimensional image is essentially ambiguous, and can be used to represent many different three-dimensional configurations (see figure 6.2). We recognize an image to be an image of something only if we have some prior acquaintance with the kind of thing that is being represented and are perceptually oriented to look for it in the image. To the extent that we succeed in recognizing an image to be the image of a particular object, we need to have known something about that object before. The subject's role—the structure of the perceiver's intentionality—is then deeply involved in determining what can be perceived and what in any instance is to be perceived. On this moreover will depend what in the environmental optical energy is counted as the relevant specific stimulus or cue.

The argument given above has led some writers to take the position that what is perceived when one looks at an image and what in the image occasions the perception of that object are just a matter of conventions. Conventions are rules that are adopted by public agreement not because of the nature of things but solely for the purpose of convenience. Perceptual intentions are learned, and thus depend on social acculturation and choice, but they are not *merely* conventional.[20] (1) They give rise to an automatic response that can even override the

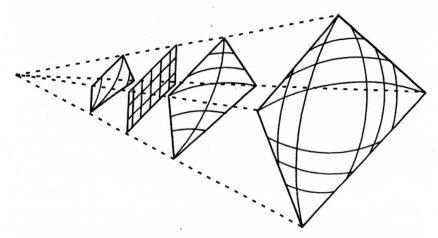

Figure 6.2: Wire ''gates'' of different shapes that give nevertheless the same projected form on a picture plane.

knowledge that one is being deceived, witness the perceptual illusions described above: a mere convention would not have this power. The persistency of perceptual illusions argues then against the mere conventionality of modes of perception. (2) The kinds of perceptual systems that can be realized are constrained by possibility conditions affecting both the subject (neurophysiologically and intentionally) and a World (as shaped by human action, as ground for alternative sets of systematic clues, etc.) but more about this below.

Influence of Picture Surface

There are subtle but important differences between the stimulus received directly from a pictorial object and that received from the image of that object constructed either by a camera or by an artist following the rules of artificial perspective. The rays that come from points in the image surface have the parallax (or convergence) appropriate for their point of origin in the image surface, not the parallax appropriate for the pictured object. Thus, a viewer of the image will tend to see an illusion of an object in depth because this is what the single rays suggest; at the same time the viewer is drawn back to the picture surface by the contrary pull exercised by the pencils of rays emanating from points on its surface.

There is more to a two-dimensional image, however, than mathematical perspective and the drawing of lines; some illusionistic images are not linear at all, but achieve their effect by size contrast, by the overlapping of motifs, by textural variation, or by shading and coloring, the painterly means used by fifteenth-century Flemish painters to create the effect of atmospheric perspective. In fact, painterly clues may run counter to linear and geometrical clues—as I shall show further on in the work of Vincent van Gogh—or vice versa. Thus, in order to create complex realistic illusions involving light and depth, the physical surface of the image which acts to provide the stimulus for a pictured or imaged object must be crafted with great skill and subtlety. As Gombrich points out,[21] an effective construction of illusionistic images is not something that can be reduced to a formula or technique once and for all. It is a matter of skill where traditional expertise has constantly to be improved and updated, in part because better ways of producing certain effects are always possible, even though traditional skills in certain areas seem close to perfection, but also because what constitutes the relevant elements of a pictorial object changes with the culture. The necessity of a "subsidiary awareness" of the pigments and canvas to establish the fixity of the representational illusion given in "focal awareness" is discussed by M. Polanyi and M. Pirenne.[22] In terms that I shall use below, the image serves as a sort of 'text' that is 'read' perceptually, and its structure leads and guides the performance of the perceptual acts.[23] Polanyi and Pirenne hold that, when the painted image is too distant to fall under even subsidiary awareness, the representational illusion, unlike those obtained from the painted surfaces of framed vertical canvases, is not fixed but changes with every motion of the viewer. The example they give is a ceiling fresco painted by Andrea Pozzo in San Ignazio in Rome.

Ambiguity of Pictorial Space

What follows from a theory of optical projection in Euclidean space does not mean, however, that—apart from past experience and the perceiver's intentionality—whatever one perceives is necessarily an Euclidean object, or that the visual space within which the pictorial object is represented is itself Euclidean. An image is essentially ambiguous: it is ambiguous, as Gombrich and others have shown in the case of "gates" or "grills" of different shapes and sizes (see figure 6.2),[24] because it could be the optical projection of any one of a family of three-dimensional Euclidean forms, but it is *also ambiguous in relation to the space of perception.*

The family of non-Euclidean spaces we have been considering as candidates for visual spaces is characterized by the condition that every object in the perceiver's World lies in the same direction from the perceiver in visual space as it does from the physical eye in physical space. That is, lines of sight from the perceiver are preserved in the transformation from physical to visual space: only distances along those lines of sight are changed by the transformation. Hence, a line of sight projection in visual space would be identical—point for point—with the results of a classical optical projection in physical space. Consequently, when it comes to perception, the same mathematical projection that has the power to mediate a variety of Euclidean forms has the power—with the reservations expressed in the last paragraph—to mediate their non-Euclidean transforms. Some perceptual illusions, such as the Müller-Lyer Illusion and the moon illusion, illustrate the fact that a more distant object can appear to be smaller in size than a nearby object despite the fact that they are of equal angular size and subtend equal optical projections on a given picture plane. The way an artist then makes hyperbolic shapes appear to the viewer does not in principle require the invention of some new kind of linear projection different from the one defined by the theory of geometrical optics: he would, however, have to rely heavily on clues of other sorts, for example, shading, coloring, brush stroke, and other painterly techniques, as well as possibly the kinds of schematic clues that seem to function in the types of perceptual illusions examined above. I do not wish to say that no modification of classical projective technique need be made, since the visual effect of such a projection may be strongly influenced by other clues in the image. If modifications need to be made, these will be a function of these other clues: they will not in principle constitute a new mathematical *system* of linear projection.[25]

Curvilinear Synthetic Perspective

Nonstandard projective techniques using curvilinear representations for straight lines have been developed and used in the past by many artists.[26] Such systems of synthetic perspective, based, as we have noted, on Euclid's *Optics*, were used in ancient Greece and Rome. Leonardo da Vinci[27] developed such a theory, in which he attempted to give rules for transcribing onto a flat surface the curved outlines of visual objects which he believed were seen as projected onto the inside surface of a concave sphere surrounding the observer. Thomas Reid[28] presented a theory of such shapes, which he called "visibles": the theory of "visibles" claims that we perceive primitively only two-dimensional forms—here spherically curved forms—to which we learn by experience to add the dimension of depth.[29]

Systems of synthetic curvilinear perspective have been developed and used for a variety of purposes among which are, for example, to represent the visibles, to compensate for visual distortions, to make a "softer" and more "living" picture, to catch "the roving quality of the painter's gaze," "to escape the limitations imposed by the picture as a framed and bounded object," as well as "to express the experience of visual reality which is only gained by introspection."[30]

Although Robert Hansen[31] has recently proposed a system of synthetic perspective like Leonardo's, the basis for his construction seems to be somewhat different. Leonardo was concerned with representing the "visibles"; Hansen is concerned with the representation of his—the artist's—experience of three-dimensional visual space. He finds that horizontals in a frontal plane bend around the perceiver, and in such a way as to meet asymptotically at points to the extreme left and right of the perceiver "on a plane with [his] forehead." Likewise, verticals bend also and curve in such a way that they too meet asymptotically at points "directly above [the] head and directly below [the] feet."[32] Neither verticals nor horizontals obey Euclidean expectations. To the extent that such an experience is an experience of hyperbolic space, as I believe it is, I have tried to illustrate its special systematic character in figures 4.1, 4.2, 4.3, 4.4, and 4.7.

The figures of chapter 4, in particular figure 4.7, sector 4 (which resembles Hansen's figures 7 and 8) are *not, however, images but kinds of maps or ideograms*. Ideograms are graphic but nonpictorial representations of ideas or meanings.[33] A map is a special kind of ideogram, which needs for its interpretation the concept of a set of projective rules, and can be read only with the aid of a legend. An image, on the contrary, delivers its message immediately, directly, and pictorially to perception. Hansen, I believe, confuses images with ideograms and maps. He errs in thinking that, because the viewer may see hyperbolic shapes, and can be reminded of this (or helped to understand this) by ideograms or maps, these maps or ideograms constitute images of those hyperbolic shapes. What Hansen is looking for is a means of constructing an image that makes the hyperbolic forms he describes so well appear to the viewer in a way "akin to hallucination," but it seems to me that he has not done so. It seems to me that his new method of perspective serves no other than the function that the figures in chapter 4 serve, namely, to help people *understand* the character of hyperbolic shapes in the hope that they may come to recognize hyperbolic perceptual episodes in their own experience. They do not themselves create episodes of hyperbolic perception.[34]

Marx Wartofsky, elaborating on the view that visual perception has a history, says that "historical changes in visual perception are in fundamental ways the results of changes in the forms of pictorial representation."[35] Using the elongated canon of El Greco to illustrate his position, he states that "canons of pictorial representation" teach us how to "perceive" and that what we perceive under their guidance is veridically a form of pictorial reality. He writes: "Though it is true that those paintings which we take to be 'realistic' are so because they most closely represent the way things look, things come to look the way they do because they are perceived in accordance with the rules of representation embodied in those pictures we take to be 'realistic.' "[36] This view is too complex and challenging to be confronted here. It involves theses that go far beyond the claims of this book in exciting but controversial ways. He suggests that the illusionary space of depicted objects comes to be projected as the space of "real" objects so that we eventually come to perceive real objects according to the canons of the depicted space. At that moment, the matching equation comes to be verified: the object-as-perceived matches the object-as-depicted, with the latter creating the possibility of the former.

It is not clear whether Wartofsky's theory is about visual perception in the strict sense of direct and immediate knowledge mediated by the optical field, or about new ways of conceiving things, or about new ways of imagining or relating things in some other sense. *Seeing* or *perceiving* in the strict sense implies the reception of an appropriately structured signal from physical objects themselves: it is then proper to ask for the clues in the physical World that support the new mode of perception. Such clues must exist apart from the images in which the objects themselves are merely depicted: for if this is not the case, then the theory is not about perception in the strict sense but about conceptions, or about imaginative or poetic creations.

My inclination, shaped by the discussions to be presented in this book, is to say that the objects as so depicted would not generally be so perceived in real life, since the possibility for so perceiving depends, not primarily on the existence of a new canon for depiction, but on the prior possibility of the new type of visual perception. This prior possibility of a new type of visual perception depends on structures in the optical fields which provide the opportunity for the establishment of a new type of perception. My position, which will be further explained below, is that hyperbolic perception depends on natural structures in the unaided Body and in the untransformed environment, and that other types of perception, such as Euclidean perception, depend on the production and pervasive presence in our environment of artifacts that play the role of what I call "readable technologies."[37] Without these, there is no direct perceptual access to objects under new perceptual forms, and it is only subsequent to the establishment of these new perceptual forms mediated by the objects themselves (and not merely by depictions of them) that a new canon of depiction will seem to be "realistic."

Among the readable technologies that have affected ways of perceiving objects are lenses and curved mirrors. Sometimes these were used by artists as substitutes for mathematical projective techniques they did not know how to use, sometimes to model an image of the World "more realistic" than that given by classical projective techniques, and sometimes to create new geometrical spaces, but not always realistic spaces.[38] It is interesting and significant that artists have used these technologies at various times in the past to generate refinements of the old canons of depiction, often in such a way as to add some hyperbolic features to the depicted object.

A. K. Wheelock has studied the use of lenses and mirrors in seventeenth-century Holland by such artists as Carel Fabritius, Emanuel de Witte, and Johannes Vermeer.[39] He concludes that the use of such instruments was influenced by new theories of optics and theories, such as that of Kepler, about the physiology of vision. The argument is based on models of the viewing subject: if the viewing subject is not the dimensionless point—the sight point—of the classical model, but has to be represented by a curved surface, that of the retina, then a "good" theory of image making should take into account the fact that the image on which the subject "gazes" is a curved one formed by light that has passed through the crystalline lens. The use of optical instruments in the seventeenth century, then, represents for the most part attempts to craft a better and more realistic image, more in keeping, as it was thought, with the best science of the time than was the image of classical (or artificial) perspective. While the psychological and philosophical beliefs underlying these attempts were certainly false, the attempts themselves bear witness at least to the artists' dissatisfaction with and distrust of classical perspective and of the pictorial space of the Renaissance. Whether they signify more than this is controversial.

As Depicting Motion or Curved Space?

Whether one can argue successfully from the use of such curvilinear systems of perspective that the artist intended to evoke in the viewer the pictorial vision of a curved World is much debated. Gombrich,[40] for example, is skeptical. He interprets a curvilinear perspective as representing not a curved space but movement within a Euclidean space: turning one's head to the left and to the right, one experiences, he says, a succession of profiles—all Euclidean—with a moving picture plane. He imagines an observer sitting in front of a wide-stretched facade:

> Naturally, as he turns right, the facade will appear to converge in one way, and as he turns left, in another . . . What we call "appearance" is always composed of such a succession of aspects, a melody, as it were, which allows us to estimate distance and size: it is obvious that this melody can be imitated by the movie camera but not by the painter with his easel. It is understandable if painters feel that the curve will suggest the movement of lines more convincingly than the straight projection, but this curve is a compromise that does not represent one aspect but many.[41]

There is a facade, Gombrich says, that "converges" now to the left, and now to the right. But what is this facade? It is not the physical facade, since this maintains constancy of height with distance from the observer, nor is it the lines of the drawing (the image on the picture plane), for this is not a facade but a set of two-dimensional lines. The facade that appears to converge must then be something else; it is, I claim, the facade as perceived in a certain way, it is perceived in hyperbolic visual space, and its lines converge in that space. A similar explanation can be given as to what we see when we see railway tracks converge. More will be said about the dynamic interpretation of curvilinear perspective in the following section.

Gombrich's fundamental reason, however, for holding the position he takes is his view that the art of perspective "aims at the correct equation: it wants the image to appear like the object and the object like the image."[42] By "image" he means (what I have called) the *object as pictured* or simply the *depicted object*; by "object" he means, however, the *object as it is physically*, not the object as it is—or would be—perceived ("how things appear to us"). But the aim of good pictorial image making is to match the depicted object as closely as possible, not with the physical object, but with the *perceived object*. Of course, the perceived object may and often does match the physical object, but neither always nor necessarily, as I have tried to show. With respect to the claim that the depicted objects should match the perceived object, Gombrich finds it "hard to see what such a claim would mean." This is because Gombrich takes the position that perceived objects are private objects. In the view put forward and defended in this book, perceived objects are not private objects, but public objects identified through the correct use of a common descriptive language, and it is meaningful to say that the depicted object must match the perceived object. In confirmation, experience tells us that it is usually not hard to discern when this aim has not been fulfilled.

If, however, it is true that we can and do perceive pictorially in a hyperbolic way, then the crafting of a good image will evoke an episode, "akin to visual hallucination," of hyperbolic space, and it is conceivable that the technique of crafting such an image is significantly different—though not systematically, as I have argued above, in the art of linear perspective—from the technique of crafting a good nonhyperbolic image.

Whether people are willing to entertain the possibility that there is an identifiable nonclassical art of image making for the picturing of hyperbolic objects depends on whether they find the theory of hyperbolic pictorial visual space plausible. People will find it plausible only if they come to recognize hyperbolic shapes in their own experience. Such shapes cannot, however, be recognized when presented, and cannot be represented to the imagination, if our memory and imagination do not already possess a broad repertory of known and tested non-Euclidean experiences. Like every synthetic concept, that of hyperbolic space emerges as an organizing principle, if it does at all, full blown or almost so.

One of the principal purposes of the first part of this book is to help one to recognize in one's own experience—artistic as well as commonplace—those familiar, often illusory, but usually unattended to features that have their origin in hyperbolic perception. Without this propaedeutic, the evidence I have presented cannot be judged: for what is referred to is a way of seeing, and either one comes to see that way, in which case one can judge the relevance of the evidence, or one fails to come to see that way, in which case one does not see the point of the claim, and other explanations for the alleged pieces of evidence arise. A perusal of the literature shows that there is no dearth of hypotheses to explain each single item of evidence that has already been introduced and discussed: there is, however, no single hypothesis, except the one proposed here, that explains all of them.

The Pictorial Space of Vincent van Gogh

The enigma of van Gogh's pictorial space has attracted the attention of scholars in a variety of disciplines, most notably art history, philosophy, and psychology.[43] Meyer Schapiro notes van Gogh's technical interest in the rules of perspective, but beyond that the enormous subjective and expressive import these had for him:

> Linear perspective was for him no impersonal set of rules, but something as real as the objects themselves, a quality of the landscape that he was sighting. This paradoxical scheme at the same time deformed things and made them look more real; it fastened the artist's eye more slavishly to appearance, but also brought him more actively into play in the world. While in Renaissance pictures it was a means of constructing an objective space complete in itself and distinct from the beholder, even if organized with respect to his eye, like the space of a stage, in van Gogh's first landscapes the world seems to emanate from his eye in a giant discharge with a continuous motion of rapidly converging lines.[44]

There are a number of paradoxes about van Gogh's forms as pictured and the space in which they are pictured: the forms are often distorted but look intensely "real"; they are created in a perspective system that is not controlled by an abstract set of rules with which we are familiar, although we sense that it is connected everywhere with the structure of van Gogh's active looking or sighting or involvement with things; the space looks realistic, organized with respect to an observer, like a stage space, but it is not objective like one constructed on scientific principles; it is one dynamically created by and for the artist with foreground objects bending towards him, and often a rapid plunge to a horizon that is nevertheless full of finite detail, a finite infinity![45]

In the course of a correspondence with Emile Bernard about painting from real models in preference to painting from the imagination, van Gogh says that, although he would modify the forms of motifs for the sake of pictorial order and expressiveness, such refinements should be consonant with a realistic portrayal. "In the matter of form," he writes, "I am too afraid of departing from the possible and the true." He adds, "I find it all ready in nature, only it must be disentangled."[46] At the time he was writing to Bernard, Vincent was already painting in a style characterized by those peculiarities of line and perspective (elements of the image) and those ambiguities of pictured forms (represented objects) with which I am concerned. For Vincent, "the possible and the true" is clearly not the realm of scientifically exact Euclidean forms.

From the first, the transcription of form was important to Vincent. Early in his career and at least on and off throughout the next ten years of his life, he used a heavy perspective frame—not the cardboard device used by many painters—to which he attached considerable importance: it was large, of wood, and was bisected horizontally, vertically, and diagonally by wires (or threads).[47] By squinting in a certain way, he was able to bring both wires and motifs into simultaneous focus. The "modern use of it," he wrote, "may differ from the ancient practice," and he stated his belief "in the absolute necessity of a new art of color, of drawing (*dessin*)—and artistic life."[48] It is this "new art of drawing" that I want to focus on. That it was concerned with the peculiar curvatures of visual space cannot be proved or disproved from his letters; evidence comes from an inspection of his works and from the general plausibility of there being such a thing as hyperbolic vision.

The linear perspective of a painting can be studied by making a ground plan and elevation of the scene represented. This can be done only if the dimensions of the motifs depicted in the painting are known. Such a plan and elevation for Vincent's *Bedroom at Arles* (figure 6.3) is given in figure 6.4; the reconstruction of the room (the original was destroyed during World War II) is based on certain assumptions about the size of the room and on measurements of furniture provided by the Musée d'Arles and other sources.[49] The reconstruction given in figure 6.4 agrees reasonably well with that given by John Ward except for the dimensions of the bed and the inferred position of the artist.[50] A mathematical projection can be constructed according to the laws of artificial perspective, taking the eye of the painter to be at the center of perspective for the painting. Figure 6.5 is such a projection assuming the picture plane to be parallel to the end wall of the room.[51]

The construction shown in figure 6.5 deviates considerably from the lines in the painting (figure 6.6) and because of the wide angle, it looks

Figure 6.3: *Bedroom at Arles*, by Vincent van Gogh (Stedelik Museum. Amsterdam).

quite clumsy.[52] Van Gogh clearly did not follow this construction. It is my claim that the changes, when read together with other spatial clues both linear and painterly, give to most viewers a visual impression that, though lopsided, is nevertheless unified, and exhibits a space that has the characteristic features of a closed, finite, hyperbolic visual space, with a near zone of which there are some foreground indications and a distant zone to which the walls seem to give impenetrable closure. The clues in the painting on the surface of the canvas that mediate this impression are more difficult to discern, but the visual impression, I hold, is inescapable. That this was the impression which the artist intended would be impossible to argue did we not share at least in an episodic way the kind of vision to which I am referring.

Among the clues characteristic of the near zone are a slight convexity of the floor in the center foreground, the convex curve of the bedboard, and the slight tilt of the chair on the left toward the viewer as well as the convexity of its curves. These are characteristic structures of rectilinear physical objects seen in the near zone of a hyperbolic

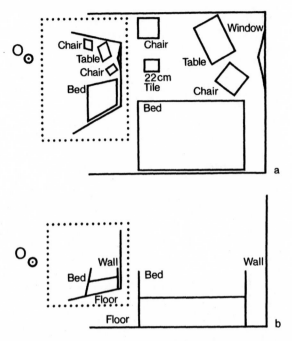

Figure 6.4: Reconstructed plan(*a*) and elevation (*b*) of Vincent's bedroom. The artist is located at *O* in the diagram. The shapes in the boxes represent one way of mapping the hyperbolic shapes of the room (as calculated from the law of transformation of the model): the shapes as drawn are drawn on a polar map with the viewer at the origin. Note that no Euclidean mapping can represent isomorphically the hyperbolic shapes of the transformed space; consequently, the shapes in the box are partially distorted relative to the true hyperbolic shapes. Arguments based on these diagrams (or maps) must use them carefully.

visual space. These features can generally be verified by the untrained viewer.

The cues for the distant zone are multiple. Some are provided by the manner in which the floor tiles are painted suggesting that, in the middle foreground, the floor slopes downward from right and left toward the center and that it rises toward the far wall. The disposition of the legs of the center distant chair—where the rear leg is on a straight line with the side legs—suggests that the floor on which the chair stands slopes upward. Likewise, the tilting of the seats of the two chairs toward the viewer suggests that horizontal planes below the line of sight curve upward. The left-hand corner of the room suggests a

Figure 6.5: Monocular projection in Euclidean space of Vincent's bedroom on a canvas subtending an 80° angle with the artist.

curved passage from the side wall to end wall, while the location and perspective of the far right corner suggests the right wall wraps around the head of the bed.[53] In summary, most viewers will see the side walls flare out, will notice that the space of the room is flattened, that depth is reduced with distance, and that the floor has an upward slope bringing it to meet the end wall in what is virtually the closure of space.

The impression that the end wall of the room bounds the pictorial space of the room is definitely strengthened by reflecting—or better, experimenting with a large reproduction of the painting—on the limits of what you see or would expect to see *if the window shutters were opened*. Having made some experiments of this kind with a large reproduction, covering the closed window of the painting with a painting of an open window displaying a scene beyond, let me summarize the outcome of such trials. You would find that certain views through the window are consistent with the internal logic of the pictorial space of the painting, while others are not. For example, if the scene displayed in the open window were of a tree filling the entire window or of the blank wall of a house, you would see these views (which block more distant horizons) as consistent with the space. If, however, the scene displayed were of the distant towers and rooftops of Arles, you would find this inconsistent with the expectations generated by the

Figure 6.6: Tracing of the linear forms of the objects depicted in the *Bedroom at Arles*. A comparison with figure 6.5 shows the "distortions" in the image vis-à-vis classical perspective.

composition.[54] From such variations of the motif, one concludes that the end wall of the room is not merely the boundary of what is depicted by the painting but also the boundary of the pictorial space in which these objects are depicted.

John Ward has a different reading of the painting: He agrees that there is "spatial distortion" in the bedroom as depicted and that the perspective is far from being accurate in detail, but for him these "curvatures" are an artifact of the technical process used by van Gogh to make a projection of the room. They are intended or at least permitted by the artist for expressive reasons, and they are presumably curvatures within Euclidean space (or a succession of Euclidean spaces): they do not indicate that van Gogh perceived his room with those curvatures.[55]

Ward argues, from a sketch of the room in a letter of Vincent to his brother, that the artist resolved his principal perspective problem—the wide angle of vision, the large canvas and the cramped quarters—by painting one side of the room with the canvas and picture plane turned one way, and the other side with the canvas and picture plane turned another way, the two directions making an angle of approximately 35 degrees. Ward concludes the painting is a synthesis of two perspectives and the "curvature" of the scene as depicted is an artifact of this process.[56]

It may well be that Vincent used a fusion of two perspectives to construct his image; it looks like a plausible hypothesis. But interesting and important as this may be, the question at issue is not how the image was crafted, but the character of the pictorial space it represents. Ward assumes that Euclidean physical space is normative for representational purposes (whatever is represented correctly, is represented for him in Euclidean space), and consequently that the only correct set of representational techniques are those of (classical) artificial perspective. Ward adds a proviso, however—if the angular field of the object is larger than the angular field of the canvas, or if the field of vision is partially blocked so that the artist has to change his viewing point in order to encompass the totality of what he wants to draw, then, he says, any compromise the artist adopts to resolve his dilemma results in the incorporation of "curvatures" in the depicted scene, though none are in the actual scene, and none are perceived when one looks at the real object.[57] What kinds of curvatures are these "curvatures," and to what do they refer? The answers are not clear. I take it that Ward is referring to distortions in the depicted scene *within* an overall Euclidean space, and *not* to a distortion of the space itself; so that the distorted scene would remain a Euclidean object and, though plastically distorted, would still be capable of being modeled in physical space.[58] This seems to be at the core of Ward's objection to my interpretation of the *Bedroom*. For his position, Ward makes two distinct arguments, (1) from the construction of the image, and (2) from the theory connecting curvilinear projections with motion. Let me consider each of these in turn.

1. If one were capable of seeing objects as *depicted* in non-Euclidean space, as I claim is possible, one would also be capable of perceiving *real* objects as construed in a space of that kind. One would, of course, have to settle the question of the ontological status of such perceptual objects, whether they should be taken as illusionary or as realistic. The fact that Ward is not concerned to debate the ontological status of non-Euclidean illusionary spaces indicates that he must want to hold that the "curvatures" affect the depicted object but *not* the space in

which the object is depicted, and possibly perceived. This conclusion agrees with the way Ward reads (or misreads?) Hansen: he interprets Hansen as attempting to describe the experience of objects seen as projected (in two dimensions) on an infinite spherical (Euclidean) field surrounding the viewer, but this is not what Hansen is doing; Hansen, by his own account, is describing an experienced three-dimensional "curvature" of visual objects, and proposing a way of representing these on a flat surface.[59]

Ward, in his first solution, assumes (1) that the artist wants to be pictorial, and (2) that the artist is forced for some reason to adopt a constrained situation in which the possibility or expediency of a correct pictorial representation is absent. In addition, he assumes that pictoriality is defined only for a Euclidean space. There are problems with this solution, even with a Euclidean constraint. It seems to me that a combination of circumstances like (1) and (2) should be rare, since the techniques of classical projection are quite mechanical and do not require that the artist ever physically occupy the center of projection for the drawing. In the case of van Gogh and the drawing of his bedroom, the canvas was large enough for a classical representation of the room; moreover, the projective outlines of that part of the room which, because of the cramped quarters, was blocked by the canvas could have been reconstructed by the artist with very little trouble.

2. In his alternative solution, which appears to have been suggested by a reading of Gombrich (see the preceding section), Ward explains the "curvatures" of the depicted object as the addition of motion—the artist's or the viewer's—to the static scene; the object-as-depicted would then be a *dynamic* synthesis of two (or more) Euclidean profiles. The motif should then be experienced as if apprehended by a lateral movement of the eyes, in which the eyes pick up successive clues to the successive profiles captured by the roving eye. Against this solution, I would make (*a*) a general argument, and (*b*) a specific one.

(*a*) In general, I would point out that a spatial synthesis of slightly differing Euclidean constructions can add up to a non-Euclidean object. Witness the non-Euclidean representation generated by a convex or concave mirror which can be looked on as the synthesis of the manifold of Euclidean representations generated by the infinitesimal parts of the mirror surface.

(*b*) Specifically, the motif of van Gogh's *Bedroom* is not experienced as a "melody," composed "of a succession of aspects." It often happens with baroque paintings that the viewer's gaze enters the picture on one side only to find itself drawn to the other. In such paintings, there is indeed "a melody of a succession of aspects," but this is not the case with Vincent's *Bedroom*; instead the gaze is drawn frontally into the room and held still in an absorbed contemplative vision. Ward himself endorses a similar reading of the painting: it was to be, in Vincent's own words, "suggestive of *rest* or of sleep in general,"[60] or as Ward puts it, to express "Vincent's longing for a home of his own."[61] It is not clear then why Ward refers to Gombrich's interpretation of curvilinear perspective, since it does not contribute to his own solution.

The key questions to be kept in mind are the following: how does one describe the spatial "distortions" (or "curvatures") of the bedroom as depicted by Vincent? and why did Vincent affirm these spatial "curvatures" (as he no doubt did)? I have already discussed Ward's presumed reply to the first question. In the final analysis, Ward's reply to the second supports my own reading of the expressive import of the painting.

The principal weakness of Ward's case is that it is formulated in a Cartesian World which van Gogh did not himself inhabit and without awareness or experience of other perceptual possibilities that van Gogh might have tried to share with us. I have given an alternative account; this works only, of course, if the shapes are *perceived* as I have described them. If they are not so perceived—and Ward says he does not perceive them in some crucial ways as I have described them—it may be because he is not interested in seeing this way, and therefore has not sufficiently tested the possibility. Alternatively, of course, my account could be incorrect, but if it is, then van Gogh's pictorial space takes its place with the other anomalous phenomena described in this book, and becomes just one of many unexplained coincidences for which there exists no satisfactory explanation separately or together.

In conclusion, I claim that, though the image Vincent crafted depends on a synthesis of traditional components which he did not hesitate to modify, the image has an expressive purpose about which there is no disagreement between Ward and myself. I want to stress in addition two points: (1) that the viewpoint is of a certain still, contemplative vision, and (2) that the closure of the space was intended. With respect to the first, I find, as I have said, Ward's argument confusing. With respect to the second, it is clear that, unlike the ink sketches van Gogh made for his brother and to which Ward refers, the *Bedroom* image has been deliberately modified in such a way that the image depicts a bedroom with a certain kind of space, closed and finite. This space is exhausted by the dimensions of the room and possessed just by the painter and his companion Solitude; it represents a total World of isolation and rest, into which nothing from the outside has power to intrude, since there is no longer an outside. The viewer should be able to verify this experience in his/her own perception by standing close to the painting—about half a meter from a full-size copy.[62]

Although I have limited this discussion to the apparent deformation of space in the *Bedroom*, similar deformations characterize enough of van Gogh's paintings between 1888 and 1890 to warrant the conclusion that the artist's aim—among others—was to record how his anomalous visual World was structured. The phenomenon I have described is not peculiar to van Gogh's later period. In the spring of 1883, for instance, van Gogh painted a view of the large rectilinear tulip beds extending from horizon to horizon in the flat countryside near The Hague.[63] In spite of the ease with which the geometrical character of the flat Dutch landscape and squared flower beds could have been drawn in classical perspective, van Gogh has recorded a spoon-shaped sweep of field rising to meet a horizon that is stretched at a finite distance from the observer. The concave force of the space as van Gogh pictures it is strikingly hyperbolic.

A reliable analysis of any pictorial space depends on the dimensions of at least some Euclidean forms being known. Thus, the inclusion in the motif of chairs, a table, and a bed—with their established sizes and shapes—makes it possible to proceed, albeit cautiously, with the analysis. In the case of a landscape, however, it is usually more difficult, even after going to the site, to develop a set of known coordinates on which to base a computation. The elements in the motif tend too often to be characterized by irregular shapes—mountain ranges, olive groves, an iris-filled roadside ditch. There are, nevertheless, some important inferences to be made from a comparative study of photographs of an outdoor motif with the motif as painted. We are fortunate in having photographs taken by John Rewald from many places where van Gogh stood or sat when he painted.[64]

A photograph of an object represents it—according to our usual way of interpreting photographs—as existing within a Euclidean space, and the two-dimensional marks on the plate by which the object is represented obey the classical rules for monocular projection on a picture plane, the rules that a painter would follow if he wished to represent Euclidean forms on a plane canvas. We have established the fact that

Figure 6.7: *The Iron Bridge of Trinquetaille* by Vincent van Gogh.

van Gogh knew the classical rules for projection in Euclidean space, and that he intended to render the *true* forms of things. Since his projections differ in what appear to be significant and systematic ways from what would be expected of a classical projection in Euclidean space, I have argued that he experienced the forms of everyday things in a different and (to him) "truer" space, and that his schema of marks on the canvas was intended to make manifest the forms of objects as experienced in this "truer" way, in this "truer" space. I have given reasons to identify the "truer" space as hyperbolic visual space.

The *Iron Bridge of Trinquetaille* (figure 6.7) and the *Railway Bridge over Avenue Montmajour, Arles* (figure 6.9) both illustrate the spatial structure of the transition to the distant visual zone and the distant zone itself. The foreground objects in both cases lie in the transition zone between near and distant, and consequently appear larger in relation to the background objects than they do on the corresponding photographs (figures 6.8 and 6.10 respectively). In the *Iron Bridge* (figures 6.7 and 6.8), the steps in the left foreground occupy nearly one-quarter of the canvas, while they occupy only one-sixth of the photograph. Likewise, in the *Railway Bridge* (figures 6.9 and 6.10), the entrance arch to the

Figure 6.8: Photograph of the iron bridge of Trinquetaille (Photograph by John Rewald).

viaduct is much larger in the image in relation to the size of the exit than in the photograph. The depicted space in both cases plunges, as it were, into a tunnel or vortex. Van Gogh also suppresses the kind of photographic detail that spells out the full sequence of receding planes, and eliminates the numerous infinitesimal objects that might ordinarily appear on the horizon. In the painting of the *Iron Bridge*, for example, the vertically oriented depth clues follow one another more closely than in the photograph. Horizonally oriented depth clues, however, such as areas of light and shade on the ground, are spaced farther apart. The total impression created is of a rising ground with distant objects closer to the observer than they are in the Euclidean experience portrayed in the photograph. In the distant visual zone, van Gogh paints a few objects that are clearly represented at a finite distance and, since the ground plane bends upward to meet the horizon, their sizes on the canvas are consequently smaller than they would be on a photograph; they are nevertheless seen as closer than their physical distance. The elimination of clear depth differentiation in the distant zone and the progressive tilting of the ground plane until it merges with the sky complete the impression of a closed and finite World.

Figure 6.9: *Railway Bridge over Avenue Montmajour* by Vincent van Gogh (Collection Erich-Maria Remarque, New York, on loan to Kunsthaus, Zurich).

My conclusion is that, as far as the spatial peculiarities of his painting are concerned, they are not *directly* the product of pathological psychology—as has sometimes been said—or *merely* the effect of strong emotion, but of a special kind of visual perception, hyperbolic in character, whose possibility has been put forward in Part I of this book. This kind of vision recreated the World as hyperbolic, and I have stated that van Gogh was able to exploit this kind of vision in his painting.

One would be curious to know whether van Gogh asked himself if this was the *true* form of the real. If it is the case that ontology is (should be) normed by common language, then the real World is Euclidean, and the hyperbolic transformation of it is a systematic distortion. If, on the other hand, it was van Gogh's persuasion that pure unaided vision has (should have) priority over cultural artifacts, then he would have cherished the new experience as the epiphany of the *real* that lies behind culture, that is usually masked by the everyday cultural overlay. It is very probable that van Gogh did believe he was the witness of a special epiphany of the real; and if this is so, his aesthetic

Figure 6.10: Photograph of the railway bridge over Avenue Montmajour (Photograph by John Rewald).

would be rooted in more than subjective emotion, it would be oriented toward a certain vision of reality and a certain kind of World.

One final question: if the formulae for linear perspective are in principle the same for Euclidean as for hyperbolic visual spaces, why does the perspective in van Gogh's paintings differ nevertheless from the classical? Whence the variations? In the first place, the variations, though often quite noticeable as in the *Bedroom at Arles*, are generally not large, leading some authors to speak as if van Gogh were just careless in his use of perspective:[65] the variations are, in my view, nonsystematic "refinements" of those rules. They work together with other, mostly painterly clues—such as the coloring and brushwork of the floor tiles in the *Bedroom at Arles*—to create the perception of flattened depth and curving three-dimensional lines. These refinements then are only part of a larger strategy for crafting images. It is very possible, however, that van Gogh was misled about the way to craft an image capable of evoking a perception of hyperbolic space and ended up with a compromise between the curvature he saw in the visual motifs and the surprisingly linear projections given by his perspective frame.[66] Finally, it is possible that the refinements were clues placed in the painting by the painter to help the viewer *understand* the painting: such clues would function as part of a legend, associated with the maplike or narrative function of the representation.

Part I, Conclusions

The conclusions of this first part are the following: (1) many common anomalous visual phenomena have structures characteristic of a hyperbolic space; (2) these structures are essential to (i.e., invariant characteristics of) visual space; and (3) these essential structures can be experienced as apodictically given, that is, the manifold of specified hyperbolic profiles in common experience can be sampled, easily and familiarly, thus enabling anybody to test and verify at will the essential structure of the experience.

These statements call for some commentary. The first of them is the least controversial: some anomalous perceptual phenomena have hyperboliclike structures.

The second is supported by the following reasoning: although we cannot look exhaustively at all possible profiles of the hermeneutical Luneburg model (here, the geometrical structure is visually interpreted), the phenomena we have examined can all be interpreted as profiles of that model. The evidence, moreover, is sufficiently persuasive that a clear and definitive possibility of visual experience is described by this model. The model is then the invariant of the profiles, the profiles are the transformation group of the invariant, and this identifies the invariant to be the essence of the phenomena. A significant value of the model is that it defines a class of anomalous visual perceptions, some of which are common and easily identified, others uncommon, more difficult to identify, and not previously associated as members of the same class.

Third, once it is understood that the phenomena as experienced are profiles of an essential structure, this essential structure becomes a *given* in perception. Moreover, its presence and structure can be verified easily and familiarly in its manifested profiles; that is, it is capable of being given *apodictically*. If new descriptive terms or new usages of old terms are introduced into public language to differentiate instances of hyperbolic vision from instances of Euclidean vision, hyperbolic visual perception can become part of everyday experience, extending the horizons of everyday Worlds.

Part II: TOWARD A PHILOSOPHY OF SCIENCE BASED ON THE PRIMACY OF PERCEPTION

7

Nature of Perception

Perception and Percept

Much of what follows is a statement of where I stand in the philosophy of perception. Some positions are simply stated: these are positions that are, generally speaking, abundantly discussed in the philosophical literature, and in these cases, I shall merely refer to some of the principal works in which they are laid out and defended.[1] However, where a novel position is articulated, or one not yet widely discussed in the literature, I shall take care to present it together with arguments that support it, and with the difficulties that it must overcome.

The approach I take to perception is through phenomenology. As explained above and more fully below, this is a first-person approach to the study of perception, where the inquirer acknowledges his/her participation in the community of perceiving subjects and uses in the inquiry the direct access this gives to the experience of performing perceptual acts.

Perception at the level of the natural attitude is a species of direct knowing of real objects or their sensible qualities, mediated by specific structures in the physical environment; by *knowledge*, I mean roughly (but in a sense to be made more specific below) warranted true belief. I take *perception* to be a completed act of knowing, and so to be a perceptual recognition, judgment, or belief capable of being expressed in a descriptive statement about a state of the World, usually by the perceiver himself/herself.[2] I distinguish the *act of perception*, which is epistemic (a form of knowing), from merely *having a percept*.[3] The latter—as I take it—is incomplete: either judgment or belief has been suspended, or no judgment has been made at all, and it is then nonepistemic. Having a percept, in this view, is possessing the appearance of a (not necessarily real) object and as such is short of being knowledge about actual or possible states of a World. It is of course not even knowledge about appearances, since having it is not sufficient to count as knowledge: it could be thematized, however, and could achieve the status of a special kind of object—an appearance or phenomenon—residing not in the World but in imagination. Appearance and reality, two objects of different kinds, use the same percept, however, although in different ways.[4]

Appearance or phenomenon, thematized as described above, is what phenomenologists call a "bracketed perception," or a "perception subject to epoche." It is an individual profile connoting an invariant perceptual essence, but considered in such a way that all judgment about its relation to the actual (here and now) situation of some World is suspended.

Contrary to the classical traditions of empiricism and rationalism,[5] and more in keeping with Aristotle, I hold that the act of perception terminates not in sensations, elementary ideas, appearances, percepts, representations, or other mental contents, but in the reality itself by intentional identity. What is perceived is a state of the World, not just a representation of that state. Unlike Aristotle, however, I maintain that the reality so reached is not any state of affairs independent of human activity, culture, or history. Since perception purports to be a direct form of knowing, it purports not to be a product of deliberate or explicit inference from or construction out of some set of prior knowns— sensations, ideas, raw feels, and so forth—serving as premises for or constituent parts of a perceptual object; we have no awareness of such inferences or constructions.[6] Perceived objects present themselves to the perceiver in the act of perception as directly given, as belonging to the set of basic, evident, and primary objects of knowing whose presence and form is in some sense imposed by the World we find ourselves in. A World being the public domain, perceived objects belong to the public and not to some private domain.

Even though acts of perception are experienced as basic, evident, and primary modes of presentation of a World, specific perceptual systems (with their specific intentionality-structures) are not, however, native and original capabilities: they are themselves achievements of human living, learned, I believe, through behavioral interaction with the environment. We learn how to perceive what we do perceive and this very accomplishment contributes to the type of World we experience ourselves to be in.[7] Moreover, in every particular act of perception, the perceiver is an inquirer, searching the environment for the presence of those things that belong to its repertory of perceptible situations. To quote Gibson, whose view in this matter I follow:

> The elementary colors, sounds, smells, tastes and pressures that were supposed to be the only data of sense (and that are indeed obtained when a passive observer is stimulated by carefully measured application of energy in the laboratory) have been thought of as an inborn repertory of experience on which the baby's later perception is founded. Learning to perceive, then, had to be some such process as the associating of memories with these bare impressions, or the interpretation of them, or the classifying of them, or the organizing of them. Theories of perception have been concerned with operations of this sort.
>
> If the senses are perceptual systems, however, the infant does not have sensations at birth but starts at once to pick up information from the world. His detection equipment cannot be exactly oriented at first, and his attention is imprecise; nevertheless, he looks at things, and touches and mouths them, and listens to events. *As he grows he learns to perceive but he does not have to learn to convert sense data into perception.*[8]

Objects of perception, as I hold, are not constructed out of more primitive and basic nonepistemic elements such as sensations, ideas, sense impressions, or "raw feels."[9] However, *learning* to perceive may involve processes of inference and interpretation, and *clarifying* what one believes one has perceived may lead to the construction of a theoretical account of perception in which sensations, sense impressions, or raw feels, are postulated as constitutive elements, but none of these, I hold, are directly perceived as components of, or necessary conditions for, the successful performance of an act of perception. In Schrödinger's words, they "drop out of consciousness"[10] when one has learned to perceive clearly. To the extent that these are discoverable in perception at all, they must then be products of a process of analysis, or "deconstruction," of the act of perception that generates these elements by the method that seeks them out.[11] They are not then part of the evidence with which any particular perceived object manifests itself as factual and imposes its presence in perception.

One of the deeper concerns of this book is to probe the primordial roots or "deep structures" of perceptual consciousness. It is these that

make possible the acquisition of such a range of intentionality-structures as one finds in cultural history or in our own experience. These
"deep structures" also account for the fact that some intentionality-
structures thought to be "logically possible" are easily and "naturally" acquired, and others not at all, though these too are thought
equally to be "logically possible." They do not explain, however, why
one finds oneself immersed in a World of the specific kind one actually
experiences; they can explain only why such a historical World is a
possible one.

Phenomenology of the Perceptual Object

Perception always takes place in relation to a horizon that has two
components, an *outer* and an *inner* horizon. In any individual act of
perception, the perceived object has an outer horizon, or boundary, or
contour, which separates it from the background against which it
appears. Each profile then has naturally a foreground-background
structure. The background too belongs to the World, but negatively: it
is that which is not part of the object.

The manifold of different possible profiles is the inner horizon of an
object (of *this* object or of an object of *this* kind). These profiles exhibit
the various facets of the object; the essence of the object is the set of
invariant structures that generate the manifolds of its profiles. In the
natural, that is, precritical, attitude, objects, but not their essences, are
recognized immediately and directly. The essence of an object (of *this*
object or of an object of a certain *kind*) is discovered only by critical
analysis and experimentation on the possible profiles of the object. This
process of critical analysis and redefinition ("prescription," in C. S.
Peirce's term) refines and purifies the commonsense meanings of
descriptive terms and gives them a philosophical status that enables
them to enter into apodictic judgments.

The term "horizon" when used below without qualification will
always mean *inner horizon*.

Apodicticity is attached only to the manifestation of an essence in
perception: it is the kind of certainty that a perceptual judgment about
the manifestation of an essence possesses. Such judgments are critical
perceptual judgments: they can be repeatedly verified at the will of the
subject, because the manifold of the object's profiles is in principle
known, and the essence manifests itself as the defining invariant—the
kind of thing—of that manifold. We have no trouble, for example,

distinguishing a table from a picture of a table, because, despite the fact that the two manifolds of profiles have a profile in common, they are otherwise different. Although an apodictic perceptual judgment, for example, of recognition, is certain, this certainty is neither infallible nor absolute; it is only as securely warranted as direct human knowledge about the World can be. As apodictic judgment, it is the expert performance of an act of perceptual knowing. The present object is constituted under the guidance of a perceptual essence, and a judgment of recognition, say, is prudentially made in the presence of its justifying epistemic condition, which is the concomitant awareness both of excellent and trustworthy subjective performance, and of the fulfillment of objective conditions imposed by the World.

I take the *truth* of an empirical statement to be its conformity to the reality that is perceived with apodicticity, and not in the Tarskian sense of conformity to an acultural, ahistorical, and atemporal measure independent of human activity and interests. Correlatively, I take falsity to be disconformity between what is stated to exist and the reality which is or could be manifested in and through perception. I take truth or falsity not to be conformity between mental representations and reality in a Cartesian sense, but to be an intentional identity between what is known as known and what is known as real. Moreover, I take truth and falsity to be not atemporal ideals, known only in the limit of perfect knowledge, but a property of empirical statements knowable when the cultural conditions are specified. A weaker notion of truth sometimes referred to as "empirical adequacy" can be defined: this is conformity to what is perceived with unaided perception.[12] I take the conditions of truth to include what I call "readable technologies" as aids to perception in the sense to be explained below.

The process of critical analysis that is responsible for revealing the manifold of possible profiles and the essence (or invariants) to which that manifold pertains is more easily described than performed. The search is always a cooperative effort of a community of critical researchers in some disciplinary area, such as the natural or social sciences, philology, philosophy, or other humanistic studies. The determination of an essence is then a historical process that may or may not be successful. It may fail because interest (or funding!) was lacking to complete the project or because sustained inquiry led to the discovery that there was no essence to be found, that the alleged profile was only a pseudoprofile, and that the essence that was sought was a will o' the wisp or illusion—a pseudoessence. Into the latter category would fall, I presume, the search, now long abandoned, for phlogiston, tetrahedral atoms, "bad blood," witches, and the "little people" that inhabit fairy raths. Thus, not everything thought at some time to be the profile of an essence turns out to be such when the critical process extended in time is complete. The extent to which an essence is dependent on a historical period and culture will be discussed below in chapter 13. Not every description that was once used but later abandoned, however, lost currency because what was described turned out to be no more than a pseudoessence: it could be that people just lost interest in a real entity. Among the topics then that a history of perception would include are (1) the story of how a presumptive essence came to be explored, recognized, and affirmed in a particular society; (2) the story of how a presumptive essence once influential in a particular society came later to be abandoned as a pseudoessence and the set of its presumptive profiles classified as pseudoprofiles; and (3) the recovery of lost essences and their associated profiles.

Physical Causality and Perception

Although it is the World-as-experienced that imposes on us with some form of necessity the objects we perceive, it does not make sense to say that these objects-as-perceived are produced in us *solely* by physical causality exercised on our bodies and sense organs by the physical reality that surrounds us.[13] In the first place, the proximal stimulus—for instance, the photons striking the retina—is rarely, if ever, sufficient to account for the character of what is perceived, and in the second place, what is perceived is not the pattern of photons striking the retina but an object or situation over there, distinct from the perceiver, and exhibiting the structure of a foreground object related to a background horizon. It is heavy with latent meanings and anticipations, because what we perceive is always perceived *within* or *against a horizon* of relationships and possibilities that function as a preunderstanding for the process of coming to perceive. This character restates the theme of the *hermeneutical circle* applied to the special materials of perception.

Nevertheless, I assume that the causality of the physical domain is in some sense necessary for perception. James Gibson, for example, maintains that the ambient array of optical energy is the necessary conveyer of perceptual *information* to vision.[14]

The term "information" is notoriously abused in much psychological literature, and philosophers are also not without fault. Information can refer to one or both of the following elements: (1) a structured stimulus field, for example, Gibson's "optical array"—let me call this "information$_1$"—and (2) a communicated content about the World—let me call this "information$_2$." *Information$_1$* is the technical sense introduced by C. E. Shannon and W. Weaver, and used thereafter by many psychologists, linguists, and others; other terms used for this are "information channel," "signifier," "(coded) cipher," and so forth. *Information$_2$* is the sense of "information" in ordinary language; other terms for it are "read or transmitted message," "signified meaning or object," "(decoded) meaning," and so forth. There is also an ambiguous usage of the term "information," in which it refers to the dual system, information$_2$ as piggybacked on information$_1$. For example, in current accounts of perception, one reads that "information"—meaning information$_2$—can be "picked up" or "extracted" "directly" from the "stimulus," meaning the dual system information$_1$ *plus* information$_2$. This dual structure of information is what I refer to as its *hermeneutical* structure.

There may be a temporary and limited advantage to treating this dual system as one piece of "information": current writers in the empirical sciences, as I have said, often use the term in this way. It is, however, confusing and tendentious; it usually assumes a one-to-one correspondence between information$_1$ and information$_2$ without consideration of conditions; and it generally carries with it a monistic (materialist) bias toward reducing information just to information$_1$. It is also an abuse of ordinary language: it tends to preempt discussion of the hermeneutical structure of perceptual information and, since fads and idealogies play a subtle role in public debate, it tends by a manipulation of vocabulary to exclude "nonbelievers" from the debate.

Gibsons's view is that within the familiar and limited context of an ordinary experiential situation in a given culture for an adult, information$_1$ is linked uniquely to information$_2$; like many other psychologists, he often makes use of the ambiguous sense of the term "information." He claims that information—information$_1$ connoting information$_2$—exists in the form of first- and higher-order structures in the optical field to which our visual perceptual powers respond. Specific information—information$_1$ connoting information$_2$, also called "specific stimulus"—exists in his view for each individual thing and feature in the

environment—information$_2$ connoting information$_1$—and so on. Information, he says, can be caused directly by the object, or it may be
present in the ambient light because it was introduced into it by artifice,
as for example, by reflecting light from a mirror or bouncing it off a
photograph or a drawing *depicting* the thing or feature that is to be
perceived. In all cases, a necessary—for Gibson, also the sufficient—
condition for perception is the presence in the ambient optical array of
that specific structure associated naturally (but produced, perhaps, by
artficial contrivance) with the thing or feature that is perceived or
depicted. The one-to-one correspondence between information$_1$ and
information$_2$ will be challenged below.

The presence or absence of appropriate specific information in the
optical array would then distinguish two classes of mental acts: (1) the
presence of a specific stimulus would be associated with the class of
perceptions—whether of X or of X-as-depicted; (2) the absence of a
specific stimulus would be associated with the class of imaginings of X,
or memories of X or hallucinations of X. In class (1), we experience
real objects that do exist—real X or some real Y (a non-X) that is an
image of an illusionary X, the difference between the two being that in
the former, the modulation of the optical array (the specific proximal
stimulus) is produced by X itself, while in the latter, it is produced by
Y. In class (2), we do not experience real objects, because X as
represented is not the product of a stimulus originating in the World,
either at X or at Y—what we experience then is only the product of
imagination, memory, or hallucination.

Imagination, Memory, Hallucination, Illusion

These causal criteria for distinguishing true perception from imagination, memory, and hallucination attempt to explain scientifically on
nonpsychological grounds why it is that such differentiations exist, and
on what grounds particular experiences should be assigned to one
category or another. The causal explanation, however, does not elucidate the original distinction: this must be analyzed within the context of
prepredicative experience antecedently to any attempt at scientific
explanation.[15] *Perception* is found in our experience of some World
together with other types of experiences, for example, *imagination,
memory, hallucination,* and *illusion,* to which it is related and from
which it distinguishes itself. An act of *imagination,*[16] for example, may

present an object very vividly, but it distinguishes itself from an act of perception by the fact that the object is not experienced as actually located in some Life World but only as possibly so located (dependent, for example, on certain conditions presently unfulfilled). An act of *memory*,[17] likewise, may present an object very vividly but is distinguished from acts of imagination and perception by the fact that the object is experienced as properly located in the Life World but at some past epoch. A *hallucination* is a "false" or "diseased" perception, not situated in the Life World of the community, but which presents itself nevertheless fitfully, inarticulately, and uncertainly as an element coherent with some tentative anomalous Life World; that anomalous Life-World, however, as Merleau-Ponty points out, fails to establish its credentials in depth even for the subject of the hallucination.[18] An *illusion* is a mistaken appearance—persistent even despite our knowing that it is mistaken—of something present in our Life World: illusions have a persistency that calls for an explanation.[19]

Perception and Past Experience

Returning to ordinary perception in the natural attitude, what one perceives is always an individual object (thing-, event-, process-, or situation-, etc., as-perceived) of a definite kind.[20] That does not imply that we cannot perceive a kind of situation that we have never perceived before, nor that our powers of perceiving are static and do not grow. Insofar as we perceive something we have never perceived before, we perceive it in virtue of those properties it has which are perceivable by us: we grasp it by and through what for us at that epoch is antecedently perceivable, that is, as related to some horizon of our World. If nothing about it is antecedently perceivable, then we simply will not recognize it, we may be systematically blind to it, and we fail to perceive even that there is something out there for which a new descriptive category needs to be formed. What for us at any epoch is antecedently perceivable is, however, among other things, a function of learning; our actual capacity to perceive, like most of our psychological powers, is a dynamic function of culture and history which is always open to future growth and change. The perceiver at any epoch then has at his/her disposal a repertory of object categories, that is, of things-, events-, properties-, processes-, situations-, and so forth, as-perceivable.

Horizons of Geometrical Perception

Shapes and spatial relationships, of course, belong to the repertory of perceivable qualities since all that is perceivable exists in space. What something that is perceived in a certain way really is, however, may or may not depend essentially on its geometrical shape or spatial relationships: it may be recognized or defined by other features, for example by its social function or purpose rather than by its precise geometrical shape or relationships. A house or a chair even when seen in a distorting mirror is immediately recognized for what it is despite the distorted geometrical percept. Distorted geometrical percepts do not in such cases prevent the identification of the correct object-category and (under certain circumstances) the geometrical form that goes with it. A pure geometrical shape, however, could be unrecognizable (perhaps even indescribable) when seen in a distorting mirror.

In this study, I have been concerned principally with the percept, appearance, and reality of that for which a specific shape, physical or visual, and/or a specific geometrical relationship is essential. A manifold of object-categories for which geometric shapes and relationships are essential diagnostic features I shall call a "horizon of geometric perception."

Realism of Everyday Life

In everyday perception, we are realists:[21] we expect things to be perceived as they are "in reality" and what they are "in reality" is by common consent in our culture what the scientific account determines them to be. Thus, in everyday perception, one expects objects to exhibit spatial properties of the kind physical measurement shows them to possess. If they do not in any particular case, we call our perception "illusionary." The structure of the ambient optical array that provides the proximal stimulus for perception, and the environmental structures that modulate these arrays are all systematically and normatively Euclidean to an approximation sufficient for the purposes of this study,[22] that is, they take their norms from the nonrelativistic and classical physical science of middle-sized objects. All of this we know, and these norms permeate the practical everyday horizon of visual perception, as well as the anticipations incorporated in common language.

Against such a background, it would be reasonable to suppose that reality is perceived correctly only when it is perceived with a Euclidean structure; and facts would be pictured correctly only when they are pictured with Euclidean percepts. This position as a *philosophical position* is that taken by Direct Realists, such as D. H. Armstrong, and Scientific Realists, such as J. J. C. Smart, W. Sellars, and others.[23]

Perceptual Judgments and Descriptive Criteria

Objects given in and through perception to individuals become public objects (and consequently accessible to scientific researchers and to others) only through the public behavior executed by the perceiver. Such behavior may take many forms: a verbal descriptive report, for example, or an elicited action made in response to a researcher's instructions such as to "match" pairs of lengths, distances, lines, or other phenomena. How this behavior is *intended by the subject*, and how it is *taken or interpreted by the researcher*, is important for establishing the usefulness for scientific purposes of information gained in this way. It is assumed that only public facts (i.e., facts that can be shared among researchers and verified independently by each of them) are useful for scientific research. Such facts *prima facie* seem to be restricted to such as have a third-person description. Nevertheless, it is appropriate to ask whether any first-person facts play or could legitimately play a role in scientific research.

As I have already pointed out, perceptual information has a dual structure composed of information$_1$ (the state of some information channel) and information$_2$ (the transmitted message that, in this case, is some state of the World). Information$_2$ is both an aspect of the World and the content of an act of perception; as the content of a subject's perceptual act, it is prima facie inaccessible to researchers except as a theoretical entity reached through the interpretation of external behavior. As an aspect of the World, however, it is, like the external behavior which it is fashioned to explain, directly accessible to each researcher in his/her own experience. Such direct first-person access to perceptual information (information$_2$) is then an inescapable condition for third-person scientific reporting, but such information is useless unless it can be treated as part of a common background to which all researchers have equal and secure access.

It will be shown below that there exists the possibility that researchers may come to 'read' the external behavior of a subject as a 'text,' and in such a way as to obtain direct—perceptual—access to some essence of the behavior. Such an essence would usually coincide with the perceptual object experienced by the subject. The essence as 'read' from behavior, however, is a scientific explanatory (though perceptual) object that is part of a scientific image, while the essence of that which is perceived by the subject is part of a manifest image; these essences may not coincide. The general relationship between scientific and manifest essences will be discussed in detail below. The possible disparity between the account offered by the experiencing subject of what is given in his/her perception—its manifest essence—and the account offered by an outside (scientific) observer of what makes sense of the subject's behavior—its scientific essence—implies that there are two distinct ways of studying perception and other intentional processes in human subjects: they are, respectively, the first-person mode of inquiry, and the third-person mode of inquiry.

A third-person mode of inquiry is one where a researcher (X) is not also at the same time a subject (Y), and the question addressed by X is: what is it for some Y not an X—that is, a *nonresearcher-subject*—to have a perception? This mode is characteristic of empirical psychology and of the social sciences in general, as well as of empiricist analytic philosophy. A first-person mode of inquiry is one where a researcher (X) is also at the same time a subject (Y), and the question addressed by X is: what is it for some Y who is also an X—that is, a *researcher-subject*—to have a perception? This mode is typical of phenomenological inquiry, methods of participant observation, "reenactment" theories of history, and "empathic" methods of social understanding.

A third-person approach assumes that there is a stable unproblematic background of perceptions and beliefs about the World that all members of the research community share. It suffers, however, from a characteristic weakness: it tends to finesse problems about differences in background World beliefs among researchers (or among researchers and subjects), which would call for a first-person and hermeneutical analysis of the acquisition and content of those beliefs. I shall return to this point in later chapters. A first-person approach such as a phenomenological one assumes that individual researcher-subjects who employ it to analyze the conditions for apodictic knowledge do so with reasonable confidence that intersubjective agreement can be reached.

I have shown in earlier chapters that first-person accounts of visual objects do not always or necessarily fall under the same set of descriptive criteria as third-person accounts of physical objects: different descriptive languages—Euclidean for the physical, Euclidean or hyperbolic for the visual—may be needed depending on whether the focus of the research is on one set of phenomena or on another. The conduct of the research will be influenced by the particular episte-mological views (and skills) of the researcher and of the subject, and by their ability to communicate unambiguously with one another.

The scientific researcher, for instance, may be committed to a scientific realism. He/she will then be disposed to accept as faithful and veridical only whatever in a particular subject's report (or elicited action) can be taken to refer with more or less descriptive accuracy to third-person scientific measures of the proximal or distal stimulus. The subject, however, may take the meaning-context of the researcher's question or instructions to include the possibility of hyperbolic visual characterizations.

I have argued that the spatiality of visual objects is a function of *hermeneutic choices*—"What makes sense here?"—among the differ-ent objects perception can have. It is the particular hermeneutic choice we make—Euclidean or hyperbolic—that determines the universe of facts that are envisioned in our calculations. These are two incompati-ble universes, though mappings of the sort to be discussed in chapter 10 exist between them.

One hermeneutic choice the *subject* may make is that of taking the researcher's question (or instructions) to be formulated exclusively within the context of a World as *physically described*; he/she would then interpret the question (or instructions) to mean: respond to what is presented in and through perception *ignoring or compensating for perceptual anomalies*. Terms like "straight," "parallel," and so forth, would then have the sense they have when they apply to physical objects, that is, a Euclidean sense. The Euclidean term "straight," for example, involves a cluster of criteria (not all independent of one another) among which are some that would not be appropriate or applicable within non-Euclidean spaces. According to these criteria, for instance, a *physically* straight edge would in general satisfy Euclid-ean but not hyperbolic criteria for *straightness*, and *mutatis mutandis* for a *visually* straight edge in hyperbolic visual space. Under the circumstances described above, the perceiver responding to what he/she believes to be the question asked (or instructions given) will address the visual objects with criteria appropriate for physical space.

Such a choice by the subject would also tend to force perception into just one of its possible channels, with the unfortunate consequence that the existence and character of other possible perceptual channels may be systematically hidden. It may well be, however, as I shall explain below, that Euclidean perception depends on the existence of a certain information channel that, in this case, is a set of clues provided by "carpentered" objects in the environment; a dearth of such clues in the physical environment would then render the Euclidean perceptual system inoperative. The attempt to achieve Euclidean vision might then be unsuccessful for want of appropriate clues, the perceived visual object would then be systematically and unalterably other than what is anticipated in ordinary life. If, in addition, the subject is "naive" and does not suspect what is happening, he/she would then be stuck with the impossible task of trying to describe phenomena (or carry out instructions) according to (in this case) inapplicable and inappropriate criteria. From such a confusion nothing will arise but radically unsystematic, ambiguous, unreliable, and confusing data, the outcome of nothing better than private guesses and individual compromises.

If the subject comes to recognize that Euclidean criteria are inapplicable and inappropriate to the spatial characteristics of the visual objects presented in the test situation, then whether or not he/she will be able to respond in a scientifically useful way to a researcher's inquiry will depend on whether both possess among their common resources the ability to use systematic descriptive criteria of other kinds.

Potential alternative horizons of perception are generally not recognized for what they are because people usually lack both knowledge of the norms that such a revision of perceptual forms should follow and the ability (including the use of appropriate language) to bring to the bar of public scrutiny and intersubjective judgment the results of such a revision. Since resources for the description of the hyperbolic visual shapes (for example, of familiar physical objects of regular shapes) do not form part of our general cultural heritage and education, one purpose of Part I of this book was to introduce the reader to the visual characteristics of hyperbolic space, to provide some intuitive grasp of a set of complex new hyperbolic visual forms, and to revise ordinary language to make it available for the description of these forms. The revised descriptive usages should then cling to the newly recognized "things themselves," more like clothes properly tailored for these hyperbolic bodies than like anything tailored according to Euclidean patterns. Reports couched in this revised language would then belong to a new domain of literal descriptive discourse. These reports, taken literally, would belong to science as truly as reports about the physical objects that serve as their proximal and distal stimuli.

Despite the confusion between two incompatible descriptive bases, it might be thought that scientifically useful results could nevertheless be obtained by some process using statistical methods. I shall argue that no ordinary process of statistical analysis will serve. To establish a probability measure, we need first to specify a common empirical base. With respect to visual facts, we have found that there is no unique common empirical base, that there are two different and incompatible universes of visual facts. We have consequently to deal with two separate probability measures, one for each empirical base. The two empirical bases are, however, *complementary* in a sense to be discussed in detail below in chapters 10 and 13. Their measures, then, are in all likelihood related in a complex way. It may just not make sense to apply a unique—single or joint—probability measure to the combined bases. The problem of using probability measures under circumstances like those described above is illustrated by quantum mechanics, where the theory tells us that there is no joint probability measure among complementary quantities; there are only separate probability measures that are complexly linked through the Hilbert space of quantum mechanical states, and these separate probability measures are obtained by a very puzzling procedure called "collapsing the wave packet." Data collected from "naive" subjects or by "naive" researchers in which the two descriptive languages are mingled or confused cannot in all likelihood serve as the basis of a statistical scientific theory of a standard sort. Perhaps something on the model of the quantum theory would be required.

I conclude then that a researcher will have the resources to acquire scientifically useful data about the shape of visual objects only if both the researcher and the subject are sensitive to hyperbolic vision and trained in the use of the new descriptive language.

Anomalous Hyperbolic Perception

Psychologists generally attribute anomalous percepts when encountered in common experience, such as those referred to in Part I above, either (1) to malfunctions of the neurophysiological mechanisms, or (2) to the use of some systematically inappropriate cognitive strategy.[24] There is, I claim, a class of anomalies that belongs to other as yet unrecognized systematic construals of human experience, and in particular, to alternate horizons of geometrical perception. Such anomalies and illusions are different from those belonging to the bulk of the two classes mentioned above because they are candidates for inclusion in a richer and more sensitive experience of reality. This is the class of anomalies that is capable of making the transition from illusion to reality once the source and origin of those anomalies is understood to be in alternate perceptual systems.

Natural vs. Conventionalist Geometry

I claim that two kinds of geometry, Euclidean and hyperbolic, have a descriptive use within human experience. Although they differ both in intension and extension, a manual of translation from one to the other could nevertheless be constructed based on the transformation between the two model geometries, with the values of σ and κ (parameters for the family of hyperbolic spaces) depending on the hermeneutical interests of the perceiver. Although Euclidean and hyperbolic descriptive languages are radically different in both intension and extension, they do not constitute a case for "radical translation" in Quine's sense, since they do not share a common "disposition for verbal behavior."[25] Prima facie, however, the case seems to be a type illustration of language *conventionalism*;[26] this is not the case, however, as a further analysis will show.

Conventionalism is a philosophical position that says: (1) that a family of descriptive languages can be created by mapping the descriptive sentences of a language on themselves in such a way that any chosen sentence can be used to describe any given empirical fact; (2) that it is a matter of indifference which of the family of languages referred to in (1) is used to describe the empirical content of human experience; and (3) that the truth of a statement is its coherence within the language used rather than its conformity with a presumed but nonexistent language-independent reality.

I shall argue below that the dual geometries—hyperbolic and Euclidean—are not conventionalist in the above sense but, on the contrary, I shall argue that, (1) both are natural, that is, easily learned and based on the manifestation of the "things themselves" in perception—the hyperbolic description, for example, is not the outcome of an arbitrary choice to permute the sentences of the Euclidean language, because genuinely new descriptive predicates are introduced; (2) each describes a possible World, given directly in perception, and not only after consultation with a translation manual, and (3) each describes a possible World with truth—that is, conformal to the way things manifest themselves in a certain public context.

8

Causal Physiological Model of Perception

Physical and Physiological Basis of Perception

In any model of perception, the sensory system (comprising the psychological and somatic processes of perception) is prepared in some way to "resonate" in response to an appropriate light structure specific to a physical object in its background in the World: the "tuning" of a "resonance" to a specific stimulus is the product of learning based on the abilities of the subject, the structures of the environment, and the cultural milieu within which perception is learned. "Resonant" response of the somatic processes—comprising the neurological network and other somatic functions—to the stimulus, being cognitively oriented, leads to the projection of a perceptual object against the horizon of a World, and thus to perception.

Thus, the emergence in the perceptual field of an act of perception involves all of the following: (a) the physical presence of the object that is perceived, (b) a specific stimulus, (c) the possible involvement of other attendant somatic and technological processes essential to the structuring of the information channel, (d) the activity of neurological networks—also a part of the information channel, (e) a specific hermeneutical circle (or intentionality-structure) oriented toward the recognition of the horizon of which the object as perceived is a profile, and (f) the persistent presence of a World to the perceiver. Resident in these are certain conditions of possibility of perception which will be a major concern in the latter part of this book.

It is perhaps more usual to focus on the conjunction between psychological and neurophysiological processes: this is called a "coding" (sometimes a "representation") of one on the other.[1] It is generally assumed that in perception there is some form of congruence between the two, loosely called an "equivalence,"[2] like that, for example, between linguistic expression and meaning. Views of this kind, usually expressed in vague terms, are often associated with identity theories of Mind and Body. For example, it is usually not clear whether by "psychological process" is meant the *kind* of psychological activity (e.g., an act of perception, or of desire, or of some other kind), or the *content* of the act; and if it is the latter, whether this, for example in relation to perception, is the percept or the semantic content of a judgment, and if the latter, whether the content—now an object—is or is not subject to epoche.[3] Moreover, there is usually no consideration of attendant somatic processes such as skilled hands, trained feet, experienced eyes, and so on, or the use of "readable technologies" about which more will be said below, some or all of which may be essential to the structuring of the perceptual object and its horizon. Any adequate account of perceptual activity must include consideration of all of the six elements (*a*) to (*f*), above.

Physiological Moment

Two inseparable moments—physical and cognitive—can be distinguished in the act of perception; in terms of the concept *information*, the physical moment is information$_1$ and the cognitive moment is information$_2$. The two moments are distinguished by the logical spaces in which they are described rather than by temporal or other sequence in some real World. The first is physical: this includes all the processes, (*a*) to (*f*), insofar as they are physically describable in the third person. They break down into (1) object in a World, and (2) somatic information processes. The latter, in turn, comprise (*b*) stimulus, (*c*) attendant somatic processes, and (*d*) neurophysiological processes. These are not always easily distinguished one from another. They constitute, however, a domain of physical and physiological causality. In the case of visual perception, (*b*) is (or includes) the appropriate structure in the incident optical array, specific to the physical object in its horizon; (*c*) may include the optical system of the human eye, as well as spectacles and other external optical devices; (*d*) is the state of "resonant" (as distinct from the random) neuronal agitation on which the content of the perceptual act (it is presumed) is finally encoded. Current thinking about (*d*) is strongly influenced by work on artificial intelligence and by computer models of information processing.[4] At the present stage of neurophysiological science, it is generally not possible to identify the character of the state on which the content of the perceptual act is encoded.

Psychic Moment

The second moment, inseparable from the first, is cognitive: this is the *content* (or *object*) of the psychic act presented to the subject. To describe it, one needs a different logical space of discourse. Relative to the somatic information processes that is a signal space (the space of information$_1$), the content belongs to a message space of perceptual horizons (the space of information$_2$). It must not be supposed that the content of the perceptual message is in any way consciously inferred from or derived directly from prior knowledge of the signal space. The signal space for a perceptual act is an unconscious element of perception, generally inaccessible to observation by the perceiver—it "drops out of consciousness"—and knowable generally only through the use of indirect or third-person methods.[5] There also exist intentional operations. These prepare us to meet our World as a set of possible objects of perception and enable us to "find" in our experience profiles of perceptual horizons which transcend the momentary impressions of any proximal or distal stimulus; these latter are not what we perceive but are at most conditions for perception. It is problematic whether what von Helmholtz calls "unconscious inference" is to be identified with intentionality. Unconscious inference, also defended by psychologists such as I. Rock and R. L. Gregory,[6] is the operation of an internalized logic or algorithm, innate or learned, isomorphic with the World and anterior to particular acts of perceptual cognition. It is clear that what is intended by "unconscious inference" is related to or some part of intentionality, but the term is most often used in an objectivist sense that does not sufficiently take account of the inseparable mutuality of subject and object in perception.

Somatic Information Channel

I propose, for reasons that will emerge, to call the "resonant" neuronal state, together with the attendant somatic processes and the specific stimulus, the "somatic information channel." The somatic information channel is then a real *physical state*: it includes neurophysiological components, other bodily processes, such as reaching for or handling things, and environmental components such as the use of instrumentation—all of these interconnected by a structured flow of energy. It is that state of physical reality (or one of those states if redundancy is permitted) which is necessary to the act of perception, but nevertheless is no part of what is perceived.

We usually do not know how to describe the elements of the somatic information channel, because we do not usually know where to draw the lines (the "cuts") between (a), (b), (c), and (d). The last three, namely, (b), (c), and (d), together make up the somatic information channel. The imagery of a line of physical causality "cut" at certain points to divide the elements of the somatic information channel from one another can be misleading.[7] It is important to remember that each chain $(a) + (b) + (c) + (d)$ belongs to a system of possibilities—a "perceptual system," in Gibson's terms—and that the determination of these systems is largely, at present, unknown.

For example, where to place the cut between the stimulus and the "resonant" neuronal state (or between the stimulus and the attendant somatic processes) is unclear, since we cannot yet identify the states of the neuronal network on which any particular class of perceptual objects is encoded: much less do we know how the process of encoding works. In particular we do not know whether, for instance, to regard the optical structures of the eye—accommodation, retinal disparity— as parts of the encoded signal and therefore belonging to the "resonant" neuronal state, or just as transducers of energy belonging to attendant somatic processes. Neither do we know to what extent attendant somatic processes external to the neurophysiological system may be essentially involved in constituting certain classes of perceptual representations.

There is evidence, for example, that the neuronal networks have a degree of plasticity enabling them to conform to recurrent features of the environment: such features would be attendant somatic processes essential to perception.[8] A blind man's cane,[9] for instance, is essential for the constitution of the blind man's World: is it a part of the structure of the stimulus, or is it an attendant somatic process essential to his perception? It could, of course, be either: now one, now the other— depending on the particular phase of the learning process, on the cultural entrenchment of the descriptive terms, and so on. In one case, the neuronal network controls the spatial dispositions of the cane, by means of a mapping within it of the domain of possible contact of the point of the cane with the environment; the blind man's cane then "drops from consciousness" and becomes an attendant somatic process essentially involved in the constitution of the percept itself. In the other case, the neuronal system is excited to a "resonant" state by physical contact with the handle of the cane. The cane then belongs to neither (*b*), (*c*) nor (*d*); it is not part of an attendant somatic process in the sense I have described, instead it is part of the physical cause of the stimulus. The two cases result in two different kinds of experiences, and consequently, in two different codings or representations in the neurophysiological system, and in the use of two different descriptive languages. We shall study in greater detail below the consequences of this analysis for the use of such technologies to which I give the name "readable technologies."

Since none of the component elements (*b*), (*c*), or (*d*) of the somatic information channel function within the act of perception as objects— not even as partial or subsidiary objects—they are not directly attainable by an analysis of the content of perception. The possible cuts between these component elements, and the coding or reencoding of one on the other, can be studied only by indirect, third-person, theoretical, and scientific methods beyond the scope of a phenomenological and hermeneutical philosophy of perception.

Ambiguity of Stimulus-Object Relation

Without appeal to what is given in the first-person act of perception as its specific object, however, there is no way of determining the relevant specific stimulus (*b*). We are aware of the fact that *without any change in the ambient optical field*, we can come to perceive different things. For example, the corner of a room where walls meet can be perceived either normally as a meeting of two orthogonal planes, or anomalously as one flat continuous plane: without any change in the ambient optical field, the colors and luminosities of the two walls in the two cases are perceived differently.[10] Or in the case of multistable spatial perceptions, such as the visual experience of the Gateway Arch at St. Louis, or of the interior of a church where the columns can be seen either as arranged in equal Euclidean ranks or as converging in hyperbolic visual space on the center of the apse, multiple perceptual objects can be evoked by one unchanging optical field.[11] Or, for instance, in the case of the Müller-Lyer and Hering illusions described above, we can read into them different spatial configurations without any change in the ambient optical field. Or again, in considering a figure such as a line drawing, we can switch our attention from the figure as object to what the figure represents as image, and so evoke multiple perceptual objects from one optical field. Perceptual *Gestalt* switches are of this kind.

The specific stimulus then is not identical with the optical field; the optical field is not an *input* field, imposing a unique response on the perceiver independently of the state of the perceiver. The optical field provides the resource in which the active perceptual system detects or from which it obtains or extracts the information (the signal structure) relevant to its perceptual concerns. In other words, the specific stimulus is a specific function of the optical field, obtained by means of an appropriate somatic-neurophysiological transformation function. Moreover, it is clear that the same optical field is a resource for a variety of different functions producing a variety of different specific stimuli *differentiated by the perceptual horizons they evoke.*

The reverse is also true: one can come to see the same objects and the same horizons, even with greatly transformed physical input. For instance, if prismatic glasses are worn which bend the rays of light as they enter the eye, or which invert the image on the retina, the perceptual systems can adapt to the new optical input in such a way as to represent one's World roughly as it was represented before the glasses were worn.[12] Thus, the perceptual object (and the World) can be kept fixed, even though the optical input and attendant somatic processes (and presumably the neuronal networks) change.

The importance of facts like these for a correct understanding of the history of science was underlined by Thomas Kuhn in the Postscript to *The Structure of Scientific Revolutions*: among the few things we know with assurance, he wrote, is that "very different stimuli can produce the same sensations; that the same stimulus can produce very different sensations."[13] To get the only meaning that I think correct—though this may not be precisely what Kuhn intended—one has to read *optical energy field* for *stimulus*, and *perceived object* for *sensation*, for what is presented to cognition is the perceptual object together with and as a function of its horizon, one profile of a family structured by an invariant perceptual essence. Kuhn, however, is correct when he asserts that "people do not see stimuli; our knowledge of them is highly theoretical and abstract,"[14] for we have no direct awareness of what our neurophysiological states are like, nor of the structures in the ambient optical array which constitute the specific stimulus. What specific structure (*b*) in the optical array mediates my perception of perceptual object (*a*)—whether it be some measure of parallax, angular size, light gradient, or repetitive textural pattern—is not part of what I perceive but —if determinable at all as a physical structure clearly distinct from (*a*) and (*d*) (or [*c*] + [*d*])—can only be determined indirectly, by a scientific inquiry that has both theoretical and experimental components. This is true even in what seems to be a simple case.

Consider, for example, three points of light in an otherwise dark background which are perceived as collinear: the specific stimulus in this case seems to be easily determinable by inspection. But not so: the determination of a specific stimulus is never obvious. A stimulus is part of an information channel: this is a system of elements, all of which need to be known in principle before any one particular element, for example a particular stimulus, can be functionally identified and correctly described. A piece of lumber may be a shelf, a support for a wall, or it may serve some other function depending on the system of which it is a part; likewise, a point of light or any particular pattern of illumination may serve a variety of purposes among which is that of serving as a specific stimulus within a perceptual information channel. The relevant information channel may, for instance, be a function just of three points of light in a dark background and of nothing else, or it may turn out to depend on additional parameters buried in the background. In either case, in terms of the relevant information channel, the three points of light may constitute three separate stimuli, or they may constitute just one complex stimulus. Moreover, the relevant parameters may be the brightness of the lights, or their color (each in some specifiable range), or their distance apart, or their angular size, or their parallax—or some combination of all of these. These are the questions that must be answered. They are not easy to answer, however, and they certainly cannot be answered merely by inspection. It has taken the host of research referenced in the Appendix and in chapter 3 to begin the process of finding the answers.

Inescapable Reference to Life World

In conclusion, the lines between the components (*b*), (*c*), and (*d*) of the somatic information channel are not presently known and probaby are not knowable *without some essential reference to the act of perception.* In the physical-physiological domain, the specific stimulus and the "resonant" neuronal state each belongs to *systems* of stimuli and neurophysiological ciphers that are in large part the product of dynamic learning processes occurring within the context of the organism's interactivity with objective environmental features of its World.[15] The pattern of that activity is related to biological evolution and cultural history. Likewise, in the perceptual domain, every act of perception selects from its World some environmental feature that is both recognizable and cognitively significant. Recognizability and cognitive significance are systematic terms related to historical and cultural horizons, and to the Life Worlds of communities objectivized through the descriptive capacities of the languages people use. As long as reality is determined by the Life Worlds of communities, these will be essential determinants of which perceptual systems are significant and worth investigating. These perceptual systems, in turn, will determine what systems of stimuli and what neuronal networks will be considered candidates, by Identity Theorists, for the impersonal, scientific, and material basis to which the psychic life of persons in their program would be reduced.

To succeed in making a reduction of any one perceptual system would be a memorable achievement in the present state of knowledge, but it would be naive to think that such a reduction is of great *philosophical* import. It would have no more philosophical import—no more and no less—than, say, the making of a dictionary, to which in fact it bears a resemblance. Like dictionaries, the significance and function of the neuronal networks to which the attempted reduction is made are historical and changing, because they borrow their horizons of objectivity from the Life Worlds of the researchers.

9

Perception as Mirroring: Realism

Perception as Mirroring of Nature

The simplest epistemological model for perception considers the Mind as a "Mirror" of Nature.[1] Its principal theses are the following. First, *the realist thesis*: this states that physical reality has elements and structures whose character and existence are independent of human perception, language, culture, and history. Second, *the thesis of objective space*: this states that space has a definite topological and geometrical structure that is *Euclidean*. Third, *the pictorial thesis*: this states that *what* is known in the act of perception is a mental content (a percept), and this content, like the image in a mirror, is not identical with that which is imaged, but in all relevant aspects is congruent with, or matches, the things of Nature which it represents. Such a set of positions entails that physical reality be essentially pictorial.

These positions call for some critical discussion and evaluation. In this chapter, I shall raise and discuss four questions relating to the thesis of objective space and the pictorial thesis; I shall discuss the realist thesis in the following chapter. The questions are: (1) Is physical reality essentially pictorial? (2) Insofar as reality is perceived pictorially, is the Euclidean structure of everyday and scientific space a privileged one, independent, as it is claimed, of human culture and history? (3) What are the subjective and objective conditions that make it possible to perceive reality in a pictorial but non-Euclidean way? (4) Within pictorial perception, is having a percept itself an act of knowing something about a World, or just a product of knowing something about the World, or merely an opportunity for knowing something about the World?

Pictorial Reality

1. Is reality essentially pictorial?

Reality would be essentially pictorial if every act of perception were to construe its object in a pictorial way, like the objects of pictorial vision.[2] But perception does not always construe its objects this way: sounds, for example, are not pictorial; they are not bounded by surfaces, and many sounds can simultaneously fill the same space and interpenetrate. Likewise, tastes and smells are not pictorial objects, although many aspects of touch and color are. The classic division between primary and secondary qualities does not quite coincide with the division of objects into pictorial objects and nonpictorial objects. I shall discuss secondary qualities below and claim that, even when not pictorial, they are nevertheless real.

Secondary qualities belong to a larger class of perceptual objects, called "manifest objects," that are given (*manifest* themselves) either as functions of the sensory response of the perceiver or as functionally related to the perceiver's interests and goals.[3] Some manifest objects also happen to be pictorial objects, like tables and chairs, but others are not, like a loving kiss, or the death of a friend, or even one's favorite recipe. For Christians, Calvary and the places associated with the life of Jesus are not just, nor even principally, pictorial objects; nor would we say that Dachau, or other scenes of human infamy, or Brutus mourning the death of his son, or the Declaration of Independence of the Continental Congress, are just, or even principally, pictorial objects. Yet all are firmly located in the space and time of the Life Worlds of the participants. Science also provides many examples: quantum mechanical objects and events fail to be pictorial in a paradigmatic way. Other examples are the structure of the physical vacuum; ordinary and isotopic spin; lepton and hyperon charge; the interaction of matter and fields; the basic properties of a Fermi gas, and so forth—none of these can be satisfactorily pictured in a Cartesian/Euclidean space, but all, as I shall hold, can be perceived.

Perceptual knowing then is not always a form of pictorial vision.

Euclidean Structure and Prepredicative Intentionality

2. Insofar as reality is perceived in a pictorial way, is the Euclidean structure of pictorial reality a privileged one, independent, as it is claimed, of human culture and history, so that all peoples, at all times, could or should picture reality in a Euclidean/Cartesian way?

The Euclidean/Cartesian character of our everyday space is logically bound up with procedures of *scientific* measurement which yield a description of reality that is, prima facie at least, independent of the cultural and historical dispositions of human perceivers.

Spatiality, as Merleau-Ponty has shown,[4] is an aspect of all prepredicative Worlds antecedent to the use of any definite geometry: that is, antecedent to any knowledge of metric geometric relations. However small may be our mathematical sophistication, any World into which people are inserted at birth is already organized spatially by top and bottom, by left and right, and in depth relative to the perceiver, and things in it exhibit pregeometrical constancies of shape and size. By this primordial spatiality, each perceiver finds himself or herself in a World organized in three dimensions.

A World has a *geometrical structure* only if there exists an unambiguous measure of distance, but no unique explicit measure is specified a priori by the common primordial spatiality of perception, or so I shall argue below. Nevertheless, one would not have thought of creating objective *measures* for distance and depth, or for the length, breadth, and height of things, had not the common universal primordial structure of this spatiality lent itself to this.

Among the possibilities of *geometrical* Worlds, there is evidently from our experience the possibility of a naturalistic World unified by Euclidean space. Such a World is that of classical science—the science of Galileo, Descartes, Newton, and Maxwell. Euclidean spatiality is the only way of giving expression both to the primordial intentionality of spatial perception and to the demands of classical science. These latter comprise a principle of atomism (physical reality is composed of small geometrically ordered rigid and indivisible parts), a principle of universality (the same physical laws and theories apply to everything), a principle of objectivity and anonymity (scientific laws are indifferent to human perceivers, to what they do and to where they are located), and a principle of geometry (there is a unique metric for all physical space).[5] Thus, if perception were also primordially accompanied by the teleology of classical science, it would always and necessarily be Euclidean.

The choice of atomism, anonymity, objectivity, universality, and a unique geometry as criteria for scientific knowing is, however, that of a particular culture, namely, of the West since roughly the fifteenth century. These criteria are not found in all cultures, at all times, and were not part of the Western tradition until about the time of the scientific revolution.

Nor is this cultural choice and its associated Euclidean pictorial space readily available to all peoples, since a Euclidean pictorial representation of the World is an option available only under certain cultural conditions that will be further explored below. Briefly, the culture will have to engineer its environment so that people will be able to "read" its Euclidean structure "off" the cues generated by the artifacts (buildings, and so forth) it contains.

Nor is this cultural choice a privileged one in the sense that it alone is capable of representing reality as it is independent of historical cultural biases. Cartesian criteria for describing reality scientifically became successfully entrenched in the pictorial perception of our culture in the recent historical past. Within the past fifty years, however, science has discovered the limitations of that classical perspective. The very large, the very fast, the very energetic, and the very small were all found to have no accurate Cartesian representation, and so fail to be pictorial objects in Euclidean space. The Euclidean/Cartesian pictorial representation of reality turns out to be the product of a particular historico-cultural choice, one based on the belief that the representation of Nature by modern science could be anonymous, objective, and universal, like a picture seen by God's eye. The many recent failures of the Cartesian scientific picture should make us aware of the cultural and historical bias that favored it as privileged, and impel us to ask whether there is any sense, apart from a merely pragmatic one, in preserving its privileged status in our culture.

Possibility of Non-Euclidean Perception

3. What are the subjective and objective conditions that make it possible to perceive reality in a pictorial but non-Euclidean way?

Riemannian geometries provide a common or universal repertory of geometrical shapes, since there is free mobility of geometrical forms in these spaces. There is no class, however, of *physically mobile but not physically rigid* things that we know of which would retain their shapes under displacements in a hyperbolic visual space. A perceiver who construes the environment in this way sees objects with variable shapes, all, however, within the repertory of what is possible in hyperbolic visual spaces; but which of these shapes he/she sees will depend on two factors: on the position of the perceived objects relative to the location of him/her, and on the conditions that determine for him/her the parameters of the space. A World so constructed is not one independent of individual perceivers in the classical sense; and so a non-Euclidean World would fail to provide the universality and objectivity required by the canons of classical science.

How could those objects we presently identify by a characteristic shape come to be identified in such spaces? Consider, for example, a plate. A plate usually has a circular shape, and this shape is, generally speaking, characteristic of the kind of thing that a plate is. In any hyperbolic visual space, however, the circular shape of a plate will be transformed into a variety of oval forms. As long as we regard visual space as analogous to Euclidean space—one and fixed—there is no unique geometrical invariant for the plate under all translations and rotations in the space which leave the *physical* shape of the plate unchanged. Consequently, either the invariant essence of a plate (that which serves to identify the plate as an individual of a certain kind) must be something other than a fixed geometrical essence (it could be, for example, the common purpose the object serves as a plate), or its invariant essence involves profiles in more than one member of the family of hyperbolic spaces. Since the parameters of visual space change with the hermeneutical purpose of vision, one could argue that if the essence of an object (say, for the sake of the argument, a plate) requires it to have a *fixed (rigid) physical shape* (say, circular), then this invariant would have to be defined in relation to the subfamily of possible visual spaces in which the object in question (say, the plate) is located at the true point. Since the "undistorted" neighborhood of the true point is of finite size, and since it is almost certain that the family of possible hyperbolic visual spaces is restricted as to the range of σ (the parameter that determines the distance to the true point) this definition places a limit on the size of an object for which a fixed physical shape can be so defined in visual terms. The visual profiles of an object of fixed physical shape would then be profiles relative to (motions among) a *subfamily of spaces* rather than—as with physical space—profiles relative to (motions in) one privileged space.

My conclusion then is that most of the objects in our Euclidean Worlds would continue as identifiable objects (or horizons) in hyperbolic visual spaces, with a possible limitation relative to objects of great size whose physical shape happens to be essential to their physical essence as we understand this in everyday life—but there are not many of them. Among them, perhaps, should be counted the heavens and celestial objects, including the moon, which, though affected by hyperbolic illusionary appearances, is nevertheless perfectly recognizable despite these illusionary variations of size and form.

Is it perhaps the case that whether we construe our visual World in a Euclidean space or in a hyperbolic non-Euclidean space is a matter merely of convention?

Poincaré-Reichenbach Conventionalism

Hans Reichenbach, in his fascinating study on space and time,[6] has argued that space as visualizable is neither Euclidean nor non-Euclidean but only a continuous three-dimensional manifold: it acquires a metric by deliberate choice of rules of congruence, rules that state under what conditions lengths are to be taken as equal. According to this view, when we attempt to construct images, we bring to the construction an antecedent logic founded on some definite rule of congruence which is not part of the image producing function itself but directs it, as it were, from outside and is a matter for choice. For Reichenbach, the internal a priori structure of our powers of visualization does not impose any privileged rule of congruence, for, as he says, *"the normative function of visualization is not of visual but of logical origin."*[7]

In Reichenbach's view, it is history, culture, developmental adaptation, and education that have taught us how to use the particular rule of congruence—transport of rigid rulers—which leads to a physical space of Euclidean structure. In his view, so deeply is this structure sedimented in our cultural experience that Kant wrongly took it to be necessary for all pure visualization, that is, visualization by the productive imagination without external sensations, and therefore, without empirical objects of visual experience.

Reichenbach holds that we have, in addition, the ability to learn to visualize physical objects according to the laws of any non-Euclidean geometry; not merely can we *conceptualize* such objects as represented in Euclidean *maps* of them, we can also *visualize* them. To visualize space in a non-Euclidean way, he says, we need only to recognize that the structure of space is specified by the mathematical form of the distance function, and that this latter follows logically from the rule of congruence chosen. Our freedom to visualize non-Euclidean forms is in his view proportioned to our ability to define non-Euclidean rules of congruence. In this, he is supported by the views of H. Poincaré and von Helmholtz.[8]

Reichenbach argues that pure visualization depends on learned empirical visualization. We are free to choose, he says, any rule of congruence for physical space; physical space will then be perceived with the corresponding non-Euclidean geometry, and the art of pure visualization according to this geometry can then be learned.

The distance between two points in space is defined not by any intuitions we may have but by the performance of a physical measurement incorporating the rule of congruence chosen. There is nothing illogical in adopting a rule of congruence which relies on rulers that are not rigid relative to a Euclidean standard, rulers that vary systematically according to that standard over the field of objects. The possibility of making such a physical measurement, however, depends on a controversial thesis Reichenbach proposed about "universal" and "differential force fields."[9] A *universal force field* is one that is all-pervasive, like gravitation, but has in addition the virtue that it would change the length of any object occupying a particular place in a fixed and systematic way independently of the physical state or chemical composition of the object occupying that place. On the contrary, a *differential force field*, like the expansion forces due to a rise in temperature, is one that acts differently on materials of different physical or chemical composition. By choosing an appropriate universal force field, his thesis claims, and modifying the laws of physics accordingly, any rule of congruence could in principle be embodied in an appropriately instrumented measuring process. This is the *conventionalist hypothesis*. Reichenbach attempts to show that we have the intuitive ability to pass beyond the consideration of Euclidean maps of non-Euclidean geometry to construct in the visualizing imagination non-Euclidean pictorial objects, an ability, he says, that is exercised particularly by mathematicians.

I agree with much of what Reichenbach and even Poincaré say, but find their conclusions problematic in ways they may not have considered. To learn to visualize in the imagination a non-Euclidean pictorial object, the imagination has to be prepared: that is, it has to acquire the appropriate hermeneutical structure for picking up such information directly; this means finding or constructing the appropriate information channel, using available neurophysiological, and possibly also technological, resources. If, as Reichenbach holds, the unaided perceptual system has no natural inclination to organize experience according to one geometry rather than another, then the new hermeneutic structure must be learned from experience; this will involve the production of known and tested exemplary non-Euclidean forms within the matrix of a special technological praxis, central to which is a suitably designed measuring process. To know what non-Euclidean shapes look like, one needs on Reichenbach's principles to have at one's disposition actual measuring processes that incorporate the new rule of congruence. Concrete experimental evidence is lacking that physics can be reconstructed according to Reichenbach's program: no one has tried, and the project itself is probably not feasible. Grünbaum calls Reichenbach's "universal force field" just a *metaphor*, not descriptive of anything, certainly not of anything like gravitation, which Reichenbach compares it to. Poincaré's illustration is a differentially heated disk in a fictitious universe where all substances have the same coefficient of heat expansion.[10] Except for such idealized domains of little practical interest, it is hardly likely that the relevant physics can in fact be reconstructed in a way consistent with an arbitrary non-Euclidean three-dimensional geometry.

In summary, Reichenbach holds that perception has no natural capacity to respond with a specific geometry other than in an illusory way. He does not recognize the possibility that specific structures of the optical field may be clues to specific perceptual geometries. He holds that the geometry or "logic" of the perceptual field can be whatever one pleases merely by applying a suitable mathematical transformation (to the Euclidean metric) and changing the laws of physics appropriately. He recognizes, however, that logic or mathematics alone cannot train perception or imagination (or Body) to recognize a particular metric structure *apart from the actual production of elementary pictorial forms that constitute the repertory of the particular geometry.* Although such productions cannot be had apart from the use of appropriately designed instrumental technology, he does not point to the existence of any technological devices that could provide for perceptual experience a training in the recognition of forms and possibilities of action based on a coherent non-Euclidean construal of the things we perceive, and without such experience and an enlarged linguistic capacity to name and identify non-Euclidean forms, there is no way to test Reichenbach's thesis on conventionality.

I shall make two comments about Reichenbach's views. (*a*) The visual experiments of F. Hillebrand, W. Blumenfeld, and H. von Helmholtz[11] with which Reichenbach was familiar demonstrated that under certain conditions the human power of visualization is internally programmed by an antecedent non-Euclidean structuring *independent of the perceiver's deliberate act of choice*, and not dependent on the reconstruction of physics or the exercise of technological devices. I have shown that independently of all instrumented measuring processes, human perception naturally (i.e., easily) reads the optical clues so as to assume—at least episodically—the form of hyperbolic geometry. Such a geometrical form is not then *conventional* for intuition in Reichenbach's sense: it is not the product of deliberate and conventional choice, since it does not rely on a "universal force field," nor does it depend on the constructability of an instrumented measurement process to provide hyperbolic congruences. Reichenbach treats such a geometry as merely subjective and illusory.

I claim, however, that this geometry results from a native ability to read the (γ, α, β) optical field as a 'text,' and to give this 'text' a reading in terms of hyperbolic geometry by judging, ordering and comparing lengths and distances as these are presented to unaided visual perception. Such a process exhibits the dual structure of information: at the level of information$_1$, its 'text' is the (γ, α, β) optical field; at the level of perceptual objects (information$_2$), its hermeneutical character consists in projecting before it that one of the family of three-dimensional hyperbolic geometries which is most appropriate to the goals of this specific vision. That choice is most likely influenced by assessments, for example, of the relation between a foreground "Newtonian oasis" and a background distant zone, where sizes and shapes of all visual objects are estimated without the assistance of any technology. I claim as a consequence that hyperbolic visual geometry is privileged for the technically unaided perceiver. Nevertheless, though privileged, it is not the exclusive geometry of perception, since our common experience testifies that we have gained the capacity to perceive, visualize, and imagine Euclidean objects and to fashion for ourselves a Euclidean World. The existence of two forms of visual geometry—Euclidean and non-Euclidean—will constitute one of the problematics of this book. The capacity for Euclidean vision, I shall argue, is mediated by the products of a universal Euclidean technological praxis, the "carpentering" of the environment in which we live.

(*b*) Though wrong in assigning the plasticity of visual perception to the power of logic alone, Reichenbach is right in recognizing the existence of such plasticity. This plasticity, as I shall explain, flows not principally from the new geometrical models invented by mathematicians but from the concrete possibility of an appropriate *Vorhabe*, or way of constructing an embodiment (in "readable technologies") "through" which we can come *in certain specific cases* to perceive according to some geometrical model, as if through a transforming lens or mirror. Our perception is transformed, then, not by the force of logic or mathematics alone but by an appropriate technological praxis molding our experience according to a particular logical or mathematical model. The principal evidence for this thesis is, I hold, our normal ability to see Euclidean objects; this is, I claim, an artifact, the consequence of being in a "carpentered" World, shaped by the classical physics of rigid bodies. The central role in such a process of "readable technologies"—here the structure of the "carpentered environment"—will be taken up more fully below.[12]

Percepts and Perceptual Judgments

4. Within pictorial perception, is having a percept itself an act of knowing about the World, or just a product of knowing something about the World, or merely an opportunity for knowing something about the World?

I have taken the position that having a percept does not of itself constitute being in the possession of knowledge; knowledge is propositional and possessed only in well-made (warranted) judgments.[13] If knowledge were according to a "representational theory of knowing," that is, if the only objects we know directly and immediately were contents of consciousness, then we could never get "outside" the Mind: the possibility of realistic knowledge entails the intentional identity of the object of knowledge with the reality that is known, and this is achieved only in the perceptual judgment. Moreover, if to know something (perceptually) is to form a percept (or "internal representation") of it, then to know this percept is, as G. Ryle and others have argued, to form another internal representation of this, and so on to absurdity.[14]

A percept—an individual profile, that is, connoting an invariant perceptual essence—is a function within a judgment. It is not a separate and identifiable stage in the process of perception; it is only a moment within a theoretically reconstructed, and therefore retrospective, account of acts of perception.[15] In this theoretical reconstruction, the process of perception is explained as a series of moments among which the formation of the percept is one.

There is not, however, just one kind of perceptual judgment. By deconstructing the judgment following the theoretical model just referred to, one finds that a particular percept, separated from its origins in the particular judgment of which it was a part, is capable of entering into many kinds of judgments. I have already mentioned a few, among them, "bracketed perception" or the judgment about appearance, veridical judgments, judgments about illusions, hallucinations, and so forth. We will need to consider some of these more closely.

Consider the percept of a tilted oval plate. This may be part of a veridical judgment, or of a judgment that what is presented in the percept is an illusion. In the former case, one judges that (the evidence warrants that) what one perceives is really *an oval plate*, while in the latter case, one judges that (the evidence warrants a different judgment, namely that) what one perceives is merely an *oval illusion of a circular plate*.[16] Whether a percept has to be veridical, in this case, structurally isomorphic with physical reality, in order to be the basis of knowledge, as the direct realists claim, will be discussed next.

We start with a consideration of the variety of perceptual judgments involving pictorial objects.

Veridical Perceptual Objects

In the first place, the most straightforward kind of perceptual judgment affirms that a percept—having typically the form of some here-and-now *A* (e.g., *this plate*) with sensible properties *B* (e.g., *of circular shape*)—is a veridical representation of reality. This judgment then connects the currently experienced profile of the plate with a family of

other possible profiles ordered—with respect to their geometrical properties—by the invariant essence of a circular form in Euclidean space.
If the plate were an oval plate, then the judgment of reality would imply
that the currently experienced profile—now perceived as an oval
plate—is a member of a different family of profiles, one ordered by the
invariant of an oval form in Euclidean space. In both cases, the pattern
of energy falling on the retina would be the same: the relevant stimuli,
however, as explained above, would be different. There is yet a third
veridical perceptual possibility which I shall discuss below: the plate
could be seen as an object (possibly of variable shape but common
purpose) in hyperbolic visual space, characterized by a set of hyperbolic profiles, some of which may be oval.

Hallucinatory Perceptual Objects

In the second place, if *A* does not exist and yet the percept persists,
the appropriate judgment is in all likelihood that we are subject to a
hallucination. Hallucinatory perception involves the nonexistence of *A*,
and its analysis by Merleau-Ponty shows[17] that it involves an attempt
by a subject to posit an alternative World. He also tries to show that
even for the subject the alternative reality does not quite work and can
only be intermittent.

Illusory Perceptual Objects

In the third place, we may be inclined to accept a percept as veridical
but nevertheless come to reject it because we judge that *A* does not have
sensible qualities *B*: we conclude then that we are experiencing a
perceptual illusion. For example, perceiving the moon on the horizon
as persistently larger than the moon at its zenith, but knowing it to be
really the same size, we conclude that we are experiencing a *perceptual
illusion*. The illusion, however, extends only to the perceived relative
sizes of the moon, not to the existence of the moon. The illusion,
moreover, supposes that we know that the moon has a fixed size, and is
at a fixed distance from us whether at the horizon or at its zenith.
Indeed, the illusion exists because the perceived relative sizes do not
agree with what we know to be the moon's normal or physical size, but
the illusion persists despite this knowledge, thereby inviting further
inquiry. What we know about the moon is what is given in the accepted
paradigm of knowledge about the moon.

Anomalous Objects: Distorted or Transformed?

In the fourth place, there is the possibility that the shape presented to perception has been *distorted*, deliberately or indeliberately, say by a distorting medium. In this case our perceptual judgment states that we are perceiving a *transformed* or *distorted* shape, one that is not the regular shape of the object. Distortion logically implies a reference standard, the regular shape, and a continuous process of variation that connects the new shape with the old. We assume that the distorted form and the regular form belong to a common family of shapes—all Euclidean—related to one another by a continuous deformation process. If what we are perceiving is the hyperbolic transform of the regular physical shape, however, then the perceived shape is not related to the physical shape by a continuous deformation process in a common Euclidean space; something more radical has taken place instead. The whole World has been transformed, foreground and background, and in the new space, the old form of the physical object does not exist *even as a logical possibility*. The repertory of possible shapes in hyperbolic space just does not include the old physical form. In this case, it would be better to speak of the anomalous form as having been "transformed" rather than as having been "distorted."

Presented then with an object in its hyperbolic transform, as, for example, the oval form of a plate we know to be circular, our perceptual judgment is challenged; what kinds of perceptual judgment have we the warrant to make? Let us consider three possible answers. (1) We could answer that we see a *distorted* circular plate, on the grounds that hyperbolic space represents things in ways other than in the way assumed by culture and convention; I think this answer—though plausible—is logically incorrect. Or (2) we could answer that we see an *oval illusion of a circular plate*, on the grounds that the scientific Euclidean representation has ontological and epistemological priority over the hyperbolic representation; this matter I shall discuss below. Or finally, (3) we could answer that we see an *alternative transformed realistic horizon*, though not one in accord with the third-person scientific account, part nevertheless of an alternative World.

The solution that I shall defend in the remaining chapters of this book is the last. Let us recall that, on the one hand, the primordial conditions of perception seem to permit both Euclidean and non-Euclidean (at least, hyperbolic) geometries for visual space and, on the other hand, that the determination of a visual geometry is not a matter of pure logic alone as the conventionalists claim. The following sections inquire into the subjective and objective conditions of possibility of this conclusion. The concrete possibilities of perceptual geometrical praxis would have to be investigated empirically.

Scientific Observer as Universal Observer

I take the term "scientific observer" to mean the physical frame of reference that carries the coordinate system; this does not have a privileged location in space, since it is identical with the normative background of space. The scientific observer comprises a virtual distribution throughout all regions of space of rigid rulers (the standard of measurement). I say "virtual" because there exist equivalent processes of length measurement which do not use actual coincidences with on-site rigid rulers but depend on ingenious applications of physical laws. A rigid ruler, however, is a kind of object definable only with the aid of physics: the norm of rigidity it embodies is only specified under conditions given by the laws of physics. Likewise all equivalent measuring processes for physical length employ physical laws in an essential way. Unlike the perceiver localized just at one spot in space, at the center of his/her World, the scientific observer is not localized; it is everywhere simultaneously in its frame of reference, it is a *universal observer*. Such an observer responds only to measures and cues that are invariant relative to the appropriate transformation group for classical frames of reference—this is the inhomogeneous Galilean group and includes all instantaneous physical translations and rotations of the observer.[18] Typical of such cues is, for example, a coincidence between marks on a thing and marks on a standard ruler located at the site of the measured thing—from these cues the length of an object is determined; another example is the occurrence of two successive events and the corresponding momentary phases of a clock—from these cues time intervals are determined.[19]

For a person to experience physical space visually according to its Euclidean metric structure, he/she must have an information channel open to receive relevant cues instantaneously from all parts of space; these cues are simultaneous virtual coincidences between distant objects and rulers and between distant events and the phases of a clock (see figure 9.1). In the first place, the ubiquitous presence of rulers and clocks dispersed throughout space must be assumed. In the second place, some means for instantaneous communication with distant

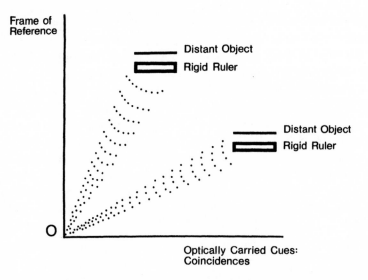

Figure 9.1: Euclidean visual space. Euclidean visual observer, *O*, responds to optically carried information about coincidences between distant objects and distant rulers.

points is needed: let us assume for our purposes that such a medium is light—the fastest medium of physical communication—and that messages encoded in light energy travel at infinite speed. A visual observer then would not be able to see the World as it is physically, namely, as Euclidean, without the ability to recognize visually the existence of (or to construct, if necessary) actual rigid rulers in space or some equivalent technology. This a visual observer could do only if he/she *already possessed the laws of physics as operating a priori in all acts of visual perception.*

Alternatively, one could argue as follows: one would not expect visual observers using the psychophysical information channel (γ, α, β) to see physical space as Euclidean, because this information channel does not share the invariance properties of the scientific (Euclidean) observers. The latter are characterized by the inhomogeneous Galilean group, and the (γ, α, β) information channel is not invariant under this transformation group. Consequently, if visual observers do actually see physical space as Euclidean, then one would have to conclude that visual observers are naturally endowed with a precognitive interpretative transformation function that maps the coordinates (γ, α, β) on the Cartesian points (x, y, z) of an infinite Euclidean space. This is, of course, conceivable. The existence of such a transformation function would mean that we are prewired, a priori to all experience, to resonate to the repertory of Euclidean shapes and relationships. Now, the significance of Euclidean shapes and relationships is, as I have said, tied to classical physics: they are the shapes preserved by inertial motions, and constitute the privileged class of those shapes that it is possible for a physically rigid body to have. Either, then, the laws of classical physics are contingent to all experience, in which case there is no reason to suppose that we are prewired to recognize and experience the constancies of classical physics, or the laws of classical physics are necessary to all experience and we come to perceive the structures and invariances we do perceive because, as many in the Kantian tradition have long held, we are prewired to perceive just these and not others.

If the laws of classical physics were necessary to all experience, then one would be led to ask why only such a limited and approximative scientific a priori? Why is not the content of the primordially given more complete? One might look for an answer in evolutionary history: it could be claimed that existence in an evolutionary context would have favored the survival of the phenotype capable of representing to itself more or less accurately the coarse scientific structure of our physical environment. If classical physics, however, were prewired genetically, why then was Newtonian physics discovered only once, in the West, and why so long after Euclid and Archimedes? Should one not expect, moreover, that the elements of biology, psychology, and other natural and social sciences, equally important for survival, would likewise be imprinted naturally in our precognition? Even granting that such capacities are especially valuable within the evolutionary context, it would surely be sufficient if we had the primordial capacity to learn to see our environment in a Euclidean way: as I suggest below, we learn to "mark" our environment so that we come to "read" its physical structures off the environment.

Contrary to the great rationalist traditions of the Enlightenment, I support the more economical and modest conclusion, that the laws of classical physics are contingent to all human experience.

Visual Observer as Privileged Observer

Consider now a visual observer. We can and sometimes do naturally perceive hyperbolic quantity, as I believe I have shown; consequently, the appropriate stimulus information must be present in the environmental optical array. Such stimulus information serves as a cue to the appropriate perceptual system of an observer; the "resonance" of this system results in the presentation of a hyperbolic horizon to imagination or perception. Such cues, the hermeneutical Luneburg theory tells us, are the appropriate visual parallax and visual orientation relative to the perceiver. A visual space tied to such cues is one in which, *given physical reality as we know it physically to be*, the perceiver has a privileged place at its center, since the visual shapes and dispositions of physical objects would adapt themselves in this space to the point of viewing of the perceiver.

The spaces of unaided vision would then most likely be the simplest geometric structures cued to visual parallax (γ) and orientation (α, β); these happen to be constant curvature non-Euclidean geometries (or Riemannian geometries). The hyperbolic visual observer would then choose among the available spaces by comparing the dimensions of familiar objects with one another and, if the neighborhood of the true point is occupied, with some local standard of length found there. The cues are given by the (γ, α, β) information channel, these of course being conveyed to the visual observer by modifications of the optical field (see figure 9.2), also assumed for the purpose of this inquiry to be transmitted to the observer with infinite speed.

This analysis raises some very puzzling questions. Generations of philosophers since Descartes have held that the structure of everyday

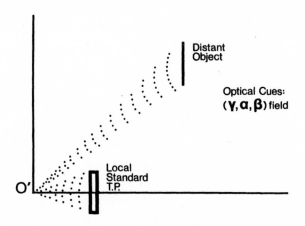

Figure 9.2: Hyperbolic visual space. Hyperbolic visual observer, O', responds to optically carried information in the (γ, α, β) field comparing distant objects with local standard at true point ($T.P.$).

Worlds is Euclidean. They have held, moreover, that this structure is essentially present in visual space, that it is an innate structure or disposition of human perception, and not an artifact. In Part I, I showed that our everyday spatial perception is penetrated by essential structures of a hyperbolic kind, though these are usually judged to be illusionary. These illusionary, essential non-Euclidean structures seem to come and go against a normative background of everyday Euclidean perception. Euclidean perception, as we have noted, is linked with a set of psychophysical parameters that are more appropriate to a universal, "godlike" observer than to one whose natural and experienced setting is to be at the center of a World. One must marvel that we have such native ability to respond like "godlike" observers in ordinary life, rather than like the kind of localized perceivers we seem born to be, irretrievably immersed in our individual heres and nows. If we are not deceived in thinking that our everyday Worlds present themselves *perceptually* according to the *conceptual* (Euclidean) norms of our scientific culture, how can this occur? How is it then that the everyday World *as experienced* is Euclidean and nevertheless has the perceiver at a privileged position at the center of a World?

One might think of various reasons why, contrary to theoretical expectation, the World as experienced turns out to be Euclidean. (1) It is possible that, moved by an urge a priori and necessary to all human experience to posit an objective and impersonal World, we have learned how to generate from the storehouse of our visual memory and imagination the only structure that makes this possible, a Euclidean space and World, *without recourse to processes of instrumented measurement*. I prefer, however, a different solution: (2) that our mode of spatial perception has been transformed by the fact that the environments we grow up with are to a great extent "carpentered environments," in which simple engineered forms paradigmatically Euclidean by physical construction are endlessly repeated. In this latter solution, we learn to compare unknown depth and distance relationships with these "carpentered" standards. Attending to these cues, rather than to others that would otherwise prevail, we learn to 'read' them perceptually, and by so doing, become aware of the presence of a Euclidean environment. Euclidean (physical) space and Riemannian (visual) spaces would, in this solution, be cued to different structural characteristics of the same ambient optical energy, that is, to different systems of stimuli.

Horizonal Realism

Scientific Realism

In a realist epistemology, the goal of critical and reconstructive philosophical analysis is generally to determine a unique and universal descriptive language, applicable grossly to everyday perception, and in a more refined way to the objects of science, in terms of which all possible empirical facts can be described. The repertory of its descriptive categories would be independent of context and natural language, as well as of personal, social, and historical bias: in a special sense of that much abused term, all the predicates of this privileged set are *objective*. I take *scientific realism* to state that the universal language is that of science, perhaps not science as it is at the present time, but in its final and complete form, as it should be, as it were, at the end of time.

Such a body of views includes positions about (1) *the real*: this is the set of structures that exist independently of human culture, language, and history; (2) *science*: this is inquiry leading progressively to knowledge of the real as so defined; (3) *truth*: this is conformity with the real as so defined. Positions of this kind have been argued, for example, by W. Sellars, B. Ellis, J. J. C. Smart, H. Putnam, and others.[1] Ellis states the strongest version: "I understand scientific realism to be the view that the theoretical statements of science are, or purport to be, true generalized descriptions of reality."[2]

The position of scientific realism starts off with the prima facie certainty that reality would exist unchanged even if there were no human perceivers. It then goes on to make two assumptions: (1) human knowers can get to know (and to describe) that reality *as totally unrelated* to human life and culture, and (2) that science is, or aims progressively to be, the form of such a knowing. One major problem that I see is in the first assumption: I take it to be highly implausible that we can come to know (or describe) anything *as totally unrelated* to human life and culture; successful as it has been, science could not be what it purports to be on this account. That does not imply that those same scientific entities whose reality is proved in the World of culture, are nothing but cultural entities; they are first discovered, however, as cultural entities. In addition, they have relationships among themselves which prescind from human cultures, and which are defined by scientific theory. Such relationships define a place for scientific entities among the furniture of the earth and universe, as *necessary antecedent conditions* for human perceptual acts and for human Worlds. Scientific entities are a part of what is or can become real, but not the totality of what is real as scientific realism would, to the contrary, propose.

In my account, *reality* is taken in a different way: it is exactly what Worlds make manifest (or purport to make manifest) to human perceivers; consequently, science to be realistic must have as its primary goal the exhibiting of reality structures not accessible to prescientific perception. Moreover, explanatory entities and processes of the scientific account—if they are genuine candidates for reality status—must have a way of exhibiting their profiles and essences directly, not, of course, to all and sundry but to expert scientific perceivers in the reflective attitude. *Reality* taken in this way does not preclude true knowledge about how things are and act in the absence of perceivers, only such knowledge that purports to be a "God's eye knowledge" of the situation. The details of this resolution will be worked out in the succeeding chapters.

Returning, however, to scientific realism, a significant point has been made by B. van Fraassen,[3] that a distinction ought to be made between the "real" and the "observable": the "real," he takes as independent of observability, and the "observable," he takes as that subset of the real which is detectable with or without scientific instrumentation. Van Fraassen argues that the goal of science is set by the observable, it is "empirical adequacy"—and it is not set by the real as such. He calls his view "constructive empiricism." It has aspects and concerns common with the view proposed below called "horizonal realism," and I applaud these aspects and concerns. I shall briefly compare van Fraassen's views with mine on a number of points. (1) On the meaning of *reality* and *realism*: *my* use of "reality" correlates with *his* "subset of the real that is observable".[4] (2) On the meaning of *theory-laden observables*: for him, theoretical terms can be used to describe what can be observed (for him, this means detected through observation), but this usage does not entail for him a commitment to the reality of the theoretical entities so described. For me, however, the goal of such a commitment—at least for the expert scientific observer in the reflective attitude—is essential to science, so that any genuine theoretical entity should eventually come to manifest itself in observation as having the profiles of an essence and so as exhibiting what it is to be real.[5] (3) On the conditions for *identifying observable objects*: van Fraassen speaks of observables as being the domain for models of a scientific theory, but he fails to notice that there are usually two stages in the modeling process—theory to abstract (or, at least, intermediate) model, and abstract (or intermediate) model to observables. The conditions for the latter mapping (conditions for correctly identifying the named objects) are more complex than he believes, and the thrust of the following chapters is to make this evident.[6]

Realist epistemologies are often called "objectivist"; they differ radically from epistemologies that see reality from the start as "horizonal." These latter epistemologies are sometimes charged with being infected with "epistemological idealism," or they may be called "antirealist."

Horizonal Realism

In a *horizonal epistemology*, perceptual facts belong to horizons of public human experience through the use of a common descriptive language. Each horizon has an *inner* and an *outer* component. In any individual act of perception, the perceived horizon has an outer horizon or boundary, which separates it from the background against which it appears. An *inner horizon*—objectively considered—is an essential structure of reality; it is composed of a manifold of possible perceptual profiles organized by an invariant essence (from which the perceptual object gets its name.) The essence is not an abstract form that exists apart from the profiles, it exists only in the variety of the manifold as its principle of organization. Thus, for example, when a table is present to perception, it is present through one of its profiles: however, that this profile correctly manifests a real table, is a function of three things: (1) that the offered profile is associated with a horizon of other profiles *of a table* (or *of this table*, if attention is directed to this individual table),

176   *Toward a Philosophy of Science*

(2) that the offered profile is connected intentionally (that is, in subjective anticipation) with this horizon, and (3) that the perceiver can by
his/her own initiative bring into evidence a representative sample of the
profiles in question. If, on the contrary, the offered profile were really a
profile of, for example, a flat cardboard painting of a table, then the
perceiver would discover the mistake when he/she tried to bring into
evidence a representative sample of table profiles; the perceiver would
find that the profiles generated by these initiatives belong instead to the
horizon of a flat cardboard painting of a table. Thus, in a horizonal
epistemology, every perceptual object fulfills an embodied subjective
intention (or perceptual system) with respect to which it is defined.

An embodied subjective perceptual intention or perceptual system
comprises a (system of behaviors or) praxis, using somatic (and
neuronal) processes and possibly also technological processes, and is
animated by an intention (alternatively called a "heuristic structure,"
or an "intentionality-structure") in which is sketched out in principle
antecedently to particular experiential acts the repertory of objects
(things, events, processes, situations, etc.) that can make themselves
present.[7] Every perceptual praxis is, of course, exercised within a
World by a community of human inquirers and uses a descriptive
language.[8] Every perceptual praxis actualizes an intentional possibility
of a Body. Moreover, every perceptual praxis is capable of making
some horizon of a World present to perceivers. Reciprocally, a horizon
of some World can impose its presence on perceivers only by causing
some appropriate function of an intentional structure of a Body to
resonate in response to its presence. Such intentional structures of a
Body mediate the presence of object to subject, that is, they mediate
both the (passive) givenness of object to subject, and the simultaneous
(active) illumination or apprehension of the object by the subject.
Noesis receives the *noema* while at the same time it gives it
significance.

The horizonal view is closely related, on the one hand, to that of
naturalistic pragmatism, for example, of W. James, J. Dewey, G. H.
Mead, A. N. Whitehead, R. Neville, and to other praxis-oriented
epistemologies,[9] and on the other, to Heideggerian or existential
epistemologies founded on the human role in the constitution of Life
Worlds. Such epistemologies are anthropological, historical, and ecological in character as opposed to the more traditional *objectivist*
epistemologies that tend to judge the validity of human knowledge by
(often implicit) standards for an absolute, perfect, and timeless (or

godlike) knowledge of the world. From the viewpoint of objectivist epistemology, a visual hyperbolic construal of the environment would have to be judged as not conforming to the basic facts about reality, and therefore as illusionary.

The horizonal view of perception assumes that plural veridical realistic perspectives are possible consonant with the plurality of different horizons of perception within a World: there is then not just one empirical basis of fact but a plurality of empirical bases. To each horizon would correspond a characteristic perceptual praxis—a kind of method of illumination—to which would correspond an intentional possibility for a Body, providing the light, as it were, for the illumination. Reality, then, would be grasped only through this network of illuminations, each one giving rise to a horizon within a World and a corresponding empirical basis of fact.

One major division between classes of horizons of perception is that between "manifest images" of reality and "scientific images".[10] The former represent objects as constituted in their essential form by a human social or cultural interest or by criteria found only in the mental life of persons, and they are functions of persons; in the words of W. Sellars, "the 'manifest' image of man-in-the-world [is] the framework in terms of which man encountered himself."[11] A scientific image, on the contrary, represents objects as constituted in their essential forms by systems of postulated (or theoretical) entities, related to one another and to some manifest World by scientific theory, and encountered only through the mediation of instruments or technology. Primary qualities in their Euclidean representation would then belong to a domain of scientific objects; but secondary qualities would belong to the domain of manifest objects, together with hyperbolic visual construals of the environment, provided of course such construals constituted, as I have proposed, true horizons of person-related *noemata*, and not just aberrant illusory episodes generated by inappropriate cognitive strategies or by malfunctions of the neurophysiological system.

The horizons of both images are sophistications and refinements of an original image, an image in the natural attitude. Each horizon is a structured domain of reality, each denotes on the objective side a set of manifesting profiles, each connotes on the subjective side a perceptual system that is capable of responding to the presence (or absence) within some World of the horizon of profiles which corresponds to it. The perceiver as subject is a (phenomenological) Body—including neural, somatic, and possibly technological processes—appropriate for the use of the perceptual system and the intentionality-structure it carries.[12]

Sellars does not argue that objects of the manifest image are any less definite, lawful, or descriptive than objects of the scientific image; it is not a "pre-scientific, irrational naive image of man-in-the-world,"[13] he says, but it is nevertheless, in his view, false as a framework for describing reality. Without accepting his terms or the consequence he derives from them, it is useful nevertheless to hear Sellars on what he regards as the significant difference between scientific and manifest images. He writes,

> The conceptual framework I am calling the manifest image, is in an appropriate sense, itself a scientific image. It is not only disciplined and critical; it also makes use of those aspects of scientific method which might be lumped together under the heading "correlational induction." There is, however, one type of scientific reasoning which it, by stipulation, does *not* include, namely, that which involves the postulation of imperceptible entities, and principles pertaining to them, to explain the behavior of perceptible things.[14]

I shall contend that the theoretical entities of science, "notoriously imperceptible and unimaginable" to the unaided senses, can never-theless, through the use of readable technologies, become fully percep-tible and as a consequence can take a place in the World of manifest objects, membership in which, in my view, gives the status of reality.

Context

To each horizon, there belongs a particular descriptive language and a corresponding *context* for its correct use. The context is the horizonal structure itself, and it has both subjective and objective components. The subjective component of context is the intentional embodiment of a subject. The objective component of context is the plurality of profiles that make up the horizon. The subjective and objective components of context are reciprocally related.

Horizons are plural, each linked to its own individual context, each connoting its own individual embodiment (or Body structure) and cognitive praxis, and each denoting objectively a manifold of possible perceptual profiles. At a particular place and time, a specific context *A* may be realized, that is, the subjective and objective components of *A* may come to be. Then a domain of real possibility is generated, and a horizon comes into existence within which descriptive statements belonging to a particular descriptive language become *truth-functional* (having the quality of being either true or false). An individual descriptive statement is truth-functional, then, only if and insofar as its context has been realized.

Perceptual statements belonging to different contexts may or may not be *simultaneously truth-functional*, depending on whether their contexts can be realized together at the same time. If the contexts cannot be realized simultaneously, the statements are said to be *contextually incompatible*.[15] It is not to be assumed that any two different contexts taken at random are compatible. To realize the contexts of two perceptual systems together, a subject must be able to share the two embodiments appropriate to each without interference from the other, and must be able to exercise simultaneously in an unimpeded way the praxis of each. Two descriptive statements may, on the one hand, belong to compatible contexts of reality and discourse, in which case they are simultaneously truth-functional. On the other hand, however, they may belong to incompatible contexts, in which case they are "complementary statements," somewhat after the usage of N. Bohr and W. Heisenberg in quantum mechanics, and taking the basic sense of their term "complementarity" to mean *context-dependence*.[16]

While compatible contexts are cumulative in a Boolean sense, that is, the resources of their respective descriptive languages can simply be added set-theoretically, incompatible context-dependent descriptive languages are generally not simply additive in this way. The next section contains the elements of a mathematical model (in elementary lattice theory) for a logic of embodied context-dependent discourse. An application of the same model will be used in chapter 13 to give an account of the rationality of scientific progress. For those who prefer to skip the more formal part, there will be a brief summary at the end of the next section.

Context-Dependency and Quantum Logic

A lattice is a set of elements, L_i ($i \in A$: A is some index set) that are partially ordered in a certain way to be defined below. In this case, L_i are descriptive perceptual languages, or the horizons which they describe; for example, L_a is the descriptive language appropriate to the horizon or context a. For the purposes of the following study, L_a, and so on will be treated extensionally as the set of its empirical contents. Each content is expressed by a potentially truth-functional sentence, and such sentences become actually truth-functional whenever the context, a, and so on exists. The set L_i is partially ordered: that is, some pairs of L_i are ordered by a binary relation that is reflexive, transitive, and nonsymmetric. If L_a and L_b are an ordered pair, then we write $L_a \rightarrow L_b$ (L_a "is ordered to" L_b or L_a "implies" L_b).

Where L_a and L_b are descriptive languages, an ordering, $L_a \rightarrow L_b$, exists under the following conditions: (1) the truth-functional conditions for L_a are compatible with the truth-functional conditions for L_b, and (2) whatever descriptive statement can be made in the (narrower) context appropriate to L_a, the same statement can be made in the (broader) context appropriate to L_b, but not necessarily vice versa. In an alternative expression of the same thing, the partial ordering "$\rightarrow$" can be given a pragmatic sense: whatever empirical question or problem can be *formulated and answered* in the (narrower) context appropriate to L_a, the same question or problem can be *formulated and answered* in the (broader) context appropriate to L_b, but not necessarily vice versa. Note that in the latter definition, problems are taken to be formulated in the same language in which they are solved, not—as in the older view—in some neutral observational language, nor—as others would have it—in a theoretical language different from those that address its solution.[17]

Consider the commonsense language of the natural attitude—a complex of inchoative, incomplete, descriptive languages that intend (successfully or unsuccessfully) descriptive horizons of some World, but not in a systematic, critical, and reflective way. Among the terms of common sense are some that refer to authentic essences and others that express intentions or proto-ideas that may or may not lead to the discovery of authentic essences but have nevertheless some role in classifying phenomena; among the latter will be some that, though in currency at the time, will be eventually discarded after critical inquiry as false, fictive, mere fantasy, and so forth. Common sense is oriented outward, objectivist in attitude, and suffers from the confusions and biases of pragmatic everyday communication: it also abounds with inconsistencies, but only when these hinder action are they faced and then as a rule merely in a pragmatic way. Common sense contains strings of descriptive languages ordered by more and more inclusive contexts: for example,

$$L_a \rightarrow L_b \rightarrow L_c \quad \rightarrow \quad \rightarrow L_A$$
$$L_m \rightarrow L_n \rightarrow L_p \quad \rightarrow \quad \rightarrow L_B$$

etc., etc.,

where L_A, L_B, and so on belong to the most inclusive (possibly ideal) contexts of each string.

Where common sense contains implicitly incompatible contexts, there will be two or more of such strings, that is, there will be L_A and L_B (and possibly more), and neither $L_A \rightarrow L_B$, nor $L_B \rightarrow L_A$. Let us assume that there is a largest contextual domain L_O common to both L_A and L_B, and let it be represented by

$$L_O = L_A \times L_B$$

where "$\times$" signifies the logical product of the two elements. Every sentence of L_O will also belong to the resources of L_A and of L_B; in other words,

$$L_O \rightarrow L_A \text{ and } L_O \rightarrow L_B$$

In the reflective attitude, the reason for the incompatibility of L_A and L_B (or horizon A and horizon B) must be that they contain some different perceptual systems that mutually interfere. This means that there are some categorial differences between L_A and L_B, giving rise at this level to a set of conceptual problems. The differences, however, are not merely conceptual; each apparently has a different empirical basis. Examining the phenomenological conditions for such a situation, one finds that the two perceptual praxes are different, bringing two systematically different horizons into focus; this implies that the perceiver is embodied differently in the two perceptual praxes. In Heidegger's language, they differ in both *Vorhabe* and *Vorsicht*.[18] Two different sets of subjective and objective conditions are involved which are not simultaneously realizable, at least not without mutual interference.

To study what kind of synthesis can be made of two such horizons, one has then first to become reflectively aware, not merely of the conceptual or categorial differences between the two *Vorsichten* but of the fact that each depends on a different set of subjective and objective conditions of possibility; that is, a different Body, a different praxis, a different horizon, and a different *Vorhabe*. All of these are implied in the context-dependent character of perceptual systems. The conclusion has to be drawn that besides empirical and conceptual problems—the only ones recognized by most philosophers and historians of science—there are in addition others, let me call them "transcendental problems," the problems that arise from the embodied character of perceptual contexts. In the following, L_A and L_B are taken to represent those languages refined and expanded to include reflective awareness of the context of embodiment within which each is exercised, and of the subjective and objective limits affecting the proper use of the descriptive terms of each.

A new, explicitly context-dependent language, L_{AB}, can sometimes be developed which includes both L_A and L_B and takes into account the fact that the two original contexts, A and B, are not compatible but interact in varying degrees. Let the combinations range from pure A, through various combinations of A and B, to pure B. If there are many possibilities for L_{AB}, let L_{AB} be the least of them in content. Then a new functor, "$+$," can be introduced, indicating a new kind of logical sum: let $L_{AB} = L_A + L_B$, where $L_A \rightarrow L_{AB}$ and $L_B \rightarrow L_{AB}$, but where, because of the impure mixed contexts of A and B, L_{AB} is richer than pure L_A and pure L_B. Interpreting the ordering relation "$\rightarrow$" as a kind of inclusion relation relative to the descriptive (alternatively, pragmatic) resources of two languages, a Venn-type diagram (figure 10.1) can be drawn to show schematically the extensional relationships between L_O, L_A, L_B, and L_{AB}, and the complements L_A' and L_B' to be introduced below. Thus, L_{AB} is more than the Boolean (set-theoretic) union of L_A and L_B (the unshaded area in figure 10.1 represents the excess).

Let us now suppose that to each context (L_A and L_B) there exist complements (L_A' and L_B' respectively) with the properties:

(i) $L_O \;\; \rightarrow L_A \;\; \rightarrow L_B' \;\; \rightarrow L_{AB}$

(ii) $L_O \;\; \rightarrow L_B \;\; \rightarrow L_A' \;\; \rightarrow L_{AB}$

(For example, *if the following sets were to constitute genuine contexts*, they could serve as complements with the requisite properties:

$$L_A' \;\; = (L_{AB} - L_A) \;\; U \; L_B$$

$$L_B' \;\; = (L_{AB} - L_B) \;\; U \; L_A$$

where "$-$" and "U" are the familiar set-theoretic operations.[19])

The six languages, L_O, L_A, L_B, L_A', L_B', and L_{AB}, constitute a lattice under the partial ordering: this lattice is nondistributive (it happens also to be orthocomplemented, but that may not be necessary) and its characteristic diagram is shown in figure 10.2. L_{AB} contains the solution of problems stemming from the variety of empirical bases of A and B; it also resolves the conceptual or categorial problems between them; moreover, none of this could have been accomplished without the

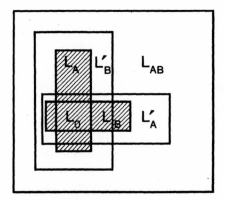

Figure 10.1: Extensional relationships between context-dependent languages of a particular quantum logic or Q-lattice.

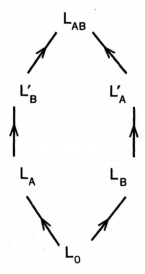

Figure 10.2: Diagram of partial orderings within a particular Q-lattice.

simultaneous solution of the transcendental problems associated with A and B.

In the lattice, the logical product of any two languages ($L_i \times L_j$, the context common to both L_i and L_j) is the greatest lower bound (g.l.b.) with respect to the diagram: the logical sum ($L_i + L_j$, the smallest context of which pure L_i and pure L_j are constituent parts) is the least upper bound (l.u.b.) with respect to the diagram.

The lattice is nondistributive over products ($\times$) and sums ($+$) provided the following conditions are fulfilled. Referring to the lattice, we find:

(iii) $L_A + (L_B \times L_B') = L_A$ and $(L_A + L_B) \times (L_A + L_B') = L_B'$

(iv) $L_B + (L_A \times L_A') = L_B$ and $(L_B + L_A) \times (L_B + L_A') = L_A'$

Thus, as long as L_A is *not identical with L_B'*, or L_B is *not identical with L_A'*, distributive relations are not uniformly observed. Consequently, as long as there are sentences in L_{AB} which are neither in L_A nor in L_B, and as long as these sentences belong to contexts that fulfill relations (i) and (ii) above, then the lattice is non-Boolean and nondistributive.

I have called this lattice a "quantum logic"; it is a nonclassical logic of *frameworks* (or *horizons*) rather than of *sentences*: it is the same kind of lattice that as I have shown is fulfilled by the "complementary" frameworks of quantum mechanics.[20] The lattice of figure 10.2 can be generalized into lattices of higher order.

In summary, embodied contexts of perception may be compatible, that is, independent of each other, or incompatible, that is, capable of interfering with one another. In the latter case, incompatible contexts give rise to three sets of problems, empirical, conceptual, and transcendental, the last mentioned including those that arise from the embodiment of the perceiver in the different contexts of perception. Incompatible contexts are joined when the perceiver finds a way to embody simultaneously the subjective conditions for both contexts. Such a joining would include the original horizons as well as the new ones, and may (or may not, of course) produce a range of new impure contexts in addition to the original pure ones. These new horizons enrich the perceptual possibilities of the pure but isolated contexts of perception by conjoining them in various degrees. I will hold that, in general, predicates of a manifest image, such as *sensed-red*, *heat-as-felt*, and *hyperbolic shapes*, and their counterparts in the scientific image, such as *(spectral) red*, *thermodynamic heat*, and *Euclidean shapes* constitute incompatible but complementary frameworks of description which are ordered by a quantum logic of the kind just described. This topic will be further discussed in a more informal manner below.

Objectivity

Traditional realists will argue that the term "reality" should be applied only to whatever is of its nature "objective" in the special sense, that is, *independent of human culture, history, and language*. They would add that it should not be used of any horizonal illumination that depends essentially on a power human perceivers may have to construe their experience in a way other than it is in scientific or physical fact, and a fortiori of a power that may well vary among particular human perceivers. In this respect, it would be argued that the framework of manifest images is useful but illusionary and subjective.

The pursuit of a scientific (or third-person) inquiry, however, depends on being able to operate with a background framework of manifest images. One must be able to set up common procedures of measurement and discriminate a common set of outcomes. There is a certain paradoxical quality about the statement that perception, which reveals (or purports to reveal) a World directly with secure evidence, does not arrive at "truth about reality," but awaits scientific knowledge to do this, while all the time scientific knowledge depends for its successful accomplishment on making (and judging accurately) discriminations among the manifest objects that contextualize the inquiry. Some "truth about reality"—even if it is only about *some of the empirical conditions* for scientific inquiry—has to be assumed. For Sellars *truth within* the framework of the manifest image would be sufficient, though ultimately he rejects the *truth of* the framework, that is, he rejects its value as descriptive of reality.[21] His reasons for denying reality to the framework of the manifest image will be taken up below.

The traditional response of scientific realists has been to appeal to two kinds of sensible qualities: primary (or original) qualities, and secondary qualities.[22] Primary qualities, like (Euclidean) spatial and temporal extensions, are taken to be intrinsic to the physical environment both *as measured* and *as perceived*. Secondary qualities, like color, taste, and so forth, are taken to be representations of something in terms of the sensory subject's response to them, and not in terms intrinsic to the physical objects themselves; *as perceived* they are taken to be no more than indications to the perceiver of some possibly not-yet-disclosed, scientifically describable source of variability among physical objects. They are not considered to be "objective," or "real," but—in the realist view—illusions of the sensory system, though useful ones.

The distinction between primary and secondary qualities has always been a cause of embarrassment to epistemology. Both seem to be natural perceptual responses to appropriate environmental stimuli! Some epistemologists, like Rorty, believe it would be intellectually more satisfying if all perceptual qualities—primary and secondary or of whatever kind—shared the same epistemological status or intent. [23]

"Horizonal realism" proposes a new sense for that overused term "objective," applicable principally to manifest objects: a World is the *objective* context of the real; whatever has a spatial and temporal location in the World is *objective*, and whatever is independent of individual perceivers who share the World is *objective*. A. Schutz calls this kind of objectivity the "reciprocity of perspectives." [24] I take it as necessary for the real that it have the kind of objectivity that is the reciprocity of perspectives.

There are few, perhaps, who would object to the objectivity of manifest objects in the sense just defined. Color terms in pictorial usage, for example, refer to a range of generally surface properties that are recognized and distinguished one from another by visual criteria learned and applied within a particular culture. To say something is green means that under standard lighting conditions (roughly, moderate sunlight outdoors or its equivalent) the normal person would judge the thing to be green and not blue, red, white, or some other color. Green (in ordinary usage) refers to a kind of property that is essentially related to specifiable standard environmental conditions and to the learned responses of normal human perceivers. Color judgments made in this way are public, testable, and intersubjective within a particular culture and language.

Color judgments have much in common with scientific judgments. [25] Like scientific terms, they are contextual; they exhibit their quality unambiguously only under standard physical conditions. Color distinctions depend essentially on a particular human culture and the descriptive language used in that culture. Scientific culture is also a particular, though not geographically circumscribed, culture, and to understand its language, its culture must first be assimilated by training: in this respect also, color terms and scientific terms are similar. Given that the boundaries of application of color predicates are imprecise, secure objective color judgments can nevertheless be made in the overwhelming majority of cases that concern the interests of persons in everyday life and within the bounds of those interests. The degree of sureness, not to say certainty, of such judgments compares favorably with determinations of predicates of the scientific image. Color terms differ from scientific terms principally in that color terms are essentially related to the trained perceptual response of human sensory systems, and to the interests and purposes of persons; and so they are not generally quantified, nor are they, therefore, precise. Scientific terms, however, are essentially related to instrumental systems of measurement quantified through the use of a theoretical model, and so can be precise, and aim at such precision.

Reduction of Manifest to Scientific Image: Problem

Realists from Democritus to Identity Theorists have proposed to redefine perceptual terms in a materialist, scientific, and reductionist way, eliminating all reference to persons, history, and culture, as well as to mental phenomena and the perceptual response to human sensory systems.

One such stipulation, for instance, is to define the color of an object by the spectral mix (of wave lengths) and intensity of the light it reflects or emits under some standard illumination. All such stipulations are doomed to failure, writes R. L. Gregory, because

> color depends not only on the stimulus wave lengths and intensi-
> ties, but also on *whether the patterns are accepted as representing
> objects*, and this involves high-level processes in the brain which
> are extremely difficult to investigate . . . The eye tends to accept
> as white not a particular mixture of colors, but rather the general
> illumination whatever this may be.[26]

While it may be convenient for some purposes to have on hand a new concept *spectral red*, defined in terms of wave length, intensity, and spectral distribution of illuminant, there is no one-to-one mapping of the sense-object *sensed-red* on *spectral red*, or vice versa, without reference to as yet unknown ''high-level processes of the brain.'' Even if these were known and taken into account, it must be remembered that the two concepts are defined within quite different contexts—one by unaided perception (sensory), the other by instrumental measure-ment (spectral). The sensory concept *red* has a structure and degree of variability that is not matched one-for-one by the account based on instrumental measurement. Its three dimensions, (perceived) hue, brightness, and saturation, correspond roughly to wave length, inten-sity, and spectral mix, but not exactly, since a variety of spectral mixes can be perceived as having the same sensed-color, and the same spectral mix (depending on the setting in which it is experienced) can be perceived to have different sensed-colors.[27] It is reported that the normal human vision can distinguish 7.5 million different colors, of which only two or three hundred are spectral.[28] Colors as sensed are not then in any simple one-to-one relation with spectral frequencies, mixes, and intensities.

This illustrates a general thesis, that there is no simple identity of denotation between manifest and scientific objects; each involves a different conceptual system ("plays a different role," in Sellars's language[29]), and each involves a different embodiment, in terms of which the object is isolated from its background and made manifest as an object of such and such a kind.

Sellars argues that one and the same ice cube, for example, is "homogeneously pink" in the manifest image and neither homogeneous nor colored in the scientific image.[30] The homogeneity of sensed-color belongs to manifest images alone; it belongs only to objects, such as ice cubes, correctly identified by a perceiver in that image, usually without the aid of instruments or technology. Turning now to the scientific image: for a perceiver to discover and identify, for example, the particular array of colorless molecules and interspersed empty regions which compose this particular pink ice cube, the perceiver has to be embodied in instruments or an appropriate technology. These instruments or this technology, I shall argue, are essential to the isolation of an object from its background and for the identification and recognition of any particular molecular array; they define what an array of this kind is. The pink ice cube remains unchanged as far as the denotation, "this pink ice cube," goes, even if (some of) its molecules are removed or exchanged with other similar molecules or permuted among themselves; the pink ice cube is an invariant denotation under a wide range of replacements and permutations of its constituent molecules. There is then no unique scientific referent congruent with the unique manifest pink ice cube. Likewise, a particular array of molecules within the pink ice cube remains denotatively the same even if removed from the pink ice cube, and put, say, into a blue ice cube; there is likewise no unique manifest referent for this particular molecular array within the pink ice cube. In addition, then, to differences in the descriptive framework, the *referent* of a particular description in some manifest image is not uniquely specified in the scientific image, and vice versa. This is a consequence of the one-to-many and many-to-one mappings between corresponding denotations of the two descriptive systems. Since there is no simple material identity of denotation between objects of the two images, the basis for asking the reductionist question is removed. The argument just given is also a basis for the rejection of Nominalism.

Manifest Objects as Real

Although clearly not an intrinsic property either of a surface alone or of light alone, nevertheless sensed-colors do enable us to identify and to distinguish different kinds of things in the World in relation to powers of sense discrimination and human significance cultivated in our society. I need hardly point out the importance for human survival, particularly in our culture, of the sense-percept *red* since it is the color of spilled blood and serves as the public, objective, unambiguous color of the STOP or danger sign. Of the two concepts of *red*, only *spectral red* is in its definition independent—but with a certain limitation to be explored below—of perceivers: *sensed-red* is by contrast, both in its definition and praxis, dependent on trained powers of human perception shared by all normal adults in our culture. The recognition of *sensed-red*, however, in any individual case is under standard conditions independent of which particular normal adult does the recognizing. Thus the concept *sensed-red* has the kind of objectivity defined above, of being *intersubjectively testable* by normal adults who share a World.

Does the use of such concepts as *sensed-red* succeed in identifying *real* differences and *real* similarities in a World? To the extent that all real differences and real similarities are in relation to the outcome of an appropriate standard praxis, we should note that some standard praxes involve essentially the use of external material instruments, and others involve only the trained human powers of perceptual discrimination. Since human perceiving and transforming agents are as much part of reality as physical instruments, functional categorizations based upon the discriminative behavior of human perceivers can also provide public knowledge of the real, contextualized by the specific horizon of the perceptual inquiry.[31]

Taking a World, as I have done, to be the context that confers reality on the objects given to perception by this World, it is not surprising that my conclusion places *sensed-red* and all manifest objects as real, objective, and belonging to possible Worlds. The argument in favor of *spectral red* is more complex: it involves the account (to be given below) of how it is possible, within the scope of the primordial intentionality of perception, for quantities, imperceptible to the unaided senses and detectable only through empirical procedures such as measurement, to come to be experienced as given directly in perception, and through this givenness, to take their place as manifest elements of the real.

Since *sensed-red* and *spectral red* have neither the same extension nor the same intention, as explained above, then both are necessary to describe what is the case and neither is redundant; we need to refer to this particular molecular array *in* this pink ice cube, or to this pink ice cube version of this particular molecular array. Only if there were a uniquely specified referent in both images for every description in either would redundancy occur. Sellars then is incorrect in arguing for the redundancy of the manifest image.[32] He also as a consequence incorrectly formulated the options for relating manifest images to scientific images. His assumption of a unique common denotation led him to suppose that there were just three possible ways of relating them: either manifest images are redundant, or scientific images are redundant, or neither is redundant because the former just deals with groups constituted of entities specified by the latter (the way a forest is constituted of its trees). It now becomes clear that none of those options is acceptable.[33] It also becomes clear why to choose "like a child" for the reality of both descriptions is not simpleminded, since each image is contextualized differently and—as I shall show in the next chapter— each has the power to make its objects present perceptually as manifest objects under its own set of descriptive terms. What then is the relation between the two images? Neither is reduced to the other. Each, however, may be methodologically connected with the other; that is, either manifest or scientific objects can be identified and its set of counterparts in the other image sought.[34]

To speak about two images, one needs, as Sellars said, an "external point of view" to compare them: is there such? Let L_A designate the descriptive language of sensed-color and L_B the descriptive language of spectral color. One view about their mutual relation is that of J. Heffner, who claims that the evidence supports the claim that the two domains of the physical order of wave length and intensity, and of the phenomenal (perceptual) order of colors as sensed, are two "mutually exclusive epistemological categories."[35] However, despite the absence of one-to-one relatedness, there exist partial one-to-many mappings and partial many-to-one mappings indicating the presence of a variety of color horizons, some based on systems of spectral variations with invariant sensed-color, and others based on systems of sensed-color variations with invariant spectral color. Referring to the lattice diagram of figure 10.2 and its interpretation, we have just proved that the two complements, L'_A and L'_B, are not identical with L_B and L_A respectively. Do they have a least upper bound? It is plausible to think so, since otherwise the use of a common set of color terms, "red" and so forth, would be simply equivocal. The framework of sensed-color, L_A, and the framework of spectral color, L_B, must then be complementary in terms of a Q-lattice of the type given in figure 10.2. The "external point of view" for relating L_A and L_B is provided by L_{AB}: this is a multicontextual language using common color terms, contextualized in use by the appropriate balance of connotations deriving from the concepts of *sensed-color* and *spectral color*. L_{AB} is of course not really external to L_A and L_B, but an extension of both.

Although the opposition and contrast between *sensed-color* and *spectral color*, and between *pink ice cube* and *colorless molecular array*, were studied to illustrate the Q-lattice of complementary frameworks and the relation of manifest to scientific images, one could equally well have used Euclidean visual space and hyperbolic visual space for this purpose. The choice of the illustration actually used was principally a function of the literature that was discussed. Illustrations of the natural way we combine Euclidean and non-Euclidean perception have been given above: for example, Arnheim's description of the interior of a church, Ballard's description of the Gateway Arch, and aspects of the visual illusions discussed above. Such examples illustrate what I called a "composite representation of quasi-stable [Euclidean and non-Euclidean] facets." What distinguishes these illustrations from *Gestalt* switches is the sense of continuity between the elements.

One central thesis has yet to be discussed without which the philosophical claims made in this chapter would not stand: this is the thesis about the direct perceptibility of theoretical scientific entities, and it will be taken up in the next chaper.

11

Perception of Scientific Entities

Scientific Entities

If perception is direct knowledge, then it would appear that a scientifically measurable theoretical quantity can only be known indirectly; it is after all a part of a substructure detectable only by instruments, and hidden to unaided perception. On that basis, I hold that theoretical scientific entities are no more than candidates for inclusion in the real order; to be accepted for inclusion in that order, they must present themselves with the proper credentials. According to the principle of the primacy of perception, this means they must be capable of being exhibited in genuine horizons of scientific observation, in which such quantities are manifestly and directly given as perceptual objects to experienced scientific observers.

The World of our place, time, and culture is structured by *horizons* of perceptual possibilities rooted in the primordial intentionality of perception. Since access to these horizons is learned together with that part of our natural language which provides descriptive terms for them, the creation of new horizons and the sharing of old is a historical and cultural process, both for the individual and for the community. No individual will actually share all the horizons of the everyday World, but everybody has the capacity to enter into possession of new horizons not previously shared. Experimental scientists—like members of most professional groups—have access by virtue of their profession to horizons that are characteristically their own. These are horizons of scientific observation in which, I claim, scientific states of affairs are given with critical absoluteness—not to all and sundry but to persons suitably prepared—in no way different from the way common and familiar objects are given perceptually in their World.

Scientific Observation as 'Reading' of 'Text'

The argument is as follows: theoretical quantities become known by attending to the response of appropriate empirical procedures, such as the use of measuring instruments. There are, however, two ways of attending to the response of an instrument: the instrumental response can be used in a deductive argument to infer a conclusion, say, the value of a quantity (provided one knows the relevant scientific theory), or the value of that quantity can be 'read' directly from this response (provided one is experienced and skillful in the use of the instrument). I want to focus on the latter. In 'reading' a thermometer, say, one does not proceed from a statement about the position of the mercury on a scale to infer a conclusion about the temperature of the room by a deductive argument based on thermodynamics; of course, one could, but then one is not 'reading' the thermometer.

To the extent one 'reads' the thermometer, the thermodynamic argument remains in the background, being merely the historical reason why thermometers came to be constructed in the first place. One can 'read' a thermometer, however, whether or not one knows anything formal about thermodynamical theory. Provided the instrument is standardized, and so can function as a readable technology, the instrument itself can define the perceptual profiles and essence of temperature.

'Reading' temperature is like reading a text: let me call it 'reading' (single quotation marks signify that the act of 'reading' the instrumentally 'written' 'text' is similar to but not identical with the art of reading texts written in natural languages). The process of 'reading' is something like this: a 'text' is 'written' causally by the environment under standard circumstances (*ceteris paribus* conditions) on the thermometer, and this 'text' is 'read' as being 'about' a presented object, here, the temperature; this piece of acquired empirical knowledge—the current temperature—is expressed in a language that uses scientific terms, such as "temperature," in a descriptive way about the World.[1] Such a process is, I claim, essentially both *hermeneutical* and *perceptual*.

Hermeneutics

The process of reading, understanding, or interpreting a literary text is guided by what Heidegger calls a "fore-structure of understanding," that is, an anticipation about the kinds of things or objects about which the text speaks.[2] This fore-structure is also called a "hermeneutical circle": in it is

> hidden a positive possibility of the most primordial kind of know-ing. To be sure, we genuinely take hold of this possibility only when, in our interpretation, we have understood that our first, last and constant task is never to allow our fore-having (*Vorhabe*), fore-sight (*Vorsicht*), and fore-conception (*Vorgriff*) to be pre-sented to us by fancies and popular conceptions, but rather to make the scientific theme secure by working out these fore-structures in terms of the things themselves.[3]

These fore-structures may be antecedently, already entrenched, devel-oped cognitive systems, or merely heuristic structures awaiting testing and deployment.

The task of hermeneutical interpretation does not have as its goal just some or any understanding consistent with the text, but a reading that attains to "the things themselves" (Husserl's term originally)—the subject matter, real or fictitious, imaginary or conceptual—about which the text speaks. Such a task is not, as Heidegger says, a work of arbitrary fancy but one controlled, on the one hand, by the totality of the text and its parts, and on the other, by the fore-structure that permits us to read the text as a text about specific kinds of things and objects. This fore-structure of understanding has three parts: (1) *Vorsicht*, or a set of common descriptive categories, a common descriptive language, as it were, (2) *Vorhabe*, a set of praxes, embodiments, skills, and so forth that mediate between the descriptive categories or terms and that to which they refer, and (3) *Vorgriff*, a particular hypothesis about the subject matter in hand.[4]

The hermeneutical task is circular, but in a peculiar "nonvicious" way, because it involves the simultaneous and mutual determination of the (meaning of the) whole by the (meaning of the) parts, and vice versa; the fore-structure of understanding—the "circle"—provides, on the one hand, a conjectured meaning for the text as a whole and for its parts; but, on the other hand, what kinds of conjectures one enter-tains about the (meaning of the) whole depends on clues scattered in the text itself. One moves from a partial disjointed set of insights or clues to an understanding of the whole and back to the not-yet-understood portions of the text, the process guided by the attempt to discover the outlines of "the things themselves."

Although the primary sense of *hermeneutics* is the interpretation of literary texts, problems of a hermeneutical kind occur also in the interpretation of empirical data, as, for instance, in the attempt to define the geometrical curve believed to link a set of variables, when only limited samples of empirical data are available. Such an investigation is guided by a series of conjectures, each of which relates points of the sample to some conjectural curve. Each conjecture opens the way to the search for and possible discovery of new coincidences in other samples; new, one hopes improved, conjectures are entertained as a result until a satisfactory solution is reached. Such a process of inquiry is hermeneutical, led by, in this case, the hermeneutical circle of functionality between the variables. A satisfactory solution in this case would fulfill the following conditions: (1) all the discovered points lie on (or *sufficiently* near) the proposed solution, and (2) one is persuaded that none of the as-yet-unsampled points lies *too far off* the proposed solution; the terms "sufficiently" and "too far off" imply reference to the goals and purposes of the inquiry. Every hermeneutical inquiry then is fraught with a certain indeterminacy. Unlike the above example, however, of a hermeneutical process, where there is a conceptual "space" common to all the conjectures, in the most general case there may be no such common "space."

When the text is imperfect or corrupt, or when the subject matter is unfamiliar, the hermeneutical task is in addition accompanied by effort and obscurity. Scholars of cultural phenomena, like Hans-Georg Gadamer,[5] are much concerned with this obscurity. The ambiguity, obscurity, and essential incompleteness of many hermeneutical tasks explains the variety and tentativeness of scientific traditions, and is connected with the possibility of alternative scientific 'readings' that were the concern of such scholars as P. Duhem and H. Poincaré, to mention two, and are currently topics of lively discussion in the sociological and conventionalist debate. In this chapter, however, I am concerned less with this aspect than with displaying the hermeneutical component in scientific observations, the presence of which is easily obscured by the practiced familiarity with which these observations are performed, and by the ideology that represents them as unproblematic because they are "observational."

All hermeneutical processes possess the dual structure associated with the acquisition or expression of information; there is information$_1$, which is the 'text,' and there is information$_2$, which is its meaning; in the case of perception, the latter is the content of the observation. Note that, in the textual analog, syllables, phonemes, or other linguistic signs, once they are read, cease to be objects in the World, such as houses or trees; they become more like *windows to a room* that by their (more or less) transparent quality give direct access to the contents of the room beyond. One does not *perceive* the syllables of a text: one *reads* them.[6] In a reading, the physical character of the text disappears from direct view, leaving no objective trace whatsoever, and one's attention is possessed directly and immediately by the *meaning* of the text. The physical character of the text drops out of objective awareness, and it becomes to that extent "nonobjective." I take that to mean "belonging to the conditions of the subject"; it has become in fact physically part of the cognitive subject as *embodied*; it has become a modulation of the somatic information channel of the embodied cognitive subject used in this particular hermeneutical activity of reading.

Although, on the whole, one is oblivious to the syllables and marks as things on paper, one is not totally unaware of them. It is through them that one is guided through the meaning of the text; the reading may be disturbed by an inability to decipher words, by typographical or grammatical errors. There is a subliminal awareness of pleasure, or perhaps of frustration and discomfort, which arises from the activity itself, from its rhythm, natural or forced, from the musical quality of the sounds, and from the resonances of heard or imagined speech. In these lie the aesthetic qualities of a text (sometimes called the "experience" of reading).

The kind of transformation I want to describe is that in which the syllables or phonemes of a strange language at first engage our attention as curious objects for possible theoretical study, and end up by being dropped from consciousness when we have become familiar with the language and have learned to read the syllables as text or to listen to the phonemes as spoken words. When one knows the language, direct access is obtained to the meaning of a text or spoken word. Such meanings, however, are expressed in judgments, but not in judgments of perception; the judgments are about the subject matter referred to by the writer or speaker, be it the history of the French Revolution or a football game. The transformation just described, wherein intermediaries (information$_1$) in the acquisition or expression of information (information$_2$) "drop out of consciousness," has been noted, for example, by Schrödinger,[7] who ascribed such transformations to processes perfected during the long course of evolution. It is a commonplace, however, that many processes perfected through the painful process of learning share this characteristic, that intermediaries drop out of consciousness. The paradigm example is reading, but there are other processes that are similar, such as playing a musical instrument, sight-reading music, driving a car, and reading an instrument: to a suitably experienced person, processes like these have "the same subjective ease and immediacy as the simplest perceptions."[8]

The hermeneutic analysis of listening to the spoken word is parallel to that of reading the written word, except that special emphasis must be placed on the context within which the word is spoken.[9] Texts are usually written for a wide audience and often seem as a consequence to be context-free. To read a contemporary piece of writing, one needs no more than the general presuppositions of one's culture or discipline, no more, for example, than the ability to understand the entries in a current dictionary. Texts, of course, are no more context-free than any form of statement, but the context of written work—particularly of work written in one's own language for wide circulation—is usually much more general than that of speech. Speech, by contrast, is highly context-dependent: group assumptions, circumstances, the cultural background of speaker and hearer, personal biographies, tone, inflection, timing are all part of the context that situates the spoken word and gives meaning to it—directly and immediately, when speaker and listener share the same context; otherwise, only indirectly and only inferentially.

Observer as Embodied

The response of an instrument, which I refer to as a 'text,' shares in the information-theoretic aspect of literary texts. A 'text,' however, is 'written' by the ambient environment on a standard instrument under standard circumstances in standard signs, and is controlled by physical causality; a text is written in a standard vocabulary, syntax, and so on by a writer, and is controlled by a causality guided by the writer's intentions. In each case, the process of interpretation involves the "resonant" stimulation of some somatic information channel of the reading or 'reading' subject, the use of some hermeneutical circle, and in both cases, when the hermeneutic task is accomplished, one is in direct possession of what the text or 'text' says, its meaning.[10]

While literary texts may speak about the World, they do not manifest, show, or exhibit states of the World. A 'text,' however, can and
usually does do this, insofar as it is usually about some state of the
World actually present and manifesting itself. Note that a scientific
observation may also be about some past state of the World, but more
about this below. Returning, for the purpose of illustration, to the
thermometer: the position of mercury on the scale functions as a 'text';
this 'text' has the character of information$_1$. Through a 'reading' of this
'text,' one gains knowledge of the current thermodynamic temperature.
The expression of this knowledge takes the form of a judgment, "The
present ambient temperature is (say) 70°''; this judgment is empirical,
direct, and uses scientific terms descriptively of the World.[11] I now
claim that this 'reading' is a perceptual process, since *it fulfills all the
characteristics of perceptual knowledge.*

Perceptual knowledge is (1) direct, not mediated by inferences, nor
is it just knowledge of an "internal representation" or "model"
constructed, perhaps, out of sensations, or in some other, perhaps,
mathematical, way. (2) It depends on the physical causality exercised
by the object on some somatic information channel, in this case, the
technologically extended Body of the subject. (3) It is hermeneutical,
that is, it acquires its meaning through the employment of a hermeneutical circle where the terms of a scientific theory are used descriptively.
(4) Its object, a state of the World, is experienced as given directly to
the knower by the World; and (5) it terminates in a perceptual judgment
of which the expression is a statement in which the terms of a scientific
theory are used descriptively, and which purports to describe what is
actually here and now existent in the World, present and manifest to the
knower appropriately embodied.

Like all perceptual knowledge, it is not apodictic in the natural
attitude; that is, though perceivers (suitably trained) usually perceive
unproblematically, they do not perceive *apodictically*, for to perceive
apodictically is to be able to sample at will the perceptual profiles of the
object. However, like all perceptual knowledge, to the extent that
profiles and invariants can be clearly articulated, scientific observation
(using instruments or other readable technologies) too is capable of
aspiring to apodicticity in the reflective attitude. It is this capacity to
become apodictic which establishes the possibility of genuine scientific
horizons in the World. This capability distinguishes the hermeneutics
of literary texts from the hermeneutics of perception and scientific
observation. While both are underdetermined, the former is about

meanings as unexhibited, possibly unexhibitable, possibilities (as states of the World), and the latter is about exhibited—possibly newly discovered—states of the World.

The hermeneutic character of a scientific observation is shown in its dependence on an accurate evaluation of the context of the observation. To perform a good scientific observation, the scientist must have the ability to use an instrument correctly (to make manifest the perceptual profiles of the scientific object), and to assess easily and accurately, often preconsciously, the *ceteris paribus* conditions on account of which the situation as presented is similar to those standard "paradigmatic" or exemplary cases of instrument use with which he/she is familiar.

For example, there are thermometers that record their readings photographically, or on tape, or with ink on paper. Such recordings, like 'texts,' can be placed in an archive and 'read' later. 'Reading' such a thermometer correctly presumes that one can judge the background context of its use, and can judge when to 'read' it as 'speaking,' say, of some *past* state of the World, and when of the *current* state. Such background judgments are part of a hermeneutical process. Suppose that the reference is to a past state of the World, then the 'text,' like a memory trace of the past event, will be 'read' as, "The ambient temperature *was* (say) 70° at such a place at such a time." 'Readings' of this kind express knowledge of past perceptual (though not necessarily perceived) events, but they are not acts of perception, for it is not the case that through them the current situation of the World imposes its presence and character directly on the knower. Such 'readings,' however, are also referred to in scientific literature as "scientific observations," but, as I have said, they do not constitute perceptions (or literally, observations). In them, the scientist (or observer) 'reads' the thermometer as making a 'statement' about a past event. Although this past event is of a perceptual kind (it could have been perceived), it was probably not one that a scientist ever actually perceived. "Scientific observations" of this kind illustrate well the hermeneutical character of such observations, as well as the fact that scientific predicates are used in them descriptively; there is, however, danger of misconstruing their character, and taking them to be literally acts of perception. Their objects are past perceptual events, a kind of "recall" that depends on the "memory trace" recorded by the instrument.

It may seem paradoxical that one can perceive thermodynamic temperature without *feeling hot*, as, for example, in the case where one 'reads' a thermometer inside that is measuring the temperature of the outside air.[12] The paradox is resolved by noting that thermodynamic heat and sensed-heat imply different contexts of analysis and consequently have different horizons (profiles and essences)—the one uses the natural heat sensors of the biological organism, the other uses artificial heat sensors, thermometers—functioning within that extension of the human subject's embodiment (in the phenomenological sense) which is appropriate to scientific observation. One perceives thermodynamic heat differently from sensed-heat, one feels the sensation characteristic of the latter, but there is no feeling characteristic of the former. One could, of course, say that one "feels" the former in some way analogous to the way one feels the latter, but here the ground of the analogy is merely the directness in each case of knowledge acquisition. There is no characteristic "feeling" of thermodynamic heat —except perhaps within the context of a standardized readable technology. Two different somatic information channels are used respectively for sensed-heat and thermodynamic heat; there are consequently differences in the aesthetic quality of these perceptions, but this is not a topic I shall develop.

Another example of the use of instrumentation to generate a new field of perceptual knowledge is the Tactile Visual Substitution System (TVSS) of C. Collins and P. Bach-y-Rita.[13] This is a device for use by the blind: it consists of a video camera attached to the temples, from which the output is fed to a ten-inch-square array of electrically driven vibrators in contact with the skin of the back or abdomen. The video image is broken up into an array of small gray dots, like a newspaper photograph; each dot is linked to a vibrator, and to each shade of gray there corresponds a definite intensity of stimulation. The device was designed to provide the blind person with essentially visual information, but in tactile form. The blind person who uses this device initially experiences only a tickling sensation, which then gives way to the experience of solid objects external to the subject and located in a three-dimensional space. These objects are experienced as having definite shapes and sizes, as being composed of patches of surface joined along edges, and as being capable of blocking the ''view'' of other (more distant) objects. This new form of experience has all the characteristics of perceptual knowledge: it is direct, it is experienced as given to or imposed by an external World on the knower, and it is the activation of a prior embodied three-dimensional fore-structure of the subject. Such knowledge naturally expresses itself in the descriptive language of perception. Morgan writes, "In general, there is little doubt that the TVSS allows the blind to see. Or to 'see,' as Bach-y-Rita prefers to say . . . as experience with the TVSS proceeds, the judgments become more and more automatic, until they are accomplished in much the same way as a sighted person perceives objects."[14]

Opacity of Scientific 'Texts'

Hermeneutic scholars have objected that the "transparency" of a scientific instrument is far removed from the "opacity" of literary texts:[15] they argue that the hermeneutic element in science is a simpleminded, highly artificial thing compared with the way an interpreter has to struggle with ambiguity and obscurity when dealing with a literary text. In reply, I say that just as there are simple literary texts easy to understand, so there are simple scientific demonstrations easy to 'read'; one would be grossly misled to believe that all empirical procedures in science are clear and transparent without shadow or obscurity. The struggle for a hermeneutic understanding is also evident in science.

Scientific 'texts' provided by readable technologies are usually opaque to students in training attempting to understand what they are experiencing for the first time, as well as to the scientifically uninitiated. The conviction that what one is seeing does make perceptual sense and that one's teachers have the key to the meaning gives confidence, incentive, and ease to the learning process.

Original research in the natural sciences often involves the effort to understand instrumental 'texts' that are only partially deciphered: such activities are clearly hermeneutical. In his often brilliant account of the historical "genesis" of the notion of *syphilis* and its dependency on the reciprocal notions of *antigen, antibody*, and *complement*, Ludwik Fleck comes to the conclusion that—in microbiology at least—a scientific fact is a "thought-stylized conceptual relation which can be investigated from the point of view of history and from that of psychology, both individual and collective, but which cannot be substantively reconstructed *in toto* simply from these points of view."[16] By "psychology," Fleck means *intentionality* or *the internal logic of the conceptual structures used for interpreting data*, rather than *behavior* or *other empirical structures of the subject*. Thought-stylization that directs and restricts action and perception is for him the characteristic achievement of mature scientific work: it enables a community of scientists eventually to see, describe, recognize, or reproduce—but with some uncertainty—a common set of basic scientific "facts." He follows the historical development of serological testing, particularly the Wassermann test for syphilis, stressing that the researchers groped half-blindly, often profiting by chance from their own mistakes and incompetence, before they ultimately arrived at the establishment of (what was no more than) a semistable set of procedures, "the irrational 'serological touch,' " which provided erratically regular correlations used in defining and then—with much uncertainty—diagnosing syphilis. "From this point of view," Fleck concludes, "the relation between the Wassermann reaction and syphilis—an undoubted fact—becomes an event in the history of thought."[17] What Fleck describes is a style of research which is fundamentally both experimental—based on 'texts'—and hermeneutical.

A 'text,' once transparent, can become obscure for either of two reasons: the 'text' has become corrupt, or scientific culture has been so radically transformed that the art of 'reading' the old 'texts' has disappeared. The great stone ring at Stonehenge once yielded a 'text' transparent at least to the 'literate' class, to the priests, astronomers, or other initiates who could 'read' the story that the heavens 'wrote' with its aid. As the centuries passed, stars and planets moved out of synchronism with its structure, the 'text' became corrupt as a consequence and more difficult to 'read,' until finally its meaning was lost. It is only within the past few years that, by following G. Hawkins or H. von Dechend, one has been able to reconstruct from astronomical laws the archaic face of the heavens, and so to recover the archaic 'text' of Stonehenge.[18] In other cases, it may be easy to recover the uncorrupt 'text,' but difficult to recover its meaning; if, for example, some piece of scientific equipment from the laboratory of Paracelsus were rehabilitated, how difficult would it be to learn to use it so as to find with its aid what it was designed to exhibit?

Observation as Theory-Laden

The claim that theoretical entities or states are observable runs contrary to a basic principle of both traditional empiricism and traditional phenomenology. This principle states that there is a hard distinction between those states or entities (observational states or entities) that can be perceived to exist and those states or entities (theoretical states or entities) whose existence can, at best, only be postulated or inferred. In terms of a theory of scientific explanation, the former enter into the *explananda* of scientific inquiry, the latter into the *explanans*, and *scientific explanation* is taken to be predictive control over observational states or entities; *observation* is taken to be *unaided* perception. An *observable*, then, is taken to be exactly something that could under appropriate circumstances be observed with the unaided senses. These usages are part of the long accepted standard or "Received" view among logical empiricists concerning the nature of scientific explanation.[19]

The standard account of scientific explanation makes then the following assertions: (1) observational states or entities exist, (2) non-observational states or entities belonging to an accepted scientific theory may or may not exist, (3) a scientific theory has exactly the aim of giving predictive control over observational states and entities, and consequently (4) a scientific theory is neutral to the reality or existence of those nonobservational states or entities whose names and descriptions enter into the logic of scientific explanation.[20]

There has been much discussion of all of these positions. It is probably fair to say that, as the result of the critique conducted by P. Feyerabend, N. R. Hanson, G. Maxwell, W. Sellars, D. Shapere, S. Toulmin, and many others,[21] few philosophers of science today hold (2), (3), or (4). The debate centered (and still centers) on two principal issues: the nature and aim of scientific theory (or explanation), and the ontological status of the theoretical states and entities of an accepted scientific theory. Subsidiary to the latter is the question as to whether, given an accepted scientific theory, any (or all) of its theoretical states or entities are *in some sense* observable: this is the question of the "theory-ladenness" of "scientific observation."

Now the position I have been defending is that theoretical states and entities are or become directly *perceivable* (alternatively, "observable," in the stipulated sense) because the corresponding measuring process can be or become a "readable technology," a new form of embodiment for the scientific observer. In this view, the term "observation" no longer means *unaided* perception. It implies that theoretical states and entities are real and belong to (what S. Stebbins called) "the furniture of the earth," *because (and to the extent that) they are perceivable in the perceiver's new embodiment*. It also implies that the nature and aim of scientific explanation is to make manifest the processes and structures of the real, the real now being taken as what is or can be given in some World.

Within such a perspective, a distinction between *observational* and *theoretical* probably no longer makes sense. It no longer distinguishes what is merely postulated from what is real, since observability *in the new sense* is the criterion of the *reality* of all physical things, including theoretical states and entities. The term "theoretical," however, can be given another sense. It can be taken to refer to one of the components of scientific activity, namely, logical and mathematical deduction, in contrast with the making of observations, where "observation" is taken in the new sense. The distinction is then of only secondary importance, for the component it identifies has no significance apart from the essential overall strategy of science, which is to make manifest the not-yet-manifest substructure (or superstructure) of prescientific experience. Theoretical activity in this sense has a separate identity and value only if the essence of science were no more than predictive control, a position rejected in my viewpoint.

There is perhaps the residue of a useful distinction between *observational* and *theoretical*, in the domain of logical usage and predication, that is, a distinction of *embodied* context. Some contexts are more firmly entrenched in language than others, and so the corresponding usages are more entrenched, not precluding, however, plurality of contextual usage.[22] A scientific instrument, always a macroscopic device, can, for example, serve an investigator in two ways: (1) indirectly and inferentially, for detection purposes—this usage is theoretical—and (2) as a "window" opening directly on the hidden processes and structures of nature—this usage is observational; for the latter the scientist must have a special training and expertise. Observations are clearly theory-laden—as Einstein[23] and others after him have noted—since "it is theory that tells what scientific observation can perceive." But this last principle may be no more than a statement about the detection, not the observation, of a theoretical state or entity. In the viewpoint I have expressed, the statement can, under suitable conditions affecting observer and instrument, also be taken to refer to a perceptual content. The presence of an essential theoretical component in the scientific description of a state does not then preclude the possibility of the state being perceived in the scientists' World; what is required, however, is the mediation of what I shall call a "readable technology."

The Ontological Intent of Science

The standard or Received view of scientific explanation argues that a good explanation does not require that the scientific structure of the postexplanatory states of affairs be exhibited in perception, nor that the scientists' World be enlarged to encompass the horizons of these new states. A post-Kuhnian argument for this view would say that scientific theories change, but it would surely be irrational to think of the "furniture of the earth" changing after every scientific revolution. Moreover, if theoretical states or entities did become manifest to perception, each prescientific *explanandum* would sooner or later produce its scientific offspring, born of its scientific explanation, and room would have to be found for that offspring in the scientists' World; adding, as A. Eddington noted,[24] to the embarrassing dualisms of that World, a scientific table to the everyday table! Surely, too much!

Going to the first part of this objection, we can say that clearly not every scientific theory that has been proposed or entertained yields genuine essences and profiles; its abandonment then does not mean a change in the furniture of the earth, but only the desertion of hopes or expectations of finding a certain kind of furniture on the earth. Not every scientific revolution, however, is a completely new beginning; scientific development generally takes place within research traditions that mediate continuity in change. These questions will be taken up in chapter 13.

Taking up the second part of the objection, Eddington's duality is not a duality of tables, but of a table and of the molecular elements of which its parts are composed. Two different embodied contexts of perception or observation are implied. To observe the table and its solid and continuous properties, is to be embodied in one's native sensory system, unaided by scientific technologies. To observe the open and discontinuous molecular structure of the table top, is to function within a Body assisted by an appropriate readable technology. These two systems of observation are complementary and incompatible.

Contrary to the standard view, it is not sufficient for the acceptance of theory, at least in the natural sciences, that it give successful nomological-deductive explanations; more is generally required, that the structures it postulates be shown to exist. Sellars says, "to have good reasons for holding a theory is *ipso facto* to have good reasons for holding that the entities postulated by the theory exist."[25] For Sellars, however, *to exist* has implications of unobservability. In physics, no theory would become entrenched unless the entities it postulates have been shown to exist; the "showing" is more than mere detection, it is being "naturalized" in the scientists' World through the use of a readable technology.

We note that the current theories about quarks and black holes are provisional and tentative, and will remain so until quarks or black holes have been unambiguously observed. This is not just a play on the meanings of the word "observe"; an explanation in terms of putative constituent parts or processes is not content merely to postulate or hypothesize the existence of those elements, it wants to *exhibit* them as part of the furniture of the earth or cosmos. One rightly distinguishes between such explanations and mere conceptual artifacts, such as cost of living indices or accounting systems, which merely classify information or present it in summary form. The former are distinguished by the existence of readable technologies that make manifest to perception the constituent parts or hidden structures of the *explanandum*: such explanations have an essential ontological intent.

This view runs counter not merely to the positivist, empiricist, and instrumentalist views, but also to the view held by many philosophers in the existential, phenomenological, and hermeneutical traditions, for whom science is merely an extension of human power to manipulate nature. [26]

Scientific Instrument as "Readable Technology"

To recapitulate the argument: from the viewpoint of a first-person phenomenological context, whenever familiar scientific instruments function in the cognitive way described above as extensions of the neurophysiological organism for the purpose of perception, they become part of a perceiving subject's Body: the subject is then said to be *embodied* in the instrument. The instrument is used in a special nonobjective way, one in which the noetic intention of the subject is embodied in the instrument joined, physically and intentionally, with the scientist. This nonobjective use is characterized by a hermeneutical shift in the subject-object "cut" so as to place the instrument and its response on the subject side of the "cut," and that response in a position of 'text' to be 'read' in the 'context' of a scientific horizon. [27] An instrument capable of being used in such a way is a part of a "readable technology."

There is also a third-person way of looking at perception from the point of view of the physical and physiological processes involved in it. Just as the analysis in terms of intentionality yields only the duality of subject and object, so the physical picture must have just two (albeit complexly structured) components, subject and object in physical contact across the subject-object "cut." To dismantle the physical picture further, say, into three separable components, such as subject, stimulus, and object, or into a chain of components each passing pieces of information in relays to the next in line (whatever this means!)[28] until a piece of relevant information reaches the central nervous system (CNS), would be making the kind of mistake I discussed in chapter 8 above. It is like trying to explain why the object on which I am writing is a table by dismantling it into a structure of molecules. Such a procedure cannot say why this object is a table, because an infinite number of different molecular complexes are compatible with its being a table. No single one of these molecular complexes is capable of explaining what kind of thing a table is, since "table" names the defining structure, or invariant, possessed by each and all such complex unities, considered as possible variants of a perceptual essence that belongs, not to natural science but to human culture.

All such chains of causality which descend from the object to the subject must function together as one articulated process of "feeling" the object. The embodied subject, then, seen in the physical picture, includes the neuronal network of the CNS, and attendant somatic processes such as hands and other peripheral organs. The subject is conjoined with the instrument (and possibly some structures of the optical energy field) as one coordinated operating system in which in some way are represented or encoded (as information$_1$) the possible states of the environment as "felt" or probed by the instrument (information$_2$). I have called this chain "the somatic information channel." The response stimulated by the object is the "resonant" causal activity descending and ascending in this chain between the subject-object "cut" and the cortex: I call this a "resonant" mode of the subject, because it is one and integral, as opposed to a collection of random neuronal firings and other miscellaneous activity. The "resonant" mode of the subject is then stimulated by the object across the subject-object "cut" and, according to current neurophysiological thinking, it should be controlled by an encoded pattern in a specific neuronal network, whose function (as an encoded pattern) is to coordinate the chain of "resonant" activity ascending and descending to and from the cortex, through the peripheral organs of the body, the instrument, and the optical (or other) energy field.

Observers and Quantum Mechanics

The account I have given of scientific observation throws light on some of the most puzzling aspects of quantum mechanics. There are two parts to the puzzle: (1) what is the nature of the "things themselves,'' the systems (quantum mechanical systems) that quantum mechanics reveals in the light of the accepted theory and praxis of physicists? and (2) what did Bohr and Heisenberg mean by the explanations they gave of quantum mechanics? I shall do no more here than sketch out the answers this account suggests.[29]

1. Quantum mechanics is the first natural science that in its explicit form includes reference to the contextual character of scientific inquiry. Referring to the account given in chapter 10 above, I take *context* to be the set of subjective and objective conditions that must necessarily *exist* if any profile of the object is to become manifest. Quantum mechanics is unintelligible without the introduction of the scientist as an observer embodied in the empirical processes of instrumentation and measurement. The term ''observer'' in quantum mechanical literature refers to the measuring system; sometimes it includes and sometimes it excludes the subject. I take the observer to include the subject; the measuring instrument constitutes a readable technology, and the subject so extended is the Body appropriate for a quantum mechanical observation. Complementary variables, such as position and momentum, are empirical horizons accessible only through contextual embodiments in complementary readable technologies. Quantum mechanics, in its Hilbert space formalism together with the interpretation of that formalism, is then the least upper bound (l.u.b.) of all pairs of complementary observable horizons (the l.u.b. refers to the kind of Q-lattice discussed in chapter 10 above). Since each observable horizon on the object side connotes an appropriate Body or observer on the subject side, the *existence* of any particular observable horizon (e.g., position) depends on the *realization* of an appropriate embodiment, that is, on the *existence* of an appropriate observer. Quantum mechanics says that such embodiments for the purposes of observing the various horizons of quantum mechanical observables can be contextually incompatible.

2. It should be clear by now that quantum mechanics in its present form cannot be satisfactorily understood or reinterpreted within an empiricist philosophy; fifty years of unsuccessful attempts have, I believe, demonstrated the truth of this statement. It is then time to return to the sources of the original interpretation of quantum mechanics in the words of Bohr and Heisenberg, and to ask whether something of importance has been overlooked in what they said. Although both Bohr and Heisenberg were strongly influenced by the new empiricist and positivist ideas stemming from E. Mach and W. Ostwald and circulating in scientific milieus, neither was committed to a systematic

empiricist construal of quantum mechanics. Both stressed that *observables* were the appropriate subject matter of quantum mechanics, and not unobservable structures postulated by some theory to explain what was or could be observed; however, the distinction between these two was certainly not understood in an empiricist sense.

Bohr and Heisenberg both stressed the distinction between measuring instruments and the observables of a quantum mechanical system; quantum mechanics, they said, applied only to the latter, the measuring instrument had to be described "classically."[30] In this respect, Bohr and Heisenberg differed from J. von Neumann, E. Wigner, and others who took quantum mechanics to apply also to the measuring instrument and measuring interaction as such, and they also differed from the subsequent conventional wisdom that quantum mechanics must be shown like its predecessors to apply to the totality of the universe. An important reason for making a distinction between *classical* measuring instruments and *quantum mechanical* observables can now be glimpsed: the measuring instrument *as such* requires an account that is capable of judging and accounting for its suitability as an extension of human sensory systems; the observable, as the potential object of empirical inquiry, is under no such constraint. Bohr and Heisenberg, in making this distinction, were expressing an intuition that only if the measuring instrument were *in relation to the subject* a classical system would it be able to fulfill its function as a measuring device as such or (in my jargon) as a readable technology. This is made plausible by an analysis of what is required of measuring instruments: a measuring instrument is a system possessing a manifold of metastable, macroscopically discriminable states that can become correlated with the states of a measured system, through an interaction the outcome of which is the stable realization of just one of those macroscopic states. Since the set of macroscopic states has to function as an unambiguous channel of information connected to the external peripheral organs of the subject's sensory systems, it appears altogether plausible, even according to our current intuitions, that a necessary condition for the existence of such a channel is that it have a classical and Cartesian description.[31] The necessity of this condition may change with increased knowledge of how our perceptual systems work and with advanced technology, but this need not have occurred to Bohr or Heisenberg since they were concerned just to state in the most elementary terms a new and fundamental insight into the nature of measurement. The nature of this insight also explains why Bohr and

Heisenberg rejected what has come to be called the "problem of measurement."[32]

The Ontological Achievement of Science

Thermodynamic temperature is one of those theoretical quantities that finds a place in the scientists' World—and, indeed, in our World also—through the thermometer; through the thermometer, according to the criteria I am following, it is a reality, a thing-for-us, and part of the furniture of the earth anticipated in the very structure of our natural language. Other theoretical scientific entities can also become part of the furniture of the earth, if an empirical process can be constructed that is capable of functioning perceptually in the scientists' World as the thermometer does for temperature.

To generalize from the case of temperature and the thermometer, we need to establish that the plasticity of the human perceiver vis-à-vis the embodiment capacity—that is, the capacity to use a prepared part of the external environment (such as a readable technology) as a "window" on structures inaccessible to the unaided senses—is a general capability and a genuine reflection of, as it were, primordial possibilities. That such a plasticity exists, follows from the account of perception given above: is it, however, a general or primordial character of perception?

The perceiver is always an embodied perceiver, a Body: this Body always includes a somatic channel that comprises: (*a*) the neurophysiological system, and in addition, (*b*) other somatic processes of the human organism, such as the limbs and sensory organs, and in addition, it may also include (*c*) a readable technology as an integrally incorporated part, and (*d*) it certainly comprises specific structures of the various energy fields (stimuli) which carry information to the peripheral organs and/or readable technologies used. A specific horizon is the objective correlate (in the phenomenological sense) of a subject with a specific embodiment, that is, a specific horizon is the perceptual response to a specific subjective embodiment. Conversely, the embodiment of the subject is reciprocally specified by the horizon to be attained. Such reciprocity of embodied subject and perceptual horizon is a necessary prepredicative structure of perception.

The presence or absence of a technological element (*c*) does not change the essential structure of the perceptual act, neither with respect to its phenomenological characteristics, for example, of directness, nor with respect to the physical and causal relationships between the embodied perceiver and the objects of the perceptual horizon. Whether the embodied subject has structure (a), structure $(a)+(b)$, structure $(a)+(b)+(c)$, or structure $(a)+(b)+(c)+(d)$, the subject's action system relates in a systematic way to the unity and integrity of the objects of the perceptual horizon. Consequently, although embodiment in readable technologies is a feature of certain special kinds of perception, this merely illustrates the general feature of embodiment which all perception shares: it is a special case of the general law. The possibility of embodiment in readable technologies then must follow from deep roots in the conditions of possibility of perception. I conclude, then, that scientific entities may come to share the reality of a World-for-us, and that they do so *to the extent that the subject succeeds in embodying itself in appropriate readable technology.*

Human Embodiment in Readable Technologies

The plasticity of the embodied perceiver referred to above would show itself at the level of the somatic information channel as the capacity of the neuronal networks to code, not merely for those horizons of a World accessible to the unaided senses but also for horizons accessible to the perceiver through the use of appropriate technological extensions of the sensory organs, even those yet to be invented.[33] Such codings in the neuronal networks would map a particular environment of the human subject as this environment is systematically contacted by the human organism, extended by technologies for the purpose of relating itself directly to that environment through perception. An example of a particular environment is that traced by the point of a blind man's cane. The existence of such technological extensions, as those described above, implies then a certain plasticity in the neuronal networks vis-à-vis *codings* for reality. It implies in addition that this plasticity can be exploited, not merely by somatic processes, for example, touching with the hands, but by readable technologies of happy design, like the blind man's cane. What makes one technology readable and another not, will remain unknown until the structure of neuronal codings is understood sufficiently to uncover that in which their plasticity lies vis-à-vis the use of technological extensions to increase human powers of perception. Up to that time, a happy design will be a happy accident.[34]

Some corollaries: two *theses* follow from this account. (1) A Body cannot be vanishingly small, since to perceive any object X it must comprise at least the physical code Y that, when 'read' with the aid of an appropriate hermeneutical fore-structure, is decoded as *X*-as-per-ceived. (2) A Body cannot be coextensive with the entire cosmos (taking the cosmos to comprise all that is accessible in all horizons of all Worlds), otherwise there would be nothing for such an embodied perceiver to know as object. Certain consequences follow from these theses concerning the limits of scientific cosmologies, and in principle certain constraints would have to be recognized if the quantum theoret-ical approach—which recognizes the embodied contextuality of empir-ical knowledge—is to be reconciled with general relativity, which does not recognize such contextuality.[35]

Furniture of the Earth and Universe

The reader may be feeling some uneasiness that the outcome of the analysis, which says that we can only come to know of the existence and kinds of scientific entities through their becoming manifest percep-tual objects, may leave us in ignorance or doubt about how to describe the earth and universe before people existed, or after they cease to exist. Is not the solution offered a form of idealism?

Scientific entities, I claim, belong not merely to the furniture of human Worlds, but also to the furniture of the earth and universe. In the first instance, they are duly accredited items in a human scientific World, that is, they exist as *things-to-and-for-embodied-human-perceivers: to human perceivers* insofar as they are the sort of thing that is defined in relation to the perceptual interests of human perceivers, and *for human perceivers* insofar as the existence of this sort of thing is recognized and asserted by human perceivers in the appropriate mani-fest image. Things of this kind belong to the furniture of human Worlds because, as I have shown, they make their presence manifest to human perceivers through appropriate readable technologies designed and constructed on the basis of scientific laws and theories. They enter human Worlds then through the readable technologies and somatic information channels that serve as "windows" through which people can get glimpses of the infrastructure of material bodies. In addition, they make possible other technologies, readable and nonreadable, which harness these entities to human cultural interests. As so revealed, they belong to one or more manifest images of scientific objects.

The existence of such entities is, however, also a necessary condition for the possibility not only of readable technologies but also of all somatic information channels, that is, of Bodies. As necessary conditions for scientific observation, for Worlds and for Bodies, each of these entities has a scientific definition that does not explicitly involve relationships to human perceivers, or to scientific or cultural institutions, but relates scientific entities of different kinds to one another implicitly through a scientific model. Such a definition does not describe explicitly the *essence* of anything (in the sense in which I have used this term): apart from human perceivers, however, it indicates by implicit definition through a model a system of interrelated "things-in-themselves"; the model alone, apart from perceivers, is no more than a metaphor of the scientific real. In this latter respect is to be found, I believe, the assertable core of scientific realism: scientific entities are indeed real, and prior to human cultures they are the necessary preconditions of all Worlds, of all cultures, and consequently, of human embodied subjectivity itself. In this respect, they belong, not merely to the furniture of human Worlds, but to the furniture of the earth and of the universe.

12

Identity Theories and Psychobiology

Monistic Realism and Identity Theories

The account I have put forward, based on *horizontal realism*, conflicts profoundly with reductionistic monisms and identity theories, *all of which would in principle reduce mental phenomena such as acts of perception to brain states or matter states with no remainder*. Such, for example, are Herbert Feigl's Identity (Mind-Body) Theory[1] and William Uttal's psychoneural (Mind-Brain) identity or equivalence theory.[2] I shall address myself to the views of Uttal, a neurophysiologist of distinction, who has given much thought to the Mind-Body question, and who calls himself a "psychobiologist."

Psychobiology

Uttal's main thesis is that "the overall pattern of information flow coursing through the nervous system is indistinguishable in all denotative regards (in Feigl's terms) from the mental states of which we are individually aware and that are reflected in the molar behavior of the organism."[3] And again, "the linguistic terms of psychology and neurophysiology denote exactly the same mechanisms and processes."[4] Uttal interprets equivalence of denotation not as an invitation to *horizonal realism*, but as a dogmatic statement of *monistic realism*. "Modern psychobiology," he writes, "is mechanistic, realistic, monistic, reductionistic, empiricistic and methodologically behavioristic."[5] He summarizes his position in this way: "the mind is to the nervous system as rotation is to the wheel."[6]

A monism of this kind—as Uttal himself admits—goes far beyond what the available evidence in neurophysiology suggests. Uttal himself warns that, despite claims to the contrary made by other psychobiologists, there exists hardly one instance of a proven equivalence between a neuronal state and a psychological state—except arguably in the most simple of sensory domains: irrelevant but analogous concomitance has often in the past been mistaken for the causal dependence sought for by psychobiology.[7] Perception, learning, motivation, and so forth, he admits, are mental functions far too complex to be able at present to be mapped on specific neuronal networks or the states of such networks.

Psychobiology then is a global program for guiding research in psychology and neurophysiology which gets whatever plausibility it has at present strictly from certain kinds of philosophical arguments associated with the following positions often taken in the Mind-Body debate.[8] (*a*) On *knowing*: what we know is an "internal representation" (of a state of the World) that we construct, rather than a state of the World itself. (*b*) On *representation*: there is no difference in "reality" between the mental state that mirrors an object (information$_2$) and the neurophysiological coding for that object in the brain (information$_1$); in other words, representation (information$_2$) is identified with a coding for that object in the brain. (*c*) On *scientific explanation*: the aim of science is to substitute for phenomena the hidden, imperceptible, and nonmanifestable structures that, it is believed, they (phenomena) stand for as surrogates. I believe all three positions are unacceptable. Instead, I propose the converse of all three. (*a'*) On *knowing*: what we know is a World, not just a mirror image of it. (*b'*) On *representation*: representation (information$_2$) is not identical with coding (information$_1$) but is a form of *decoding* a coded cipher with the help of intentional hermeneutical structures brought to bear on it by a language-using subject. (*c'*) On *scientific explanation*: that science is essentially realistic and aimed at manifesting to perception hidden structures that are necessary and antecedent conditions of possibility of Worlds and of perception itself.

Metaresearch Principles of Psychobiology

It is of course also the belief of those who follow its guidelines that the metaresearch principles of psychobiology or similar identity theories will better help the design and execution of neurophysiological research than will metaresearch principles based on opposing philosophical considerations. This belief I also oppose. The set of principles used, for example, by Uttal, (1) are vague; (2) are unnecessarily global vis-à-vis present possibilities of scientific research; (3) are not consistent with a philosophy of science sensitive to hermeneutical and phenomenological issues; (4) restrict the design of research programs unnecessarily; (5) prevent the possible emergence of contrary empirical evidence; and (6) contribute to the fallacy that scientific achievement would provide unique evidential support for their truth.

In most identity theories—and Uttal's is no exception—the notion of psychological state is ill-defined; no clear or persuasive answer is given to such questions as the following: are psychological states differentiated by their kind, such as, for example, desire, perception, purpose? Or are they differentiated among themselves by their objects, for example, perceptions *of* a table, or perhaps *of this table*, from perceptions *of* a chair, or perhaps *of this chair*? If they are differentiated by their objects, then is the object merely the exhibited content, the *table* or the *chair*, or is it the semantic content, that is, is it the content of a statement about what is or is not the case? It is not clear in Uttal's identity (Mind-Brain) theory, and in most identity theories, in what sense psychological states are encoded in states of neuronal networks.[9]

The identity theory, moreover, does not permit a many-to-one, redundant, possibly stochastic, mapping of several psychological states, however they are defined, on one neurophysiological state, because such codings would deny the determinism of the brain state, as well as the absolute priority of matter over mind. An analogy can be made with language, where words, sentences, and larger and more complex linguistic structures are, like neuronal states, material vehicles that function as carriers of meaning. If we define the *picturing use of language* as one where there is a one-to-one correspondence between descriptive sentences and the reality they designate, then the program of psychobiology looks very like the linguistic program of reducing all use of language to a *picturing* use. This early Wittgensteinian thesis, rejected by the later Wittgenstein, no longer receives much support from philosophers or linguists.

I want to point out that since the occurrence of many-to-one reverse mappings between neuronal and psychological states would falsify the metaresearch principles of psychobiology, the identity theory is not immune to possible contrary empirical evidence. However, it can be made immune to falsification by a methodological choice, and this is apparently what psychobiology has chosen to do.

Psychobiology claims that there is no aspect of mental life which is not determined in its contents *uniquely* by neurophysiological activity. Such a claim exemplifies a point made forcefully by the later Heidegger and by Gadamer, that modern science *as practiced in our time and culture* is intentionally antagonistic to the interpretive role of persons through which persons establish a meaningful perceptual World in which to live and to do research. All third-person research, as I have said above, assumes the existence of a background World common to members of the research community, and given to each in first-person perception; common sense and the presumptions of social life also support the claim of an inescapable background of shared first-person experience, implying—among other things—a priori fore-structures of understanding. The comprehensive claims of psychobiology—paralleled by similar claims in other disciplines, such as linguistics—work then against good and essential virtues of open empirical scientific inquiry, not only because a priori fore-structures of understanding are denied, contrary to evidence, but because these claims go far beyond what is necessary for the research in question and virtually constitute a scientifically endorsed philosophy or metaphysics of the human person.

Psychobiology, moreover, encourages the fallacy of asserting the consequent; that is, it asserts that successful achievements in this kind of research give support and credence to the premises of that research, namely, a materialistic (reductionistic) philosophy of the person. We ought to remind ourselves (1) that the premises stated by psychobiology have redundant philosophical elements and that the materialistic (reductionistic) philosophy of the person is such an element; (2) that mental acts are not identical with their necessary and antecedent physical conditions; and (3) that even if there were no redundant premises, logic warns us that a true conclusion is quite compatible with false premises. To illustrate the last, one may, for example, deny the vertical dimension and refuse on that account to look up or down, one may even discover a great deal about the surface on which one lives by looking only to the right or left, but one will never discover the sky overhead or the flowers underfoot. The kind of attitude referred to above is the one that Merleau-Ponty castigated ceaselessly as being both naive and dishonest.[10]

The particular form of identity theory proposed by Uttal—Mind-Brain Identity—has a special weakness in relation to perception. For Uttal, the correlate of a percept is an activity state of the neuronal network, *exclusive of other somatic structures or of technological structures in the external World*. I have argued above (1) that the environment is perceived as the correlative of the perceiver's Body; (2) that the neuronal system is only a part of the somatic and other physical processes of that Body; (3) that the Body may include technological extensions; (4) that the purely neural codings are not intelligible, that is, not known to be significant or worth investigating, except as a function of the Body as a whole; and consequently, (5) that the horizons of reality coded for in neural codings suppose a structural apparatus on the side of the subject—whether these be attendant somatic processes or technological processes—through which possible states of the environment are shaped, defined, and contacted. In summary, because of its exclusive preoccupation with the brain to the exclusion of other somatic processes that shape natural horizons, because of the exclusion of readable technologies that shape new horizons of a more sophisticated kind, and because it plays down the World as perceived, Uttal's Mind-Brain identity theory is a poor bet for the understanding of anything but the simplest sensory responses.

Not all identity theories would identify the perceiving subject with the brain alone; some would include in addition other organic parts and systems (some, perhaps, might even include energy fields or other parts of the physical universe) in the materialistic reduction of the subject. What I have shown is that a perceiver's Body is complex, and that the variety of components that comprise it depends on the character of the perceptual system that it is using. Since the human subject then uses a variety of perceptual systems, the Body of a human subject is not unique and invariant for all the perceptual systems the human subject uses. It is not possible then to reduce the human (perceiving) subject to just *one single* materialistic system.

Perceptual systems are distinguished from one another by the fore-structures of understanding (hermeneutical circles), including the various embodiments that they use. There is then no way of describing or explaining the human subject without taking into account the existence of the hermeneutical dimension, and the potential multiplicity of material embodiments that human perception involves. If, as I suppose, the essence of the human subject as perceiver—its invariant in multiple perceptual roles—is to make possible the potential multiplicity of its own perceptual systems, then the essence of human perceptual subjectivity is inseparable from the unique hermeneutical presence that differentiates among these systems while revealing each as belonging to the same subject. The conclusions arrived at are the following: (1) the human subject as perceiver is Body but not a fixed

Body; (2) the diverse Bodies of a human subject are self-identical, that is, express essential identity of subject, exactly through the common hermeneutical presence that in all perception opposes subject to World; and consequently, (3) the human subject as perceiver designates primarily this hermeneutical essence, its denotation—insofar as it is the generator of many Bodies—connoting its multiple perceptual roles.

13

Hermeneutics and the History of Science

Hermeneutical Circle

Hermeneutics is the science and art of interpreting texts and textlike materials in order to arrive at the "things themselves" about which the text speaks.[1] The task of interpretation is led by fore-structures of understanding, suggested by the cultural traditions—the "biases and prejudices"[2]—we share, and by clues in the textual material itself. "A person who is trying to understand a text," Gadamer says, "is always performing an act of projecting. He projects before himself a meaning for the text as a whole as soon as some initial meaning emerges in the text."[3] This search for the holistic understanding of a particular text is always situated at a definite time and place in history: the text is not assumed to have one timeless meaning, it functions within many different cultural milieus, perhaps with many different meanings, related, however, by the text. The holistic search for meaning, though circular, is not a "vicious" circle in the logical sense; it is a "hermeneutical circle,"[4] which, we are told, is a certain kind of "virtuous" circle.

This search is guided by appropriate fore-structures of understanding: these comprise (1) *Vorsicht,* or the resources of a common descriptive language; (2) *Vorgriff,* or a hypothesis about the sense of the materials being investigated; and (3) *Vorhabe,* or the culturally acquired skill and practices we need to understand, recognize, and name the objects in our World.[5]

We are warned, however, that "all correct interpretation must be on guard against arbitrary fancies and limitations imposed by imperceptible habits of thought."[6] Our "first, last and constant task," to quote

Heidegger, is to direct our gaze "on the things themselves,"[7] that is, on the objects spoken about, indicated, or described.

What kinds of things are the "things themselves"? In the case of literary texts, these are manifold. They may be, for example, formal theories of a mathematical sort, metaphysical arguments about Matter and Mind, historical or fictional stories, programs, projects and plans. They may be serious or comic, presented for their content or merely for the play of words and associations, or a thousand other "things": none of these "things," however, are perceived states of the World, they refer to unexhibited, often unexhibitable, World possibilities. They always presuppose, moreover, a background World, necessary for interpretation, and against which they are understood. In the case of scientific 'texts,' the "things themselves" can be new horizons of new perceptual objects, which come to be named and recognized in scientific observation. Scientific observation is, as I have shown, both hermeneutical and perceptual.

Let me recapitulate the senses of *hermeneutics*, and in what way perception in general, and scientific observation in particular, are hermeneutical processes. The descriptive term "hermeneutical" is applied to the acquisition and expression of information, whenever the duality in the structure of information (in terms of information$_1$ and information$_2$) is connoted. Three principal kinds of activity are referred to as *hermeneutical*.

(1) "Hermeneutical" describes the conscious and deliberate *process* of research into the possible meanings of textual and other symbolic materials where the choice among alternative systems of meanings is deliberate and conscious. The background of such research is an established World. This is the usual and primary meaning of the term.

I have shown, however, that the term "hermeneutical" extends to include (2) the *process* that tries to gain information from experience by searching for the perceptual horizons made manifest—or capable of being made manifest—in experience; essential to this, as Gibson affirms, is the presence of structures in the ambient energy fields which cause "resonances" in available somatic information channels open to the (potentially) skilled or experienced subject. Although this process is guided by fore-understandings (or intentionalities) that in some sense are passed in review and among which one chooses, the review and the choice may not be deliberate and conscious, as they would be in the interpretation of literary material. There is, however, often a pre-conscious choice, for example, in the constitution of that geometrical structure (information$_2$) that satisfies both the optical clues (information$_1$) and the purpose of visual perception (its intentionality). One reason for the lack of deliberation and conscious choice is that the subject may not ever have practiced or become skilled in using available somatic information channels (or learning how to construct them). On the one hand, then, the physical structure of the somatic information channel (its states are possible 'texts') shapes what it is about the object that is revealed—thus information$_1$ specifies information$_2$. On the other hand, the existence and structure of an available somatic information channel may not be known to the subject in advance of the successful performance of perceptual acts—thus information$_2$ specifies information$_1$. What counts as an available information channel is the fact that its 'text' can be 'read' as conveying empirical information$_2$, about an object capable of 'writing' that 'text'; that is, of producing that information$_1$. Examples of such perceptual hermeneutical processes are the geometrical structure of visual space, or a perceptual 'reading' of a 'text' provided by a readable technology—but more about this below. The hermeneutical process I have described is a horizon- or World-building process, it is reality in the process of constitution.

In addition to hermeneutical processes, there are hermeneutical outcomes. "Hermeneutical" then also describes (3) the successful *outcome* of a hermeneutical inquiry, or its *expression*, when this results in gaining direct access to meaning, perceptual or nonperceptual. Such, for example, is the structure of visual space as given to a perceiver, the meaning of a text as given to a reader, or a scientific fact as given to a scientific observer through the 'text' provided by some instrument or other readable technology. Note: A meaning (a piece of information$_2$) is "given," if it is presented to the experienced knower directly—not mediated by inferences or "internal" constructions—through the causality of an ambient structure stimulating a somatic information channel to "resonate" in the appropriate information$_1$ state.

Natural Science as Nonhermeneutical

Contrary to my claim, is the well-entrenched tradition in the sociology of knowledge as well as in hermeneutical philosophy that hermeneutics has nothing whatsoever to do with natural science; hermeneutics is taken as a method characteristic of the humanities and social sciences by which these are distinguished from the natural sciences. K. Mannheim expresses what was once the prevalent view of sociologists of science: the natural world (the subject matter of the natural sciences) is "static and timeless." Like the empirical concepts that describe it, its empirical laws are universal and unchanging, and its characteristic method of observation and measurement is cumulative in a linear way; above all, its truth criteria are constant and uniform, and not subject to changing and historical social interests. Thus, for Mannheim, natural science would have no need for a hermeneutic of perception, of theoretical models or research traditions; hermeneutic methods apply only to the study of sociohistorical objects—cultural objects—not to the study of natural objects. For sociologists of science, the only legitimate interest of sociology in the study of natural science is in the social goals of the pursuit of this kind of knowledge, and these they hold to be entirely pragmatic.[8]

Hermeneutical philosophers hold a very similar view of the natural sciences, their subject matter, methods, and goals. For hermeneutical philosophers, the aim of the natural sciences is essentially pragmatic, ordered to the control and manipulation of people and things; and the characteristic method of the natural sciences is the construction of theoretical model systems as surrogate "descriptions" of nature, since their components are radically imperceptible and do not, and cannot, have a place in any historical World. (It is this last assertion that I have contradicted.)

It is not surprising that hermeneutical philosophers of our culture and sociologists of science agree in their description of the phenomenon of science in the West, and in the denial of any role in its internal development for social, historical, and hermeneutical factors. This view is after all the one suggested by the particular cultural history of Western societies. Empiricism in philosophy and a pragmatic bent characterized the rise of science. One finds both these attitudes, for example, espoused as a public ideology by the Royal Society at its foundation in 1660. Empiricist philosophy at that time was, it must be remembered, as much a political as a scientific ideology. Nonhermeneutical empiricist science made an early ally of the Enlightenment, and soon became identified as the political tool of bourgeois liberation from the "authorities" of the *ancien régime*, and from the philosophical (mostly hermeneutical) traditions that were identified with princes, popes, and divines. "Conservative" religion and "liberal" science were born within this sociopolitical scene.[9]

It is evident that even today—long after the demise of the old political order—our culture supports science more for pragmatic reasons, mostly economic and political, than for contemplative, theoretical, or religious reasons.[10] Our social scientists, our professions, our bureaucratic office holders all subscribe to nonhermeneutical third-person behaviorist models of society which underplay the role of human values, the sense of personal destiny, and the critique of historical traditions.[11] Finally, to cap it all, the reigning schools in the philosophy and sociology of science overwhelmingly take the content of natural science to be beyond examination by hermeneutical methods.[12]

There are those who, like Gadamer and the later Heidegger, have experienced science only as functioning within the historical matrix I have described, and who have not performed the kind of critique of historical prejudice both those men so strongly recommend in their own work. It is understandable, then, that they would accept the view that since the goal of natural science in our culture is control of natural phenomena, the content of its theories need be no more than a reconstruction of nature according to a system of models or surrogates that permit the achievement of this goal.

Role of Hermeneutics in Natural Science

The real core of the hermeneutical problem in natural science, however, is that at diverse times, places, and circumstances, diverse and presumably authentic scientific horizons become manifest. The problem must be understood in light of what has been said in the preceding chapters.[13] The role of hermeneutics in natural science is not restricted to the study of literary and graphic materials (textual material in the strict sense), for these alone do not bring the reader to confront the "things themselves" about which science speaks insofar as these are *perceptual* objects. These objects are confronted only in and through empirical procedures, that is, through 'textual' materials (in the special sense of that term) made available by the use of appropriate readable technologies.[14] In the case of an old scientific theory, these materials could be provided by the performance, say, of the classic exemplary demonstrations of the theory. Empirical procedures in science are not pieces of Nature, pure and simple, but each provides a 'text' for the trained scientist to 'read,' that is, each is a humanly contrived phenomenon in which Nature is made to 'write' in conventional symbols a 'text' through which it 'speaks' in a particular descriptive language of science. The "things themselves" come to be understood, recognized, and named through a 'reading' of these materials.

'Texts' of this kind can be 'read' only by the initiated. They give direct and immediate access to a domain of scientific entities and processes—referents for the scientific account; that is, provided the scientific entities and processes are well-understood theoretically and clearly manifested in the empirical procedures adopted. The transparency and clarity of such 'texts' varies; in well-established areas, such as physics, the degree of transparency may be great, but in areas such as microbiology the degree of transparency may be small, since what is presented to perception through microbiological techniques is often incompletely understood and obscurely presented to experience. Fleck has given a brilliant account of this kind of groping and obscurity.[15] A similar experience of obscurity and travail will even accompany periods in the history of well-established disciplines, where the necessity for further theoretical development is painfully recognized by researchers confronted with empirical anomalies. Such experiences are closest to those of philologists in the interpretation of archaic texts or texts from other cultures.

The hermeneutic process described above affects both the science that is coming to birth as well as science that has passed into the archive of history; and in both cases interpretation would be indeterminate without the inclusion of 'texts.'[16] To illustrate further what I mean by authentic scientific objects possessing perceptual essences, consider current speculation about "quarks" and "black holes," accounts of which can be found in the literature of physics. The essence of such radically new and speculative entities cannot be understood, I hold, from a study of literary and graphic materials alone. On their own, these speak only of the conceptual or imaginative models as abstract objects, and not of the *real* essence to which such models apply through the appropriate metaphorical use. The *real* is that which would be manifested (were quarks and black holes to exist) as an invariant structure of a system of perceptual profiles accessed by experienced observers through the embodied use of an appropriate readable technology. Appropriate readable technologies may be difficult to construct, but this remains in my view a necessary means to authenticate the reality of such postulated entities.

Hermeneutics and the History of Science

Thomas Kuhn, whose work has had such dramatic impact on so many fields, spoke about the transformation of his sense of the history of science which took place when he understood that Aristotle in his *Physics* was not making an unsuccessful attempt to do mechanics, but was concerned with the larger question of general qualitative change.[17] To interpret Aristotle's *Physics* correctly, he found, one had first to experience change (all, or the most part) as charged with intentionality; he describes this discovery as a "hermeneutical" experience. Ludwik Fleck[18], Paul Feyerabend[19] and N. R. Hanson,[20] to name a few, have pointed to similar phenomena: perceptual transformations generally accompany the widespread adoption of new scientific theories leading to the reformulation—hermeneutic in my language—of the original problem in theory-laden language. The problem comes to be perceived, as it were, with a *Gestalt* switch. Thus, a particular set of archaic perceptions, governed by an archaic set of interests, could have comprised in their own time authentic essences and authentic profiles that were lost to our culture perhaps for no other reason than a change of theoretical and cultural interests.

The importance and necessity for an aspiring scientist to be apprenticed to the experimental and theoretical life or praxis of an ongoing scientific community is stressed by many, among them Fleck, Kuhn, and Polanyi.[21] Learning the *Denkstil* (or "thought style"), as Fleck puts it,[22] or learning to operate within the "disciplinary matrix," as Kuhn calls it,[23] does not, however, consist solely of the conceptual ability to construct mathematical models, but in addition comprises for the trained researcher the ability to use these models in empirical procedures to transform the content of empirical experience. This ability to "empiricize" (H. Törnebohm's word) mathematical models is learned "by some non-linguistic process like ostension" in which mastering exemplary or paradigmatic studies plays a large role.[24] Fleck puts it differently: the researcher becomes "experienced" in the performance of scientific observation; this training of the perceptual powers is a sharing of a culture. "In science, just as in art and life," he writes, "only what is true to culture is true to nature."[25] I have explained this training as one of coming to be embodied differently as a perceiver: this special aspect of *Vorhabe* is what is missing from the discussions of Quine, Reichenbach, Goodman, and other conventionalists.

The History of Science and Progress in Knowledge

According to a very influential view presented by Kuhn,[26] science according to its *internal* history falls into two patterns: "normal science," which is progressive and linear in its development within a fixed and dominant "paradigm," and "revolutionary science," which is science in a more or less discontinuous mode, involving competition between paradigms that are more or less mutually incompatible (they use different descriptive categories). A paradigm is "the entire constellation of beliefs, values, techniques and so on shared by the members of a given [research] community": at its core are a "disciplinary matrix" composed of "ordered elements of various sorts requiring further specification," and classic exemplary demonstrations of what constitutes good science.[27]

Revolutionary science, in Kuhn's view, is initiated when a formerly entrenched paradigm is queried persistently by a significant group of scientists because of intractable empirical anomalies to the paradigm. During periods of revolutionary science when one paradigm vies with others, there is, he holds, no sufficient rationale internal to the discipline itself to justify a particular replacement: a rationale to be sufficient will comprise in addition external factors, coming from the social and political milieu. On the problem of the reference of scientific accounts, he writes, "There is, I think, no theory-independent way to reconstruct phrases like 'really there'; the notion of a match between the ontology of a theory and its 'real' counterpart in nature now seems to me illusive in principle."[28]

In Kuhn's account, we note the significance of the distinction between a paradigm and the progressive developments supported and inspired by the paradigm, each more successful in problem solving than its predecessors. Kuhn's statement/principle of continuity in "normal science," as well as his denial of a comparable statement/principle of continuity among paradigms, has exercised a powerful influence on the present generation of philosophers of science. One of his legacies has been a series of problems about research traditions: (1) about "intertranslatability" between the descriptive languages of different research traditions (closely allied in philosophical literature to the problem of "analyticity"); (2) about the fixity of "reference" (the warning quotes are used because these first two terms have not been properly introduced, but borrowed from analytic philosophy); (3) about the criteria for rational choice among more or less incompatible research traditions; and (4) about relating the internal history of science (taking account only of criteria internal to the discipline) to its external history (where factors external to science—social, political, religious, and other—are also taken into account).

It should be clear from what has been said above that all of these problems appear differently in hermeneutical and phenomenological philosophy than they appear in analytic philosophy. In the former, for example, concern will focus on such aspects as the authenticity of the perceptual horizons (about which more below), on conditions for embodiment (an aspect of *Vorhabe*), and on the descriptive use of metaphorical language (particularly in science) to describe perceptual objects. In analytic philosophy concern will focus on problems of "conceptual meaning," "analyticity," "translation," and "reference," and on the interpretation of the logical connectives of classical logic, and of the metalinguistic concept of *truth*.[29]

The problem of intertranslatability, for instance, has been addressed in many ways by philosophers. Quine, Duhem, Rorty—to name a few—resolve it in a conventionalist sense (different categorial systems are possible, they are not in general intertranslatable, and none is the privileged bearer of a language-independent ontology).[30] Sellars and other scientific realists give privilege to scientific languages and to the continuity of description and reference assumed by such languages.[31] Others (positivists and instrumentalists) give privilege to prescientific observational languages, while some few, like Hesse,[32] try to resolve it on the basis of nuanced human consensus, or *convergent realism.* Although all of these classes use some concept of "truth," they divide as to whether science (and all knowledge) needs a fixed reference independent of the scientific and descriptive frameworks people use to describe it, whether it has one, or whether chasing a (fixed) reference is indeed chasing a will o' the wisp.[33]

Looking at the problem of diverse incompatible descriptive languages from the point of view of phenomenology and hermeneutics, we are led first of all to ask which among perceptual objects are authentic, that is, possess well-determined (or determinable) essences. Without such an essence, the presumption that a classifiable perceptual item corresponds to a phenomenological horizon fails. Are all perceptual objects authentic? Without going deeply into the matter, I take it for granted that some of those experiences we have in the natural attitude are in fact no more than pseudoprofiles of pseudoessences, and are destined to be supplanted at a later date by other, more authentic perceptual structures. This line of thought suggests that some "revolutions" in science (the abandonment of *phlogiston, humors,* and *signatures*), like revolutions in common experience (the abandonment of belief in *leprechauns, witches,* and *monsters*), were of this pseudo class. I see this process as one of critical focusing on the contents of perception, and the winnowing of authentic from inauthentic contents.[34]

In contrast with Kuhn's account of the history of science is the view of Popper, and his students or followers.[35] For them, science is *normatively* the effort to supplant old theories and old research traditions by new ones of greater explanatory power. Kuhn's idea of "normal science," in their view, describes a jaded caricature of living science, living science being always, as it were, in a "revolutionary" phase, such being the Popperian view of its essence. Defining science as rational knowledge, the Popperians define rationality as the search for excess explanatory power. Despite his earlier indifference to the questions of truth and reference, Popper came later to see the enterprise of science as a search for truth in a quite traditional correspondence or Tarskian sense: the search for what is the case independent of "external" criteria (since these are "irrational"—not part of the internal rationale of knowledge development), and independently of the constructive activity of human inquiry and research (since these are "subjective"). Scientific research, for Popper and the Popperians, has an "internal" history. This history is identified normatively with the pursuit of rational inquiry, rationality being the progressive choice among competing theories or research programs of that which is (or at least promises to be) the more comprehensive in empirical content or problem solving.

Perhaps the most ambitious account in the Popperian tradition of science (in fact, of all knowledge) as rational to the extent that it is progressive is given by Larry Laudan.[36] For Laudan, science—and all knowledge—is "essentially a problem solving activity."[37] Problems are of two kinds, empirical and conceptual; conceptual problems may be internal to a theory or research tradition (coherency, ambiguity, etc.), or they may be external (with metaphysics or theology, with world views, with rival theories, etc.).[38] An unsolved problem, he holds, is irrelevant to science until solved by some theory; then it becomes relevant and tells as an anomaly against other theories that fail to solve it.[39] A problem (solved by at least one theory) can be stated in terms (generally theory-laden but) neutral with respect to the variety of theories offered for its solution.[40] Theories solve problems by the character of the underlying structure of entities and processes which they postulate.[41]

The adequacy of a theory or research tradition is to be assessed, Laudan says, by its problem-solving capacity alone, and that capacity always is evaluated comparatively to other proposals: "the choice of one tradition over its rivals is a progressive (and thus a rational) choice precisely to the extent that the chosen tradition is a better problem solver than its rivals.''[42] He defines an "appraisal measure" for a theory: "the overall problem-solving effectiveness of a theory is determined by assessing the number and importance of the empirical problems which the theory solves and deducting therefrom the number and importance of the anomalies and conceptual problems which the theory generates.''[43] (The principle that rationality is essentially dependent on making progressive choices among theories has, for Laudan, the character of a transcultural, quasi-transcendental principle.)

An important component of Laudan's view of science is that it is pursued within the general matrix of a culture that also includes nonscientific domains that generate their own share of problems to be solved.[44] The rationality of science is then part of the rationality of life and is influenced by world views, and social, economic, and religious values; the history of science witnesses to this influence, Laudan says, and to the fact that this influence can be and has been exercised in a rational manner (that is, by maximizing problem-solving capacity in the culture).

Laudan is unhappy with most of the standard models of (scientific) rationality found in the current literature of the philosophy of science, all of which, in his view, fail in many ways. They do not ring true to the story of science as told by historians of science, and they are hobbled by the notion of truth as correspondence (in the Tarskian sense), and so are unable, he believes, to make sense of continuity and change. Laudan consequently breaks with Popper and with the main current of today's philosophy of science in stating that truth and falsity (in the Tarskian sense)—consequently, also reference—are irrelevant to scientific knowledge; in this respect he is in agreement with Feyerabend.

Since Laudan offers his model as being both *normative* for rationality and *descriptive* of what was going forward in the history of science and culture, his work contains a fascinating assortment of arguments based on the rich ecology of past and present forms of scientific thinking.

Hermeneutical Model of Rational Progress: Linear Part

The contributions of Kuhn and the "new" history of science he represents are impressive, as are the contributions of Popper and his students to the study of the progressive imperative of scientific rationality. Yet their analysis overlooks that dimension of empirical knowledge stressed in previous chapters, and this incompleteness leaves important aspects of the problem of progress, continuity, and change more obscure and indeterminate than need be.

I offer in this section and the next the outlines of a normative model—linear and dialectical—for the progressive aspect of scientific knowledge in keeping with the hermeneutical/phenomenological analysis of perception and science given above. Perceptual criteria are made basic and fundamental for the acceptability of a theory or research tradition—though it may appear they are not always sufficient for its retention. The model will be presented sketchily, and the evidence to support it will be very much abridged; it is presented more as an illustration of how my analysis would affect the study of the history of science, both in the large and in the small sense, and for its potential value as a research-guiding principle with similarities to those already in the field.[45] For readers who prefer to pass over the formal parts of the model, a brief summary is presented at the end of the following section.

Descriptive frameworks (scientific and nonscientific) comprise a historical set, the members of which are prima facie at least partially ordered;[46] that is, the historical set comprises historical sequences that are prima facie ordered. What is the significant structure of this ordering? We are looking for an overall ordering that is both *descriptive* of and *prescriptive* for rationality and progress, particularly for science. I make a threefold claim: (1) that all knowledge is progressive in intent; (2) that its progressive intent is witnessed by the history of science and culture; and (3) that this progressive intent is expressed in the internal history of science by a goal to construct Q-lattices (see chapter 10) among research traditions. I shall not present arguments in favor of (1) or (2), since these positions are well-entrenched in the current literature of the history and philosophy of science, some of which has been referred to in the preceding section.[47] I shall explain what I mean by (3), and offer some plausibility arguments in its favor.

Let L_{A1}, L_{A2}, L_{A3} . . . be a sequence of descriptive languages representing historical stages in the development of a particular research tradition A.[48] We are concerned principally with the great research traditions of the history of scientific thought such as Aristotelianism, Cartesianism, Atomism, Newtonianism, and evolutionary theory, to name a few, but the model also applies to smaller research programs,

such as atomic theories in a particular period. The symbol *A* represents
the general context of the tradition, which I take to be the invariant
(fixed, constant) elements, vague and even contradictory at first, which
take on specific forms as the tradition develops and are seen in retrospect as constituting the particular noetic intention of the tradition. This
is a composite of: *explicit elements* such as the terms of a descriptive
language containing general assumptions about the kinds of entities and
processes involved, a rudimentary conceptual, mathematical, or
imaginative model, standard operating procedures, and guidelines for
the development of research, and so on; and *implicit, tacit, or sedimented elements* rooted in bodily or instrumental expertise, in subjective habitualities or in that limit-mode of sedimented cultural consciousness which Husserl called "the unconscious."[49] While in certain
accounts the latter (implicit) elements would be called "irrational,"[50] I
agree with those who would extend the rationality of science to cover
them to the extent that they serve knowledge and its growth.

I shall consider first of all whether it is plausible to assume that there
is an ordering of a logical kind among them and what it could be. Let
"→" represent the ordering that is sought among the historical stages
of an evolving tradition. Putting aside for the moment the fine structure
of historical development, the ordering will be presumed to be linear,
that is,

$$L_{A1} \rightarrow L_{A2} \rightarrow L_{A3} \ldots \rightarrow L_A$$

where L_A, the last of the sequence, is conceived as some *ideal limit* of
complete disclosure of the horizon of the tradition.

What possible logical relationship, we ask, could the ordering repre-
sent? It might be suggested that L_{Ai} "implies" L_{Aj} (or in notation, $L_{Ai} \rightarrow$
L_{Aj}; $i <$ or $= j$), whenever L_{Ai} is a sufficient condition for L_{Aj} (though
perhaps not necessary). Such a choice suggests that traditions develop
entirely from resources within the human subject, a position I find
unacceptable, as did Husserl.[51] Alternatively, it might be suggested
that L_{Ai} should be a necessary condition for L_{Aj} (though perhaps not
sufficient); in elementary mathematics, for example, one must first
learn to count and then one learns to add, subtract, and multiply. But
who is to say that every tradition develops along a unique and necessary
trajectory? It is at least conceivable that the present stage, say, of
nuclear science, could have been reached by a path other than the
actual historical one. The ordering in question, I take it, is then not
necessary ordering.

I propose instead to interpret the ordering as equivalent to that defined in chapter 10 for a quantum logic but with an explicit reference to the continuity of the historical tradition: $L_{Ai} \rightarrow$ ("implies") L_{Aj} ($i <$ or $= j$) exactly if (1) they have a common set of truth-functional conditions (that is, if they have compatible contexts), and (2) whatever descriptive statements can be made truly and appropriately in the narrower context of L_{Ai} *in retrospect from* L_{Aj} can be said truly and appropriately in L_{Aj} but not necessarily vice versa. Alternatively—as discussed in chapter 10 above—(2) could read: whatever empirical question or problem can be *formulated and answered* in the narrower context of L_{Ai}, *in retrospect from* L_{Aj}, that same question can be *formulated and answered* in the broader context of L_{Aj} but not necessarily vice versa. Let $L_{Ai}*$ be the *current contemporary version of* L_{Ai}, that is, the form L_{Ai} takes when historically reinterpreted in retrospect from L_{Aj} to give the sense and reference of L_{Ai} for the later time, L_{Aj} being assumed to be contemporary with the historian (and therefore identical with $L_{Aj}*$). The basic ordering then takes the form,

$$L_{Ai}* \rightarrow L_{Aj}*$$

where the stars indicate that the current contemporary versions of the descriptive languages are used, in relation to the standpoint of the historian.

Some comments on these definitions are called for.

1. Each language (whether starred or unstarred) is a *descriptive language*, and as such is (or was) accepted on the basis of the evident perceptual objects that it brought into public and manifest presence; such a language implies a specific form of embodiment for the perceiver.[52]

One might think that a descriptive language is not ever a candidate for replacement, but only perhaps for development. That is not so; one has only to think of the descriptive possibilities in its time, say, of the Aristotelian system. A descriptive language once well entrenched could apparently be abandoned for a variety of reasons: (*a*) because theoretical and cultural interests changed, or (*b*) because these were better fulfilled by other perceptual horizons. The model I am offering applies solely to transpositions satisfying either of these conditions, (*a*) or (*b*). A third possibility has already been mentioned, where perceptual objects fail to qualify as authentic horizons of perception: in this case, the line of development terminates at some point in time.

2. Each (unstarred) language named has a *current contemporary*

version (starred). The ordering "→" implies a certain translatability of L_{Aj}^* into L_{Ai}^*, or vice versa.

3. The ordering "→" is an extensional mapping of one domain onto another: however, it does not place an absolute measure on the extent of the empirical domains of L_{A1}, L_{A2}, and so on—there would be no sense to such phrases as "the number of empirical problems in L_{A1}," or "the number of separate empirical contents in L_{A1}," etc.—but it places a comparative measure permitting only the analogous use of phrases such as, "L_{A1} and L_{A2} have equal (larger or smaller) empirical contents."

4. The frameworks so ordered linearly are retrospectively compatible with one another; that is to say, if the transition was once beset with "internal conceptual problems"—to use Laudan's phrase—these have been resolved by the hermeneutic of historical interpretation. Contrary to Laudan's model, the ordering includes then a comparative assessment not of conceptual problem-solving capacity but solely of empirical contents or empirical problem-solving capacity. The reason for this is that the resolution of internal conceptual problems is, I believe, essentially hermeneutical, and therefore, to some extent indeterminate like all textual interpretation; witness, for example, Einstein's or Heisenberg's own accounts of their discoveries, which contain many elements of a historical hermeneutic of science.[53]

5. "External conceptual problems"—to use Laudan's phrase—are associated with the existence of incompatible frameworks within science and in relation to the World; these generate problems for development, but this problem will be taken up below.

6. Besides empirical and conceptual problems—the only ones recognized by Laudan—there is a class of problems connected with the subjective and objective conditions of possibility of perception, for example, as to whether there exists an embodiment for a certain theoretical proposal, or how to bring it about so that the entities and processes postulated by a theory can be made manifest to perceivers. Since no descriptive language can become entrenched unless its subjective and objective conditions are fulfilled, it is a matter of primary importance whether or not there exist satisfiable conditions of this sort. Whether or not such conditions exist is not something, however, that can generally be resolved by argument alone.

7. Empirical scientific problems are always theory-laden: I take this to mean not merely that they can both be formulated and resolved in the same framework, but that they must be. While there are aspects

associated with an empirical problem (e.g., the spur or motive of the inquiry, conflicting appearances, etc.) that can be expressed in some other (possibly theoretical) framework or context, whatever happens to be experienced in this way is not experienced as such as a scientific problem, that is, as a problem for a particular descriptive scientific framework, say, for L_{A_I}, until it has been resolved in that framework. Laudan, on the contrary, seems to hold that a problem can be determined (*univocally*?) outside of the framework in which it is solved.

8. The standpoint of the speaker is always contemporary with the historian, earlier stages having undergone a hermeneutical reinterpretation to become their respective current and contemporary versions. This latter condition asserts, (a) that there is no absolute history, only history for a situated historian, and (b) that a past form of human activity, like research within a particular tradition, may not have been conscious of its meaning and eventual significance for the future history of that tradition, such meaning and significance being gained by a later historical insight into what was "really going forward" at that time in the past. The historian then makes decisions about what from the past belongs and what does not belong to the historical materials needed for coming to understand a movement, and how these materials are to be understood in order to derive the contemporary version of the old theory. It is through this current and contemporary version of the old theory that the "things themselves" of the old theory are made present in their continuity with the past, despite changing theoretical forms and despite forms of mimicry and disguise that can conceal the meaning and significance of the past from the present.[54] Necessary inclusions in the historical materials used for this essentially hermeneutical task are empirical materials or 'texts,' through which the profiles of the old essences—assuming these to have been authentic—become present and manifest to present-day perceivers.

That a "hermeneutical gap" exists between present scientific practice and its account of past practice (that one cannot take past accounts as simply literal, if one wants to understand what was going forward at the time) has long been recognized *in some sense* by most philosophers and historians of science, often, however, without their being able to give an account of its conditions and necessity.[55] Fleck, for example, writing in the thirties in the heyday of logical positivism, found that its account of science failed to note and explain the hermeneutical gap in the history of medicine. He found that even scientists who were involved in the development of a theory later changed the sense and intent of the original observations to make them coherent with the picture provided by subsequent, more mature scientific accounts. Such scientists, he showed, are often blind to the confusion and fumbling even of their own early work and 'read' later theoretical developments into their earlier observations.[56]

9. The historian's charge to simplify and order the complexity of past events presents the danger of covering up a rich and diverse fine structure. For certain purposes details of the fine structure of a historical development may be needed. How did atomic theory at the beginning of the century (L_{A1}) become atomic theory at the end of the century (L_{A2})? The historical narrative would be helped by having a series of fine structure models to refer to. Figure 13.1 below illustrates some lattice structures (all nondistributive, and therefore enriching in the sense explained in chapter 10) of the transition between L_{A1} and L_{A2}. The nodes (small circles) in the figure stand for the different intermediate frameworks of the fine structure, each presented in the light of the historical synthesis L_{A2}.

Hermeneutical Model of Rational Progress: Dialectical Part

We turn now to the paradigm nonlinear model of a cumulative dialectical sort.

Consider traditions, *A* and *B*, which have developed more or less in contextual isolation from one another, and are incompatible in the sense discussed in chapter 10. The reason for incompatibility is that each involves a different perceptual system, and these perceptual sys-

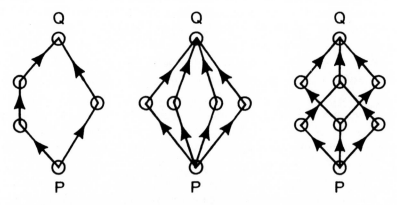

Figure 13.1: Examples of nondistributive lattices.

tems interfere with one another; not merely at the *logical* level of concepts, but also at the *performance* level, where the perceiver finds that the appropriate embodiment for one precludes the unhampered exercise of the other. Such perceptual frameworks are *complementary* more or less in the sense intended by Bohr and Heisenberg. Not merely then do the two frameworks conflict at the empirical and conceptual levels, but they also conflict at the level of subjective and objective conditions of possibility for perception, that is—as explained in chapter 10—of Body, praxis, and horizon.

Conflicts between research traditions A and B fall into two classes: (*a*) where the perceptual objects the older tradition (A) purports to identify turn out to be no more than pseudoessences with pseudo-profiles: in this case, A may simply be abandoned in favor of B (see above for a brief discussion). I assume that the defeat of the *phlogiston theory* by Lavoisier was such a case, as was also the gradual abandonment of the theory of *humors* in medicine, and of the theory of *signatures* in pharmacology.

(*b*) It can happen that the conflict between A and B is resolved within a new and more comprehensive research tradition, $A + B$. In this case, the logical structure of the development is diagrammed in figure 10.2, which is a Q-lattice of partially ordered traditions. L_O (the g.l.b.) comprises all the traditions jointly presupposed by A and B. $A + B$ (the l.u.b.) is the least tradition composite in character that is capable of subsuming each of the older traditions, exhibiting them to have been but partial perspectives of the more comprehensive horizon. I presume that special relativity is the synthesis of mechanics and electromagnetics in just this way. The ordering, "→", is the same as that described above, except that all the languages (L_A, L_B etc.) have to be taken as in some sense ideal languages or limit languages ordered to the disclosure of the full potentialities of their respective research traditions. In figure 9.2, L_B' names a tradition intermediate between L_A and L_{AB}, L_A' names a tradition intermediate between L_B and L_{AB}. L_{AB} is the l.u.b., or logical sum, or the synthesis of L_A and L_B.

Some comments on the interpretation of the Q-lattice are called for.

Nondistributivity: the nondistributivity of the Q-lattice is associated with non-Boolean enrichment (the sum is greater than its parts). Non-distributivity in this case implies that L_A is not identical with L_B' (or L_B with L_A'). In the Venn type diagram, figure 9.1, the ideal horizon of each tradition is represented by a rectangle, and the partial ordering, "→", is interpreted as a relation of inclusion: thus L_O is included in L_A,

L_B, L'_A, L'_B and L_{AB}; L_A is included in L'_B and L_{AB} but not in L_B and L'_A, and so on. The nondistributivity condition then entails that L_{AB} *must* be more comprehensive than L_A and L_B taken separately as pure traditions. Thus, nondistributivity of the lattice signifies that the horizon of the synthesis of two traditions A and B is more extensive than the horizons of the pure traditions A and B, taken separately.

Complements: The complements, L'_B and L'_A, refer to the largest subhorizons of L_{AB} which are *internally* independent of L_B and L_A respectively: they will, however, usually contain some external reference to L_B and L_A respectively. L'_B may be taken, for example, to represent a certain extension of L_A including what Kuhn describes as that part of the A-tradition which can be "translated" into the B-tradition.[57]

Illustrations abound in which it is helpful to apply, as a research-guiding principle, analysis according to the underlying Q-lattice of frameworks. For example, if L_A is taken as "physicalistic" language (language descriptive of physical objects) and L_B as "mentalistic" language (language descriptive of mental or phenomenal objects), then L'_B would represent a reductionist language like that of psychobiology, and L'_A a reductionist language like that of phenomenalism. Or again, if L_A is taken as a perceptual language of the manifest image and L_B as the counterpart language of the scientific image, then L'_B would represent the attempt to reduce the "explanatory" framework of science to the "observational" (or perceptual) and L'_A would represent the attempt to reduce the "observational" (or perceptual) framework to the "explanatory" one of science. If a Q-lattice is the norm of rationality, as I claim, the clash between physicalism and phenomenalism, or between those who favor scientific images exclusively over manifest images or vice versa, would be in principle dialectical, that is, oriented toward the development of a synthesis $L_A + L_B = L_{AB}$.

Cumulative dialectical development: While the interplay between subjective anticipations (molded by a tradition) and experiential encounter with reality is often described as dialectical (so that even the development of a single tradition is in this sense dialectical), still, I believe, a dialectic is better conceived as a clash between opposing intentions in the subject. Since a prepredicative encounter with experience can have no purchase on the subject except by means of some habitual intentionality that gives it sense, no such encounter has the power to shock the perceiver unless two or more intentionalities are actively competing to give sense to the encounter. Such a situation could arise, for example, if the subject, puzzled by unanticipated and insoluble problems—anomalies to the research tradition—turns to reconsider critically horizons that were taken for granted, and searches for possible alternative construals of experience. Anomalies then bring up for reflective questioning the "cut" that separates the subject from the object of science, and invites a new interpretation of experience based on a new embodiment.

Conflict resolution: The new tradition enters on stage in conflict with the old. Sometimes the old tradition turns out to belong to class *(a)* above, the battle between old and new ends as a consequence with the definitive defeat of the old tradition, and little, if anything, is carried over for the enrichment of the new tradition.

Conflicts of class *(b)* are resolved by a transforming synthesis, which is the essence of a dialectical move. Such was the conflict between the complementary frameworks (in the Bohr-Heisenberg sense) of mechanics and electromagnetic theory and its resolution by Einstein in the special theory of relativity; the current and contemporary versions of all three will be found in any textbook of physics. The path of reflection for Einstein, as naturally happens in such cases, turns on the historical roots and path of development of the research tradition in an effort to find the negativity that has been overlooked; this latter is often contained in an existing but minor tradition, incompatible with the former. The accounts given by Einstein in his *Autobiography* and by G. Holton in his studies of the origins of relativity describe the process of discovery in many respects as a hermeneutic recovery of past oversights. Heisenberg's own account of the discovery of quantum mechanics similarly illustrates this process of historical recovery and reorientation.[58]

Historical accounts: A corollary of the above model of rationality is the recommendation that the history of science (or any other cultural system) be presented in terms *(a)* of those traditions that were simply abandoned because they were inauthentic; and *(b)* of those elements that, on analysis, comprise the (completed or, as yet, uncompleted) Q-lattice of its development, as seen from the hermeneutic viewpoint of a historian situated historically in the present and culturally at the point of synthesis (if there is one, otherwise at the highest level of the lattice). When this is done, the structure of rationality in the historical narrative is displayed. A consequence of this is that the problem of distinguishing the large-scale structure from its fine structure is in principle resolved. The decision of how to abbreviate (or expand) a historical account without substantially falsifying the account is also in principle resolved—since the narrative threads are traced by the arrows, one need only include as many of the nodes as serve the purpose of the account that is being given. Diagrams with the form of nondistributive lattices have often been used in art history and in the history of ideas to display the network of historical influences among styles or periods; such diagrams, moreover, would seem to be well suited to structuralist approaches. What makes the present model distinct from all others is the fact that the elements are interpreted to be descriptive languages hermeneutically reappropriated from the past, and not just timeless ideas or structuralist metaphors.

A diagram or "tree" of a historical development will then be a combination of an "evolutionary tree" with terminating branches, and a "dialectical tree" with networks of branches that join above as well as below; these latter are the Q-lattices. Figure 13.2 is a sample of such a structure.

In summary: the linear development of a research tradition depends on the ability *in retrospect from the present* to order the sequential

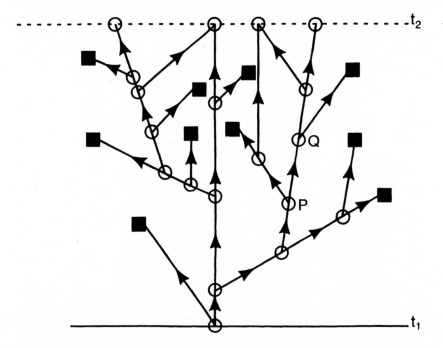

Figure 13.2: Diagram of a "tree" of historical development.

stages of the tradition—the main theories through which it has expressed itself—in a sequence of greater and greater empirical content (or empirical problem-solving power). In order for older theories to become comparable with later theories, it is assumed (1) that the older theory in its own time was authentic (designating, that is, through appropriate readable technologies, perceptual profiles of scientific essences), and (2) that the older theory has a *current and contemporary form*. This latter is the outcome of a historical and hermeneutical reappropriation, not merely of the literary and graphic materials relating to the theory, but also of empirical materials (or 'texts') in which the theory is currently applied to a range of its classic exemplary problems.

Cumulative dialectical development, whether among research traditions, in the history of science at large, or in general cultural history, is taken to be the enriching product of a historical process arising out of a clash between incompatible context-dependent and context-limited intentionalities, where the resolution or synthesis is achieved within a larger structure of mutually interacting contexts. In the case of perceptual horizons (scientific or nonscientific), these contexts are those in which the perceiver is embodied.

The synthesis of two dialectical contexts generates three classes of problems: empirical, conceptual (both internal and external), and that class concerned with subjective and objective conditions of the possibility of perception; these last are transcendental problems. The synthesis preserves the original systems as limited possibilities within a complex lattice structure of systems partially ordered (roughly, by inclusion) in which the least upper bound is the synthesis. The formal property of nondistributivity guarantees, on interpretation, genuine enrichment in the synthesis.

The rationality of a progressive science—part of the progressive rationality of progressive World development—can be defined as the imperative of growth according to the norm of a Q-lattice, where its nodes represent descriptive scientific (or possibly nonscientific) languages. This kind of growth possibility is exhibited in our own culture, for instance, by the clash between various polar reductionisms, such as physicalism and phenomenalism, manifest and scientific images, Euclidean and hyperbolic vision, and so on. The principle of rationality in knowledge then dictates the search for—not necessarily, however, the discovery of—inclusive upper bounds for these polarities. (Among the purposes of this book is that of studying the conditions of possibility, subjective and objective, of the synthesis of polarizations of this kind.)

Finally, some corollaries are drawn from the model about the writing of the history of science.

Modes of Rationality Compared

There are areas of agreement and disagreement between the model I have just proposed and the models proposed, say, by Popper, Lakatos, Feyerabend, or Laudan. All agree that both texts and ancient instruments are included among the historical materials of the history of science, and that some reinterpretation of these materials is inevitable and necessary, if sense is to be made of actual history. Not all agree with me that the systematic performance, say, of classic exemplary demonstrations may be necessary to come to understand, recognize, and name correctly the "things themselves" about which an archaic scientific descriptive language spoke.

All, except me, agree for the most part that a scientific problem (or *explanandum*) can be defined unambiguously without reference to the theory (or *explanans*) within which the solution can be given or proposed. Empiricists by their tradition would attempt to define a scientific problem in primitive observational terms alone; others, like Laudan, take the position that a scientific problem could be unambiguously defined in theoretical terms provided these did not belong to any family of solutions of the problem. On the contrary, I take the position suggested by the hermeneutical character of explanation in natural science, that a scientific problem is categorically indeterminate until a theory (a hermeneutical circle) is proposed capable of resolving the problem; in this view, the *explanandum* and the *explanans* are simultaneously transformed by the discovery of a solution.

All agree that the empirical categories of science are theory-laden; not all agree with me on the reasons for this, that they become perceptual categories through changes in the embodiment of the scientific observer. Scientific objects become public and manifest objects to "experienced" researchers through this reembodiment. Consequently, when the objects of science exhibit their authentic character as perceptual essences (as I suppose they regularly do), they provide direct evidence for the acceptance of a scientific theory.

All agree that knowledge—preeminently scientific knowledge—is progressive; for Laudan that is the sole criterion of rationality (rationality, he says, is parasitic on progress). For the other Popperians and for myself, other criteria also enter or are implied by the rationality criteria themselves; for myself, the most important other criterion is the perceptual 'reading' of empirical 'texts.'

All agree that the decision to accept a theory or research tradition involves more than criteria and values internal to science; it involves world views, social and religious values, and so forth—a World, as I call it, in fact a scientific World. Some (e.g., Scientific Realists) hold that there is a fixed reference for scientific accounts, with specific though possibly unknown properties independent of the scientific frameworks that attempt with greater or lesser accuracy to specify what these are. On the contrary, I hold that there is no sense to such a claim, since the reference of descriptive scientific statements is itself a function of the *context* of the perceptual system used to make the scientific object manifest.

All agree that the principle of rational progress justifies the abandonment of a theory or research tradition that is less progressive *relative to current historical and cultural interests*. Not all would agree with me that, if the abandonment is due merely to a *change in cultural interest*, the permanent loss of a theory or research tradition on this account would be an impoverishment of knowledge and, consequently, *irrational*; nor would all agree—though some would, such as Feyerabend and Kuhn—that the resuscitation of such traditions—aspects of Aristotelianism, for instance—could then become a *rational* project for a future generation.

All agree that scientific theories articulate physical reality, and that scientific theories are historical. Not all would agree with me that physical reality is also historical—with a history like that of the theories that are used to describe what is real. Reality, as I hold, is what belongs to Worlds and is given directly and apodictically in perception to one trained by experts in its performance; it then shares the historicity of the horizons of human perception. Physical reality is just that part of reality which is found to be among the universal conditions of possibility of both Worlds and (acts of) perception; it is roughly that part that is mediated by readable technologies.[59]

Models and Their Descriptive Use

M. Hesse, applying her convergence formula ("accumulating data plus coherence conditions ultimately converge on the true theory"[60]) to a historical case, in fact, to the sequence of atomic theories since Lavoisier, found that ''each successor theory in this sequence interprets its predecessor in its own terms as an approximation to itself.'' She concluded that, as a consequence, ''there is no obvious sense in which convergence of [theoretical] *concepts* can be maintained.'' This raises the question of how or whether conceptually discontinuous scientific models can express a continuous research tradition. Hesse, unwilling to give up convergent realism, resolved the semantical problem of continuity and change by a theory of the metaphorical use of scientific models in which a certain continuity of reference is possible despite the changing metaphors afforded by a series of scientific models. The solution I offer to this problem is similar in fundamental respects.

A *model* is a structure of ideal or empirical elements which is capable of entering into an appropriate semantical use to represent or elucidate a significant structure of something else, the *modeled*. There are various usages of the term ''model'' in the philosophy of science.

The formalists[61] make a distinction between a *theory* and *models of a theory*. A theory is (syntactically and ideally) an axiomatic system with undefined primitive terms; a model is any of its semantical interpretations; a semantical interpretation of a theory is any set of elements— *abstract*, as for example geometrical points or lines, or *empirical*, as for example measured values of physical variables—which can be represented by the theory (onto which the theory can be mapped); a model then possesses a structure isomorphic with the theory. No theory in the above sense is exhaustively represented by one model, and any theory in this sense is underdetermined by any finite set of its models. In applying this account to ''scientific observables,'' there is often deep confusion about how the theory comes to be applied to identifying empirical subject matter. In the working out of virtually all sufficiently elaborate scientific theories, it has been noted that the process of modeling involves two stages: *(a)* from (axiomatic) theory to abstract (say, geometrical) model (model$_1$), and *(b)* from abstract model to observables (model$_2$). Formalists often lose sight of this distinction and treat the process of modeling as if model$_1$ and model$_2$ could be collapsed into one model. The distinction is important because different kinds of considerations are appropriate for each stage of the modeling, from theory to model$_1$, and from model$_1$ to model$_2$.

I propose to take a *model*—or a *scientific model*—to be a theoretically-structured set of elements, usually mathematical objects (a model$_1$) purporting to be useful for the description or explanation of an empirical domain (this would be a model$_2$).

To illustrate the distinction between a scientific model and its empirical domain, consider Newtonian point-particle mechanics. The *model* is a system of dimensionless mass-points governed by the equations called "Newton's laws of motion"; a mass-point is an abstract entity, imperceptible and therefore unreal. Sentences formulated in the model are then used *in a special semantically descriptive way* to make particular statements with realistic intent about the World. For example, the Sun-Earth system (to which this model for certain purposes can be correctly applied) is not a system of mass-points, but an identifiable set of empirical objects which this model can be used to *represent* or *model*. The point of my emphasis throughout this book on empirical or perceptual acts in science is to insist that the principal aim of science in making models is none other than to apply them descriptively to the World through the use of appropriate readable technologies. The connecting link between the model and the horizon of the World that it describes is the special praxis within its research tradition in which the subject embodied in the appropriate technology comes to recognize the presence (or absence) of the modeled structure.

The elementary objects of the abstract model are not *in any literal sense* like the empirical objects they are used to describe; the model elements are (usually) mathematical objects, while the elements of its descriptive horizon are physical objects, but there is a certain likeness of structure. As the model elements are structurally related, so *in significant ways* are the empirical elements of that which is modeled. Aristotle takes the stating of such a likeness to be the function of metaphor. However, the scientific model does more work semantically than merely stating this proportional likeness; we have no access to an understanding of the physical horizon (the "things themselves") except through the use of the model functioning through a readable technology. The model then is more than an Aristotelian metaphor, it is for us a necessary (and perhaps sometimes unique) way of identifying and recognizing what it is the metaphor refers to; we do not know its referent apart from the metaphor, since the metaphor is *needed* to disclose its referent. Hesse, Black, and others have held that this function too is a characteristic of metaphor; accepting this account of metaphor, I conclude that models in science are metaphors of their objects.[62]

Differences between models are then like differences between metaphors: two models/metaphors may be different as abstract objects (or in their character as model$_1$) and nevertheless point to the same thing (model$_2$) *within* the intentionality of a particular research tradition. It is for this reason that continuity and convergence of a historical tradition may be preserved despite the fact that different scientific models replace one another in a given research tradition. The continuity that is sought need not be in syntactical congruences between successive theories, nor even in semantical congruences between successive models (model$_1$), but in the greater or lesser inclusivity of their empirical horizons (model$_2$). This is represented by the ability to order them under the arrow relation, "→", used in this chapter and in chapter 10.

Science in Cultural Perspective

To people in our culture today, nature presents itself with an impersonal majesty before which people stand in silent awe; science is nature's sole (or at least chief) accredited interpreter. To place a hermeneutic of perception at the core of scientific observation, however, is to give science a historical and an ontological essence. The current cultural/historical profile of science in the West would then appear as just one of its possible profiles, one historical route science can take in human culture. The science that we have come to know is a creative historical dialogue between a society in search of freedom and its World in process, and consequently is full of human choices and prejudices, of political fights and religious feeling: such is the picture that the new historians and sociologists of science paint of Western science.[63]

The distinction between the profiles and essence of science opens the way for the "redemption" of science from its Babylonian captivity to this particularly Western cultural path.[64] If science has an ontological essence, then new historical possibilities can be glimpsed for nature as the revelatory goal of science, for human Worlds as the natural setting for human life, and for other interpretations of scientific activity and research.[65] With such freedom, I expect that science would also come to acquire what it often seems so sorely to lack, an element of playfulness, which, as Gadamer states,[66] is at the center of aesthetic experience.

14
Euclidean Space as a Scientific Artifact

The Everyday World as Artifact
of the Scientific Imagination

A hermeneutic of common descriptive language reveals that our everyday Worlds have a basically Euclidean normative structure. Insofar as common descriptive language approaches everyday Worlds *pictorially*, objects in these Worlds can be analyzed into or synthesized from a repertory of elementary geometrical forms, such as Euclidean lines and surfaces, and surface properties, such as color, texture, and light gradients, that are either largely scientific redefinitions of primitive perceptual qualities, or compromises between these two. This is in contrast with the naturally hyperbolic form that unaided human perception takes.

The readable technologies necessary to convert primitive terms into scientific terms are ready to hand in our culture.[1] That such technologies are ready to hand particularly, but not exclusively, to the culturally advanced levels of our society, makes plausible and, to some extent, inevitable the substitution of World horizons that depend on technology for those that do not. Indeed, Western society seems to be driven by a deliberate teleology to replace the primitive horizons of perception with new horizons accessible only through technology, thereby replacing common naturalistic descriptive terms at a primitive level in everyday language with a new repertory of scientific terms for the more cultured users of everyday language. This is, of course, a substitution that is fraught with dangers, particularly if the process by which it comes about is not understood, and is not critically and deliberately evaluated.

The process I have described leads to the replacement of *nature* as pictured in prescientific language by a World where common descriptive norms refer—not to the use of the unaided senses—but to contexts of standard scientific and technological intervention ready to hand in a "carpentered environment." Such a World is, in a special sense, *artificial*: it is, in a special sense, an *artifact* of human culture. E. Cassirer speaks of different Worlds, created by the different myths and religious symbols that have given meaning to human cultures.[2] N. Goodman also speaks of many Worlds, or versions of Worlds, which are constituted by processes of composition and decomposition, weighting, ordering, deletion and supplementation, and deformation: these are artifacts that presuppose common individuals grouped in different ways by convention.[3] But there is a special and more radical sense in which Worlds are or tend to become artifacts—the categories within the World depend or come to depend on a special class of referential artifacts, the class of readable technologies. The contents of such Worlds are or come to be divided into two classes of things: (1) readable technologies, and (2) a repertory of events, things, situations, and so forth, organized by horizons, defined in relation to contexts (of comparisons or interactions) employing artifacts of the first class. Elements of the second class then take on descriptive properties that are functions of readable technologies. It is in this radical sense that I speak of *everyday Worlds as artifacts*: their horizons imply essential reference to a special class of artifacts.

The horizons in question are not simply attainable by unaided perception, but only by such as can 'read' the way the referential class of technologies responds to or is modulated by its surroundings. The environment we experience tends to become the surroundings of those technologies that interface between us and our World. World becomes a meaning of a 'text,' a 'text' that is itself both a part of that World—since it is a set of objects or modulations of objects—and not a part of that World—since it is a part of 'language.' Such a World can be possessed only by those who can recognize its characteristic 'texts' and can 'read' their meaning, or—in popular figurative language—who are capable of experiencing its "vibrations."

The elements of the first class, then, possess a special ambiguity: as objects they belong to a World; but as functioning within the embodiment of human subjects they cross the "cut" between Body and World and come to be part of the subject for whom the World exists. These artifacts are often spoken of in scientific literature as "observers," or "instrumental scientific observers," or "frames of reference," but apart from a subject, they are no more than material systems, information channels capable of producing signals or marks but not capable of producing knowledge. As constitutive of the Body of a subject, however, such a channel becomes a 'text' that is capable of producing knowledge by being 'read' perceptually.

Euclidean Visual Space as a Scientific Artifact

Visual space is Euclidean when it matches physical space. Apart from a subject, a frame of reference for such a space is no more than a virtually infinite set of rulers (or virtual rulers) dispersed throughout space; to have a *visual* Euclidean space, a visual subject must be added. A visual subject, however, is always localized at a definite place, say, the origin of the coordinate system, and must receive through some medium of communication the spatially dispersed pieces of information (about coincidences between rulers and objects) generated in the dispersed parts of space. If such information is communicated instantaneously, the distance between physical things as defined above is additive, nonnegative, and obeys the triangular inequality for a metric distance function. It gives, moreover, a consistent geometry that turns out, not surprisingly, to be Euclidean.

The physical measurement of distance depends on the availability of rulers that are *transportable, physically rigid bodies*. The notion of *rigidity* is different from visual shape and size constancy. The latter, as we saw for the hyperbolic model, agrees with the physical criteria only in the neighborhood of the true point and deviates from it beyond in progressive and systematic ways, most markedly in the distant zone.[4] Thus, physical shape and size constancy—necessary for scientific rulers—does not imply visual shape and size constancy, nor does visual shape and size constancy—necessary for visual rulers—imply physical shape and size constancy. The difference between the two measures is least in the near zone directly in front of the perceiver, and greatest in the distant zone. Recall Ballard's thesis: perceptual distance is characterized, in general, by *ambiguity* in depth, shape, and orientation, while perceptual closeness is characterized by *clarity of form*—both of these I interpret as relative to *our* everyday World in which ambiguity and clarity are in great part taken with respect to the constancies of physical expectations.[5]

Although the spatial structure of unaided visual phenomena is hyperbolic, we are not accustomed to notice the anomalous behavior of visual objects. In the first place, the anomalous behavior is not seen as following an easily understood pattern—the character of hyperbolic shapes is not part of our general background knowledge: consequently, the anomalous behaviors remain consistently outside any ordinary system of classification. In the second place, the anomalies contradict our well-founded expectations regarding the physical behavior of objects. The geometrical properties of physical objects possess the kind of objectivity, universality, and precision around which can gather a community that wants to see the World in a *pictorial* and at the same time a *scientific* way—according to the goals of classical science. Hyperbolic phenomena appear as outlaws to the realm of scientific order, and are consequently dismissed as perceptual illusions. In the third place, the significance we attach to them is a function of the intentionality we bring to bear from our everyday Worlds: this instructs us to hold that depths and distances are depths and distances *as measured* by processes of a scientific kind which include as an essential element the physics of rigid bodies.

Is the general conviction we all share, that the depths and distances we directly perceive are *essentially* Euclidean, fundamentally deceptive? In chapter 11 above, I showed how the structure of our perception and embodiment can change so that scientific entities—reached through empirical (usually measurement) procedures—can come to be the direct objects of perceptual acts by a 'reading' of these procedures; these empirical procedures can become "windows" into the recesses of Nature. In this way, *provided an appropriate readable technology exists*, Euclidean depth and distance—the structure of the physical environment—could also come to be perceived directly, and become an essential structure of everyday Worlds.

It would surely be strange if no such readable technology existed to enable us to perceive our everyday World in a Euclidean way. There is broad and consistent testimony by philosophers, scientists, artists, naturalists, and others, that the structure of empirical space is intuitively grasped as being essentially Euclidean. So persuaded, for example, was Immanuel Kant of the apodicticity of Newtonian physics, that he proposed as a self-evident truth that the space of empirical objects and intuitive experience is Euclidean. Unable to rebut the testimony of so many authorities, we need then to ask about the grounds for such an intuition. What are the conditions for intuiting in human spatial experience the essential form of Euclid's geometry?

Husserl attempts to answer this question in his essay "The Origin of Geometry."[6] Euclidean geometry follows, he says, from the technical praxis of grinding edges, smoothing surfaces, of trying to realize preferred shapes, and so on, from which eventually "the rough estimate of magnitudes is transferred into the measurement of magnitudes by counting the equal parts."[7] He then continues: "Here, proceeding from the factual, an essential form becomes recognizable through a method of variation." Thus, Euclidean geometry emerges for Husserl, as we would expect, as the essential form of a *technical and scientific praxis*—and not of visual phenomena as such.

If visual phenomena in our everyday Worlds are also Euclidean, then we must look for the ready to hand artifacts from which we 'read' the Euclidean structures of everyday perceptual space. The artifacts we look for—the 'text' to be 'read'—must be widely distributed in our environment, and must be such that they can serve to establish for the observer the fact that Euclidean criteria are in possession of space. Such artifacts, I believe, are the simple engineered forms of fixed markers, such as buildings, equally spaced lamp posts, and roads of constant width, as well as mobile markers, such as automobiles, trains—and, though not the product of engineering but also of relatively stable size, the human figure. A name has been coined for this—"the carpentered environment."[8] Each of the objects just mentioned brings to its immediate neighborhood an exemplary local standard that works in fact in a Euclidean way, and serves as 'text' to be read by everyday perceivers. One thinks of the beginnings of our modern era, and of the way Perugino and Raphael deliberately painted a pavement of regular square marble slabs from which the measure of the horizontal extent of pictured space could be 'read.' The standard "carpentered" markers referred to above are dispersed widely, and provide a visually accessible 'readable' technology from which measures of length and distance can be obtained by 'reading' coincidences between their known dimensions and the unknown dimensions of nearby objects.

Works of engineering, such as the rectilinear facade of a large apartment building, or the shapes and sizes of cars in motion, must be included in the 'texts' that the carpentered environment provides; from their paradigmatic Euclidean geometric forms is 'read' the fact that things of different sizes or in different locations can be similar, a property characteristic of a Euclidean space. Though often inflected with illusory hyperbolic "distortions," these works of engineering nevertheless by and large provide an existing 'text' from which the Euclidean character of the local spatial domain within which they occur is 'read.' In untouched Nature, among mountains and in wild places, in sea and sky, beyond the domains populated with visually accessible Euclidean standards, the geometrical structure of visual space may become indeterminate, or more likely may tend toward the hyperbolic.

The movement of our culture, however, is such that more and more frequently, although we find ourselves in the kind of situation that has a scientific description sanctioned by use and convention, we do not genuinely perceive that situation in that way, because the 'text' through which its presence would 'speak' to us is simply not present. Our culture tells us *what* is there, but this *whatness* remains hidden, because its manifestation is a function of what happens on such occasions to be absent; our culture tells us to ignore as illusion the perceptual forms actually present or striving for presence in unaided perception. Under such circumstances, our recognition of the scientific horizon can be no more than indirect. We must reconstruct in imagination what we see according to what we would perceive were we appropriately embodied, or to what we would 'read' if the appropriate 'text' were there. Such imaginative reconstructions or projections are generally neither conscious nor deliberate, nor are they, of course, contrary to fact; but the fact is known not by genuine perception (since the appropriate 'text' is lacking) but by some other process of cognition which we can easily and systematically mistake for perception. If we do, we have failed to discriminate the truth that is given directly in perception from the truth that is reconstructed by the imagination.

The quality of human subjectivity is likewise affected; we can fail to discriminate genuine perception from a pale and anemic surrogate. In genuine perception, we embody ourselves in the technology that "opens the window" to direct experiential contact with scientific reality. If that technology happens on occasion not to be ready at hand, or if we fail to learn the skills that permit us to use such technology as extensions of our sensory organs, then we have no option but to use the descriptive criteria of our culture as if it belonged to a foreign language; we are forced by our culture to translate what we genuinely perceive but may fail to recognize, into a form of "reality" that is reconstructed by thought according to norms for the scientific imagination. Such a "reality" must then be experienced as pale, anemic, and ambiguous, since it is not the presence of an authentic essence manifested through its profiles. [9]

There is then the danger of believing that we possess direct experience when we may only have a pale substitute for it; and that as subjects and embodied perceivers, we are what in fact we may not be or may only rarely be, but of course what we should be if we wanted to speak in the new scientific terms with the authority of direct perceivers. When exercising a surrogate ''perception'' according to norms for the scientific imagination, we are not connected to the reality that we perceive by a real connection, for the 'text,' which is the real connection, is missing. The 'text,' however, is the source of those full-bodied, colorful aesthetic resonances that accompany the manifestation of the real, the ''existential joy,'' that S. Florman speaks about ''at the heart of engineering.''[10] With an anemic sense of the real and a detached or alien sense of the self, the contemporary stage is set for failure to respond critically, or with appropriate existential or aesthetic sensitivity, both to the old but genuine horizons condemned by our scientific culture to shadow and illusion and to the genuine but new—the scientific, which, for the generality, are experienced as pale, remote, and alien—metaphors of the real.[11]

15
World Possibilities

Hyperbolic Worlds

Worlds are potentially multiple and differentiated by communities, historical epochs, and dominant interests[1]. What constitutes the unity of a World? What is the root of possibility of multiple Worlds? What would be the characteristics of a hyperbolic World, with or without access to scientific Euclidean horizons? We ask further: were the Worlds of premodern societies hyperbolic? Are Euclidean and hyperbolic Worlds complementary to one another and dialectically related?

A World is the public home, the reality in which persons live as social beings. That which gives ultimate and final unity to the World as a whole and to its parts is a cosmology.[2] The principle of unity of a cosmology may be religious, or it may be naturalistic.

Premodern Religious Worlds: Premodern societies in the West and primitive societies everywhere organized their reality by religious myths that projected the space and time of human life against the background of mythological events occurring in sacred Space and Time. The profane World, that zone of activity, work, recreation, commerce, or whatever was disconnected from the sacred, and had its meaningful setting within the one sacred Cosmos, with which it had in some mysterious way common boundaries.[3]

Profane space was local and well structured: within it, the sacred had its place, in sacred temples, places, and rituals; through these, connection with the sacred Cosmic environment was maintained. Local spaces, on the one hand, were often highly structured, like the cities of ancient Greece, China, and the civilizations of Central America. Far-away distant space, on the other hand, was all part of sacred space. The sky and heavenly bodies, the sources of rivers, mountain tops, the seas beyond the horizon, distant, inaccessible places, such as the bottoms of caves, or the depths of forests and deserts—all belonged to the gods or the demigods. Distant space was unlike local space in that it was metrically and pictorially ambiguous, like the mythological places that were located there and the mythological events that happened there. Moreover, there was no clear geographic frontier between local, mostly profane, space, and the sacred space of distant regions; the transition zone between the two was systematically displaced as one journeyed toward it, always tantalizingly near, yet never reachable, for sacred space was where one could not be.

This difference in the characteristics of local and distant space connects them naturally with hyperbolic visual space, where the near zone and the distant zone are linked to the perceiver, the near zone is a "Newtonian oasis" of definite, public, quasi-Euclidean structure, and the distant zone is characterized by ambiguity, relative to form and distance, and by inaccessibility, since it is defined in relation to the observer.

F. Cornford[4] has shown that the commonsense space of the ancient Greeks before Plato was spherical and finite—like the Being of Parmenides. This space, consecrated by Aristotle, became the cultural space of the West for two millennia until the scientific revolution, when the infinite Euclidean space of the Void of the atomists was rediscovered and applied, first to the astronomical heavens, and then to the explanation of terrestrial phenomena.[5]

There even seems to be some explicit evidence that Dante conceived his universe, the universe of medieval religious cosmology, as a space with the characteristics of a finite hyperbolic space. Dante's universe is divided into two zones or "hemispheres," each roughly "mirroring" the structure of the other, and each organized about a pole. He established a symmetry between the nine celestial spheres of moon, sun, planets, stars, and *primum mobile*, surrounding the Earth, with Satan at its center, and the nine angelic spheres of the Empyrean spinning with ever-increasing ardor around the antipodes on which "depends the heavens and all that nature holds."[6] From his point of observation on the *primum mobile*, Dante was able to look down at "this our threshing floor,"[7] the Earth, far beneath him and then look upward through the angelic choirs to the Godhead, "a point of radiating light, so piercing, that the eye on which it smites must close perforce by reason of its glow.''[8] Above, he saw in imagination the radiant Sun or *Gloria* of the Godhead, from which rays of light emanated as from a source, crossed the heavens, and rejoined one another at the dark antipodes of the Earthly orb. Each pole, one bright, one dark, reflected the other in counterpoint, and each exhibited the hyperbolic structure of the radiant rising or setting Sun.[9] These symmetrical counter-universes are described in Cantos 27 and 28 of the *Paradiso* as joined along the *primum mobile* as if this were their equator. I think it is plausible to suggest that Dante, in proposing his structure, may have been influenced by the properties of hyperbolic visual space.[10]

Modern Naturalistic Worlds: Modern cultures are distinguished from premodern cultures by the fact that their cosmology is organized on naturalistic principles instead of being based on a religious cosmological mythology. In a naturalistic cosmology, whether scientific or not, the heavens no longer speak of the glory of God, not necessarily because naturalism of its nature is hostile to religion, but because the naturalistic principle takes its stand—against those who would oppose it, whether on religious or secular grounds—that there is no part of space, no matter how distant, that is not like local space, and that is not metrically continuous with local space, and consequently, that is not in principle profane. This is a principle with which modern religious consciousness is quite at ease.

The integration of cosmological space according to naturalistic principles was begun in the late fourteenth and early fifteenth century with the invention—or reinvention—of pictorial reality, and completed in the seventeenth and eighteenth centuries with the definitive establishment of classical scientific cosmology.[11] The space of classical scientific cosmology was an infinite space of Euclidean structure, traversed by the straight line paths traveled both by light rays and by inertial bodies in force-free motion. Such a cosmology was brilliantly successful in its account of physical objects. The new science set itself to construct a pictorial model of reality which was quantitative (through appropriate measuring processes), universal (for all parts of space), and objective in the strictest sense (that is, exact and comprehensive as seen by putative ''godlike'' observers). Add to these principles, as the Greek atomists did, that all bodies are made up of small physically rigid parts, and the only geometry that satisfies these criteria is Euclidean. Thus, given the naturalistic principles that led to the development of modern science, space and the naturalistic World of which it is the unifying principle had to be Euclidean.

Modern science, however, in its classical form had no place among the objects of its inquiry for Mind, and was unable consequently to account for the phenomenon of human persons. Perceiving and purposing were displaced from the physical realm to a nonphysical mental realm that was not a part of cosmology, though it was a part of nature in some larger sense. Mind so disembodied was none other than the pure "godlike" observer of the new science, and the objects it contemplated were none other than the ideas it possessed. These were objective, geometric, pictorial representations of physical reality, images rationally reconstructed in the Mind so that they matched or "mirrored" physical nature, or so it was thought. Mind for the new scientific culture was the disembodied spectator of an infinite physical universe spatially integrated by the laws of classical physics and Euclidean geometry.

The everyday World of modern culture—still only lightly touched by the discoveries of relativistic cosmology and quantum mechanics— is one normed conceptually by the laws of classical nineteenth-century science: from its artifacts, one 'reads' off its essential structure, which is Euclidean, and its essential ontology, which is Cartesian. Such a World has much to commend it, particularly its capacity to transform Nature in the service of human purposes and goals. There is also much to deplore, for example, the inability to 'read' from it anything about the dignity of human persons, or about our relationship with God, or an elenchus of worthwhile purposes and goals, for such things are not contained in the 'text' that was 'written.'

The everyday World of modern culture also promotes the false and dangerous assumption that a person is no more than a detached and disembodied Mind capable of contemplating no more than its own ideas, reflections of things—in their scientific constitution—outside the Mind. In opposition to this view, I have claimed throughout this book that persons are embodied, purposeful perceivers and agents fully naturalized in physical nature, and that the functions of Mind have no existence apart from the life of persons. Mind is an artifact of a certain mode of analyzing the way persons have come to create and experience a World of a particular kind, a World whose meaning is 'read' in the carpentered World of the scientific Enlightenment. Persons, however, are not Minds: they are natural entities with a special power, that of constructing Worlds and horizons to live in. Their being, then, is to be beings-in-Worlds.

By this criticism, I do not mean to suggest that classical nineteenth-century science is wrong; it is, however, relative to a particular historical cultural project. Nor do I mean to say that the spatial synthesis it has constructed is not a significant achievement, but only that, for historical reasons, the kind of World we live in is dominated by the perspectives of classical nineteenth-century science and is fraught with dangers to the personal dimension of human life. By uncovering the latent structure of Worlds as based on readable technologies, and of our (Western) World as based on those derived from classical nineteenth-century science, I have tried to show how the modern person takes an ontology and a cosmology from the works of Newtonian science and engineering dispersed in our culture. Emancipation from the prevalent inauthentic way of living with these artifacts does not involve or require a repudiation of science, or technology, or the naturalistic principle, but rather a determination to explore the new possibilities that such an emancipation makes possible.[12]

Though normed both conceptually and perceptually by Euclidean space, it is not at all evident that the space we perceive is everywhere at all times Euclidean. Enough evidence has been presented in Part I to show that daily experience would have instructed us well in the recognition of hyperbolic visual forms, had we a language with which to describe them. Our ability to perceive a domain as structured in a Euclidean way depends on the pervasiveness in that domain of engineered figures that we can 'read' or that can serve as local standards for size, depth, and distance. Outside of the domain of engineered figures, if we are to experience genuine perception, a different rule of congruence must dominate founded on a natural capacity to estimate size, depth, and distance by visual comparison. Such a rule of congruence leads, as I have shown, to a hyperbolic visual space. Regions such as the open sea, virgin forest or countryside, and the heavens above, which lack the presence of engineered artifacts or natural standards of length, will then appear to be hyperbolic. Such perceptual structures, though they constitute an essential part of the phenomena of our everyday Worlds, will nevertheless in our culture be judged to be illusionary.

Naturalistic Hyperbolic Worlds: One of these new possibilities is a World structured by hyperbolic visual space. In the first place, such a possibility is, as I have shown, primordial,[13] and so is capable of grounding alternative Worlds. Second, despite its probable association with premodern mythological cosmologies, such an organizing principle is thoroughly naturalistic. Meditating on its origins and mode of constitution, one thinks a hyperbolic World is likely to possess the following characteristics.

(1) A hyperbolic World would be person-related. Space would have different parameters for different visual purposes. A hyperbolic World then connotes the activity and decision-making capacity of the personal embodied knowers who use these spaces.

(2) A hyperbolic World would be locally structured. Its standard for size, depth, and distance is local, and visual comparisons with such standards provide a clearly articulated foreground space, but an ambiguous distant space. One set of principles of spatial organization would govern local space—whether home, neighborhood, or city—and another set, or none at all, would govern the space beyond the local space. Since shapes in the distant zone are shallow and much distorted compared with those experienced in dealing with local objects, we would probably consider distant space chaotic and uninteresting. Moreover, the distant zone—without science—would be mysterious and inaccessible, since its distance is defined by its separation from the perceiver. If visual space is a finite space, then, a hyperbolic World would be of finite size, differentiated qualitatively between near and distant zones. Such a World would have a center in the local community, and its World View would be anthropomorphic and tribal.

(3) Without science, a naturalistic hyperbolic World would have little technology and few readable technologies. To have a standardized technology, one must have, besides energy, materials and capital, the "thinking cap" that makes possible the mass production of standardized equipment. Mass production of that kind requires that many people, in many places, in many occupations, understand in a uniform way the specifications for the design, production, servicing, and operation of equipment. Such information is generally communicated through the use of plans and blueprints: these are diagrams and projections of the assembled and disassembled equipment, represented in the space appropriate to physics, that is, Euclidean space. Without the ability to visualize a piece of machinery according to the geometry appropriate to its construction and operation, there is no possibility that such plans would be understood and used in a uniform way. Consequently, being at home in Euclidean horizons is a prerequisite for those whose business it is to make and maintain a plethora of readable technologies for the service of the general public.

To be at home in Euclidean horizons, is not, however, necessary for the general public, except insofar as the general public is also the pool for possible technological talent. But such talent could be nurtured in special environments and special schools. It is said that the modern industrial revolution would have come into existence and gone on steadily even if not a ton of coal had been mined in England, and not a single iron mine had been opened. It was sufficient that a new "thinking cap" had been adopted by the peoples of Northern Europe in the seventeenth century which saw the possibility of a World where things were all mechanisms.[14] Mechanisms in turn provided the 'text' from which people 'read' the basic facts about reality, and 'taught' them how to convert Nature into an industrial resource. The Mechanization of the World-Picture could not have followed from a purely hyperbolic World.[15]

(4) A hyperbolic World complemented by a full range of scientific horizons would be very different from a purely hyperbolic World. We have seen that complementary manifest and scientific horizons can and do exist side by side, and that by dialectical development they can form a new and reflective synthesis. The synthesis is a least upper bound of a nondistributive Q-lattice. Such a synthesis would be achieved within the context of a dominant World. If the dominant World were hyperbolic, what effect would such a synthesis have?

In the first place, more distant places would be reachable directly or indirectly by instrumentation or by readable technologies. There would be less mystery about distant parts of the earth or distant parts of space: all would be recognized as we now recognize the surface of the moon, as localities just like other localities we know, but as possessing perhaps more extreme properties.

In the second place, the introduction of sophisticated technological communications media, such as telephone and television, would change the meaning of *closeness* and *distance*.[16] Any object within the actual or potential reach of any person anywhere could, with these media, be made present in one's immediate neighborhood by a "live" *re-presentation* of the object in one's local space. This re-presentation would, of course, conform to the metric of the local space. Would the re-presentation of distant parts on local space by means of the new communications technology exert a subtle pressure to convert visual space into the universal, objective space of Euclidean geometry? I do not think so. There would, of course, be a strong tendency to redefine *closeness* and *distance*, by analogy with ease of access and clarity of re-presentation; but these are functions (1) of the design and use of the technological system and (2) of local space within which the re-presentation takes place, rather than of relationships within a universal, objective space. It seems to me that the redefinition of *closeness* and *distance* would not of itself force a reconceptualization of space, or if it did, not necessarily as a Euclidean space.

Consequently, apart from the needs of designing and servicing such new technologies, for which a Euclidean intuition would certainly be required, the prevalence of the more sophisticated communications technology in our environment would seem to be compatible with a lived hyperbolic World. This would be the tendency more surely in any population given predominantly to social, political, and personal service occupations where significant human action on the World takes place in the near zone of space, either on objects themselves or on their re-presentations there.

Finally, the dialectical question on which this book will end: could scientific Euclidean horizons that presently constitute our cultural Worlds be complemented dialectically with more primitive hyperbolic horizons to constitute a single, multicontextual cultural World? Although it is assumed that a cultural World is a singular totality and the source of reality-schemata, it is, nevertheless, the case that Worlds are not fixed and permanent conditions; they change historically. Prior to the fourteenth century, we have surmised that Western Worlds were spatially unified in a primitive way by hyperbolic geometries. Since then, modern Western Worlds have adopted the scientific norms of Euclidean geometry. Could these two great synthetic principles, Euclidean and non-Euclidean, coexist in one multicontextual cultural World?

If the answer is yes—and this is the answer I am suggesting—the multicontextual synthesis should have the formal structure of a Q-lattice. Let us construct it: let W_A designate some Euclidean scientific World, and W_B some hyperbolic visual World. By a move that is critical, reflective, and dialectical, like that described for complementary scientific and manifest terms, each World can come to manifest itself apodictically according to its essential structure; what is to be 'read,' and how this is to be 'read,' now becomes plain for each, and the different principles of spatial unification come to be recognized as incompatible. This process of reflection is accompanied by an intent toward a dialectical synthesis suggesting a complementary relationship between the two. Complementarity is suggested by the ''smoothness'' with which Euclidean or near-Euclidean perceptions mesh with hyperbolic or near-hyperbolic perceptions of the same motifs. Ortho-complements, W'_B and W'_A, would be formed of W_A and W_B respectively, with W'_A providing hyperbolic profiles of fixed physical shapes and W'_B providing Euclidean profiles for fixed hyperbolic visual shapes. The synthesis, W_{AB}, would then be the least upper bound of W_A, W_B, W'_A, and W'_B. W_{AB} would represent a World for a society culturally enriched by both scientific and visual horizons.

What would the furniture of such a World as W_{AB} be like, and how would its subjects be described? W_{AB} would have to be equipped with *two* sets of 'texts': each set would comprise artifacts such as, for example, works of art, architecture, or city planning which 'speak' of the geometry of space. One set is a 'text' 'written' in a scientific 'language,' and the other, a 'text' 'written' in a visual 'language.' One is decipherable only with the aid of Euclidean space, the other only with the aid of hyperbolic visual space. The subjects in W_{AB} would have to be equipped with hermeneutic capacities to 'read' both series of 'texts' in the way appropriate to each.

Can we point to examples of such 'multilingual' Worlds in the West? Not yet, as far as I can tell. But anthropologists and sociologists, such as M. Griaule, P. Berger, H. von Dechend, and others, who are sensitive to the hermeneutical dimension of art and culture, have already pointed the way. Moreover, historians and philosophers of science, such as T. S. Kuhn, G. de Santillana, H. Marcuse, N. R. Hanson, and P. Feyerabend, have often contributed in spite of themselves to a positive solution by undermining the imperialistic hold of science on our cosmology. With the recovery of our premodern roots in cultures that once flourished and no longer exist except perhaps here or there outside the mainstream, we might expect dialectical development toward 'multilingual' Worlds to take place.

Among non-Western cultures, Japan and China have assimilated Western scientific culture while in many respects they have succeeded in keeping Western social perspectives at bay. Islamic countries and the countries of the Third World long for the benefits of modern science and technology, but fear Westernization. While coming to share the benefits of a scientific technology, they hope nevertheless to be able to maintain their traditional social forms and rituals, often incompatible, though possibly complementary (in the quantum logical sense) with the perspectives of a scientific Westernized society.

General Conditions of Possibility of Worlds

A tentative conclusion of the main line of argument of this book is then that insofar as reality is grounded in the general conditions of possibility of perception, the structure of this ground has (possibly) the historical and dialectical structure of a nondistributive quantum lattice. This would admit of many contexts, some incompatible with but complementary to others and, therefore, mutually enriching. The capacity to create new contexts exists through, among other things, the ability to design and use readable technologies, even those that are non-Cartesian, such as those constructed on the basis of the quantum theory. The ontological dimension, therefore, is historical, cultural, and never closed.

In the natural attitude, we accept the givenness of the real; we acknowledge that our World is the successor of past Worlds and we take this succession to mean that our World is better than past Worlds, and that part of the improvement is in the addition of new horizons of truth and the elimination of old errors and ancestral "myths."

In the reflective attitude, however, we come to a different conclusion. We find sedimented in our present World 'texts' appropriate to both Euclidean and hyperbolic horizons; these "texts" are artifacts comprising works of art, architecture, and other products of readable technologies 'written' in the past for a variety of perceptual languages, scientific (Euclidean) and hyperbolic. Some of the 'texts' are decipherable only with the aid of Euclidean space, others only with the aid of a hyperbolic visual space. The World then reveals itself as a dialectical synthesis of many contexts, constituting a Q-lattice of past horizons and past Worlds, the origins of which have been largely forgotten. If we are ever to live authentically in the presence of our past we will need to be able to 'read' the stories our ancestors have left us in many media about their historical Worlds. This is a hermeneutical task, and a work of philosophical cosmology.

16

Retrospective

Hermeneutical Character of the Study

In a study of this kind, we can distinguish the general philosophical principles ("premises") that are used from the outcome (or "conclusions") of using these principles in a particular domain. However, "premises" and "conclusions" are linked in this study—and in many others that share the same character—in a peculiar way: the evidence for one is part of the evidence for the other, that is, the two parts relate to one another generally in a *hermeneutical* way.

The "premises" in a hermeneutical inquiry are used neither deductively nor inductively but in a special and characteristic way. They serve to confer sense, meaning, and significance on material signs that are then "read" as "conclusions"; note that the significance does not fall on the material signs as such, but only on what these signs come on interpretation to "say." Thus, the progress of a hermeneutical inquiry is circular—more accurately, *spiral*—from "premises" to "conclusions" to assessment, and then from refined "premises" to refined "conclusions" to a new assessment, and so on, until a refinement of the original proposal is reached which fulfills the conditions for a satisfactory solution.

In contrast with both deductive-nomological and inductive models of explanation, there is in the hermeneutical model no categorically definite question, problem, *explanandum*, or "conclusion" to start with, that is, prior to the "premises" that serve as the sought-for explanation; that is why I have used warning quotes around the words "premises" and "conclusions." Once the explanation is found, the question or problem becomes definite and determinate; in retrospect, it can only be seen to have been categorically incoherent before the hermeneutical circle was invoked which gave it progressively sense and significance by supplying (constructing?) the absent categories.

The "conclusions" then are like a continuous curve fitted to empirical data that gives a normative sense to the data by defining what about the data is significant.[1] On this imperfect analogy, general philosophical principles, when used in a hermeneutical way, are refined by the attempt to make sense of philosophical problems, while, at the same time, it must be admitted that what constitutes a philosophical problem is specified only by the general philosophical principles one entertains to begin with; it is they that tell us what to look for in a problem, and how to recognize and name it. The philosophical problems that will be addressed are the general conditions and structure of perception, and most particularly, the spaces of visual perception, scientific observation and, to the extent that Worlds are based on perception, World-building.

In writing this book, however, I presented in Part I certain "conclusions" about the spaces of visual perception, *before* presenting the "premises" or general theory about perception which made it possible for me to present the "conclusions" in the way I did in Part I; this general theory is presented only in Part II. This order inverts the logic of the hermeneutical circle, since in the hermeneutical circle the general theory, which informs (gives meaning to, provides descriptive categories for) its concrete applications, has a certain logical priority over its conclusions, which cannot even be stated coherently without it. The justification of the order used in the presentation relates simply to the rhetorical or pedagogical goals of this work. There is then room at the end of this book for a retrospective that reestablishes the logical order.

Note: to the extent that an explanation—in philosophy or in science—is hermeneutical, the relation between "premises" and "conclusion" is peculiar and characteristic. For example, unlike the *deductive-nomological* model of explanation, the same categories are used in the "explanandum" (or conclusion) as in the "explanans" (or premises). However, neither does the hermeneutical model agree with the *inductive* model of explanation for, despite the fact that in both the same categories are used in premises and conclusions, in the inductive model, they are first exemplified in the conclusion and then used in the construction of explanatory premises; in contrast, in the hermeneutical model they arise simultaneously, with, however, logical priority given to the premises. The hermeneutical model of explanation then agrees neither with inductive models nor with the deductive-nomological model of explanation.

General Philosophical Theory

The general philosophical theory or method I am using is to be found in a specific hermeneutical reading of the works of Husserl, Heidegger, Merleau-Ponty, Ricoeur, and Gadamer. In this study, however, I am not concerned with the theory specifically as an interpretation of these authors, but only with its value as a philosophical theory, that is, with its ability to make sense of a cluster of problems relating to the existence of and conditions for perceptual knowledge, mostly in the sciences. The theory can be summarized, though not easily or well, in the following principal points.

1. About *hermeneutics*: the process of reading, understanding, or interpreting a literary text is guided by what Heidegger calls a "fore-structure of understanding," that is, an anticipation about the kinds of things or objects about which the text speaks. This fore-structure is also called a "hermeneutical circle"; in it is

> hidden a positive possibility of the most primordial kind of knowing. To be sure, we genuinely take hold of this possibility only when, in our interpretation, we have understood that our first, last and constant task is never to allow our fore-having (*Vorhabe*), fore-sight (*Vorsicht*), and fore-conception (*Vorgriff*) to be presented to us by fancies and popular conceptions, but rather to make the scientific theme secure by working out these fore-structures in terms of the things themselves.[2]

The task of hermeneutics (or interpretation) has as its goal not just some or any understanding consistent with the text but a reading that attains to "the things themselves" (Husserl's term originally) about which the text speaks. Such a task is not, as Heidegger says, a work of arbitrary fancy, but one controlled by those elements that constitute the fore-structure of understanding: (1) *Vorsicht*, or a set of common descriptive categories and a common descriptive language; (2) *Vorhabe*, or a set of praxes, embodiments, skills, and so forth which mediate between the descriptive categories or terms and that to which they refer; and (3) *Vorgriff*, or a particular hypothesis about the subject matter in hand.

The hermeneutical task is *circular* in a peculiar but "nonvicious" way, because it involves the simultaneous and mutual determination of the (meaning of the) whole by the (meaning of the) parts, and vice versa. On the one hand, the fore-structure of understanding—the "hermeneutical circle"—provides a conjectured meaning for the text as a whole and for its parts, but, on the other hand, the kinds of conjectures one entertains about the (meaning of the) whole depends on clues scattered in the text itself. One moves from a partial disjointed set of insights or clues to an understanding of the whole and back to the not-yet-understood portions of the text, the process guided by the attempt to discover the outlines of "the things themselves."

A satisfactory solution to a hermeneutical inquiry would fulfill the following conditions: (1) all the clues lie on (or *sufficiently* near) the proposed solution; and (2) one is persuaded that none of the as-yet-undiscovered clues lies *too far off* the proposed solution; the terms "sufficiently" and "too far off" imply reference to the goals and purposes of the inquiry. Every hermeneutical inquiry then is fraught with a certain indeterminacy, and all are radically incomplete.

When the text is imperfect or corrupt, or when the subject matter is unfamiliar, the hermeneutical task is accompanied by a special kind of effort and obscurity.

All hermeneutical processes possess the *dual structure* associated with the acquisition or expression of *information*. The duality is between information$_1$, which is the text, and information$_2$, which is its meaning or content. Note that, for a linguistic text, once syllables, phonemes, or other linguistic signs are read, they cease to be objects in the World, like houses or trees, they become more like *windows to a room* which by their (more or less) transparent quality give direct access to the contents of the room beyond. One does not then *perceive* the syllables of a text: one *reads* them. In a reading, the physical character of the text disappears from direct view leaving no objective trace whatsoever, and one's attention is possessed directly and immediately by the *meaning* of the text. The physical character of the text drops out of objective awareness, it becomes to that extent "nonobjective," and I take that to mean "belonging to the conditions of the subject."

Although, on the whole, one is oblivious of the syllables and marks as things on paper, one is not totally unaware of the fact that through them one is guided through the meaning of the text. There is then a subliminal awareness of pleasure, or perhaps of frustration and discomfort, which arises from the activity itself, from its rhythm, natural or forced, from the musical quality of the sounds, and from the resonances of heard or imagined speech. In all of this lie the aesthetic qualities of a text (sometimes called the "experience" of reading).

The kind of transformation I want to describe is that in which the syllables or phonemes of a strange language at first engage our attention as curious objects for possible theoretical study, and end up by being dropped from awareness when we have become familiar with the language and have learned to read the syllables as text or to listen to the phonemes as spoken words. When one knows the language, direct access is obtained to the meaning of a text or spoken word. Such meanings are not, however, expressed in judgments of perception, but in judgments about the subject matter referred to by the writer or speaker, be it the French Revolution or a football game. The transformation just described, wherein intermediaries (information$_1$) in the acquisition or expression of information (information$_2$) "drop out of consciousness," has been noted, for example, by Schrödinger, who ascribed such transformations to processes perfected during the long course of evolution.[3] However, it is a commonplace that many processes perfected through the painful process of learning, such as, for example, reading, playing a musical instrument, driving a car, and so on share the characteristic that intermediaries drop out of consciousness.

2. About *perception*: primary ontological intent is carried by perception; that is, reality is exactly what is or could be manifested through perceptual essences and profiles as horizons of Worlds. This is the *principle of the ontological primacy of perception.*

Perception is epistemic (that is, a form of knowing). It is not just having a percept or "internal" representation of the object; it terminates directly at its object—a state of the World—as manifested in perception; the form of that knowledge is a descriptive judgment (capable of being stated in a language, usually by the perceiver, but at least by others of his community).

Perception is hermeneutical. Perception is like reading, but the meaning picked up—'read'[4] from the physical clues that constitute perceptual information$_1$—is an exhibited state of the World, and may in fact constitute an enlargement of that World. Perception is a process that tries to gain information by searching for the perceptual horizons made manifest—or capable of being made manifest—in experience; essential to this is the presence of structures in the ambient energy field which cause "resonances" in available somatic information channels open to the (potentially) skilled or experienced subject.

An act of perception then starts with a physical resonance (information$_1$) in some available information channel caused by an interaction between subject and object (or more precisely, between what later will be more precisely specified as the Body of the perceiving subject, and the perceived object). In a characteristically hermeneutical way, the subject searches for the clues that can be 'read' systematically as manifesting horizons (systems of perceptual profiles organized by objective perceptual essences) and Worlds (systems of horizons organized by, among other principles, space—and, of course, though hardly mentioned in this book, time); such meanings in the hermeneutical process belong to information$_2$.

Reading, as I have said, is a kind of cognitive activity that does not require attention to the descriptive specificity of the phonemes, syllables, and so forth that serve as clues for a reading. In fact, it is not necessary that these clues be cognitively known before they are interpreted. After successful interpretation, the clues, whether or not they were antecedently known, "drop out of consciousness," and become "transparent" to the new objects of perception; the new perceptual system, once it has been established, 'reads' the new clues *perceptually*, that is, has direct access to perceptual meaning. The Body of the subject as perceiver then comes to be defined correlatively with the new horizon of perception; physically the Body of the perceiver is the new information channel (its states comprise information$_1$) for perceiving a new repertory of World states—the profiles of a new perceptual horizon (these comprise information$_2$).

To the extent that perception is hermeneutical, its contents are never unique, definitive, complete, final, absolute, ahistorical, or acultural. Unlike the meanings of literary texts, they do, however, enjoy the status of being (or becoming) possible states of the World. Through such essentially hermeneutical processes, limited, however, by what *Vorhabe* will allow, perception builds cultural and historical Worlds.

3. About current debates in the *philosophy of science*: in the first place, my position is neither that of *Scientific Realism* nor that of *Instrumentalism*. I call my position *"Horizonal Realism."* Horizonal Realism holds that science has the intent of describing the elements and the structures of reality; these may be hidden to (theoretically and instrumentally) unaided perception, but have nevertheless genuine perceptual essences which can be made manifest directly in perception

with the aid of theoretically structured instruments used as "readable" technologies. Theoretical entities made present in this way are not just detected through a process of inference, they are directly perceived.

In the second place, contrary to many theories of scientific reference, I hold that there is no *identity of reference* between individual objects of a manifest image (e.g., this patch of sensed-color) and individual objects of the relevant scientific image (e.g., this spectral mix of wave lengths), only many-to-one and one-to-many mappings of perceptual objects contextually defined within mutually incompatible but complementary contexts.

In the third place, my position is *neither* that of *Conventionalism* nor that of *Cultural Relativism*; I admit, however of plural, incompatible, empirically descriptive frameworks, some of which are complementary. I differ from both Conventionalism and Cultural Relativism by insisting on the fact, necessity, and limitations of *Vorhabe* (in this case, trained bodily expertise) for linking descriptive categories to their appropriate empirical objects; the plasticity of *Vorhabe* places limits on the descriptive frameworks that can be conventionally chosen or used in a culturally relative way. I surmise that these limits can only be known through empirical, historical, and cultural studies.

Specific Applications and Conclusions

The general philosophical theory or method summarized above is used in a variety of projects to study the subjective and objective conditions of possibility of apodictic (critically established, systematic) perceptual knowledge. The first project is related to the geometry of visual space, and is twofold: (a) to exhibit, through the use of an appropriate descriptive model, the existence of hyperbolic visual perception, side by side with Euclidean visual perception, and to show that such descriptions can be apodictic; and (b) to inquire into the subjective and objective conditions of possibility of both Euclidean and hyperbolic vision. The second project is also twofold: (a) to show that there can be direct perceptual knowledge of theoretical scientific entities, and that it can be apodictic, and (b) to inquire into the subjective and objective conditions of possibility of such knowledge. The solutions to the first two projects match one another. The third project concerns the criticism of identity theories. The fourth project studies hermeneutics and the history of science and perception. The fifth project is an inquiry into

World-building, that is, into the way scientific and technological horizons affect our Worlds, and change our reality schemata.

Visual Spaces

In this book, the general theory (or hermeneutical circle) outlined above is applied principally to the domain of visual spaces and their structure. Such a use has a manifold character. In the first instance, it directs an *empirical* (in this case, descriptive) study; in this respect, the general theory provides an outline of the relationship between hypothesis and evidence, between model and reality, between continuity and development and change, and it states the hermeneutical and perceptual criteria for the acceptability of a theory. In the second instance, applying the general theory to visual spaces has the character of a *test* for the general theory; for if it cannot make sense of the phenomena, it may have to be revised or set aside. And in the third instance, its application to visual spaces is an *illustration* of the general theory, serving, if successful, as a paradigm for other, similar studies. In all of these respects, the general theory serves as a philosophy of science vis-à-vis the empirical inquiry, but hermeneutically related to it, so that it is clarified and strengthened by a successful outcome of the inquiry, and rendered ambiguous and weak by an unsuccessful outcome.

I take the usually Euclidean character of visual space to be uncontroversial. My task then is to determine whether the phenomenon exists of perceiving physical objects not as Euclidean visual objects but as construed (or reconstrued) in a hyperbolic visual space. The general theory of perception is used to construct (or refine) a model of the hyperbolic vision of physical objects; the model is called "the hermeneutical Luneburg model," after R. Luneburg, who proposed the geometrical prototype. The model is applied to empirical material hermeneutically—that is, there is a circular relation between alleged or possible instances of hyperbolic vision, and the shapes and structures derivable from the spaces of the model. The model includes not only the descriptive framework for hyperbolic shapes (part of the hyperbolic visual intention), but also the specific kind of embodiment in the optical energy array which makes such vision physically possible.

Chapter 3 sketches the history and rationale for the hermeneutical Luneburg model. Chapter 4 studies the transformed visual shapes and structures of physical objects "carpentered" to simple paradigmatic

Euclidean shapes and structures; it discusses many everyday anomalous phenomena, such as the appearance of the sea and sky, of churches, highways, railway tracks, and so forth, which are prima facie exemplifications of these hyperbolic visual structures. Chapter 5 studies many two-dimensional visual illusions, such as the Müller-Lyer and Hering illusions, and other three-dimensional illusions, such as the moon illusion and various rotational illusions, and shows that the illusionary effect can be described as episodes of hyperbolic vision. Chapter 6 argues that the pictorial spaces of many works of art in the past exhibit characteristic structures of hyperbolic space, implying that the space perceived and intended by the artist in the portrayal was hyperbolic; examples are found in the "refinements" of Greek Doric architecture, in Pompeian wall paintings, in the paintings of Cézanne, and especially in the paintings of Vincent van Gogh. The empirical conclusions of the study are briefly summarized in chapter 2 as well as the end of chapter 6 of this book.

Manifest and Scientific Images

Contextuality of Perception: Acts of perception are always contextual. The context of a particular act of perception is the set of subjective and objective conditions of possibility of this act of perception. On the side of the subject, there is the perceptual system which comprises (1) an antecedent *intention* or conceptual framework, expressed in the determinate use of a descriptive language (cf. Heidegger's *Vorsicht*); (2) an antecedent *embodiment* for the subject, the constituent physical parts of which are multiple: neurophysiological networks, somatic processes (e.g., sensory organs), structures of external (e.g., optical) fields, and possibly, technological processes (e.g., "readable" technologies); the activity of the embodied subject is functionally (that is, physically and intentionally) indivisible. The latter in the form of trained bodily expertise is related to Heidegger's *Vorhabe*. The antecedent embodiment of the subject is designated a "system" only by reason of its intentional function; it is only because of this connection that it becomes of interest to the psychologist and the epistemologist.

The analysis of the contextuality of perception is pursued in relation to: (*a*) sensed-heat versus thermodynamic heat, (*b*) sensed-color versus spectral color, and (*c*) hyperbolic visual space versus Euclidean visual space. In each of these cases, the first element belongs to some manifest image, and the second belongs to the relevant scientific image. The structure of the analysis leads to a general theory concerning (*d*) the relationship between a manifest image and the relevant scientific image.

In all of these cases, the differences between the manifest and the scientific image are not merely in intention (descriptive categories), but also in extension (reference). It is not the case that one and the same manifest object (e.g., a patch of sensed-color, a feeling of sensed-heat, a hyperbolic shape, etc.) has a unique scientific description (a unique spectral mix, a unique thermodynamic state, a unique Euclidean form, etc.). The reason for this is that the two descriptions refer to different contexts of the perceiving subject. Each context has the capacity to identify a range of individual referents, but each individual of one image is invariant under a variety of permutations and changes within the other image, and so the mappings are one-to-many and many-to-one. Consequently, there is no descriptive redundancy between the two images. This analysis places limits on what is or can become a conventional framework of perception, and undermines the kind of Conventionalism defended, for example, by Duhem, Poincaré, Reichenbach, and Quine.

Incompatible but Complementary Frameworks: The root of incompatibility between two descriptive predicates may or may not appear to be a theoretical inconsistency. It would not so appear, for example, if the two predicates simply did not belong to the same theoretical picture; the root of incompatibility is the practical inability to identify and isolate individuals having simultaneously the properties believed to be incompatible. Quantum mechanics provides a paradigm of how predicates can be both theoretically and practically incompatible. For quantum mechanical systems, it is not possible, either in theory or in practice, for each of a pair of "complementary quantities," such as position and momentum, to have a precise value; each, however, can have simultaneous *imprecise* values regulated by Heisenberg's Uncertainty Principle (according to one of the original interpretations of that principle). The logical structure they constitute is that of a lattice of complementary descriptive languages, where the lattice is nondistributive and (usually) orthocomplemented. I call it a "quantum logic," or "Q-lattice." A similar structure is found in the complementarity of manifest and scientific images.

About *perception* and *scientific observation:* there are genuine perceptual horizons of scientific observation, in which theoretical scientific entities are or could be manifestly and directly given as perceptual objects to experienced scientific observers. The argument for this thesis is based (1) on an analogy between reading a text and 'reading' an instrument—both are hermeneutic, and contextual acts—and (2) on the conditions for a perceptual act, which involve the "picking up" of "information" about the present and actual state of the World from the environment directly via the Body.

Theoretical quantities become known by attending to the response of appropriate empirical procedures, such as the use of measuring instruments. There are, however, two ways of attending to the response of an instrument: the instrumental response can be used in a deductive argument to infer a conclusion, say, the value of a quantity (provided one knows the relevant scientific theory), or the value of that quantity can be 'read' directly from this response (provided one is experienced and skillful in the use of the instrument). I want to focus on the latter. In 'reading' a thermometer, say, one does not proceed from a statement about the position of the mercury on a scale to infer a conclusion about the temperature of the room by a deductive argument based on thermodynamics; of course, one could, but then in this case one is not 'reading' the thermometer.

To the extent one 'reads' the thermometer, the thermodynamic argument remains in the background, being merely the historical reason why thermometers came to be constructed in the first place. One can 'read' a thermometer, however, whether or not one knows anything formal about thermodynamical theory. Provided the instrument is *standardized*, and so can function as what I call a readable technology, the instrument itself can define the profiles and essence of temperature.

The process of 'reading' is something like this: a 'text' is 'written' causally on the thermometer by the environment under standard circumstances (*ceteris paribus* conditions). This 'text' is 'read' as being 'about' a presented object, here, the state of temperature; this piece of acquired empirical knowledge—the current temperature—is expressed in a language that uses scientific terms, such as "temperature," in a descriptive way about the World. Such a process is, I claim, essentially both *hermeneutical* and *perceptual*.

The response of an instrument, which I refer to as a 'text,' shares in the information-theoretic aspect of literary texts. A 'text' is 'written' by the ambient environment on a standard instrument under standard circumstances in standard signs, it is controlled by physical causality; a text is written in a standard vocabulary, syntax, and so on by a writer, it is controlled by a causality guided by the writer's intentions. In each case, the process of interpretation involves the "resonant" stimulation of some somatic information channel of the reading or 'reading' subject, the use of some hermeneutical circle, and in both cases, when the hermeneutic task is accomplished, one is in direct possession of what the text or 'text' says, its meaning.

While literary texts may speak about the World, they do not manifest, show, or exhibit states of the World. A 'text,' however, can and usually does do this, insofar as it is usually about some state of the World actually present and manifesting itself. Returning, for the purpose of illustration, to the thermometer; the position of mercury on the scale functions as a 'text.' Through a 'reading' of this 'text,' one gains knowledge of the current thermodynamic temperature. The expression of that knowledge takes the form of a judgment, "The present ambient temperature is (say) 70°"; this judgment is empirical, direct, and uses scientific terms descriptively of the World. I claim that this 'reading' is a perceptual process, since it *fulfills all the characteristics of perceptual knowledge:* (1) it is direct (not mediated by inferences, nor is it just knowledge of an "internal representation" or "model" constructed, perhaps, out of sensations, or in some other, perhaps mathematical, way). (2) It depends on the physical causality exercised by the object on some somatic information channel, in this case, the technologically extended Body of the subject. (3) It is hermeneutical, that is, it acquires its meaning through the employment of a hermeneutical circle, where the terms of a scientific theory are used descriptively. (4) Its object, a state of the World, is experienced as given directly to the knower by the World. And (5) it terminates in a perceptual judgment of which the expression is a statement in which the terms of a scientific theory are used descriptively, and which purports to describe what is actually here and now existent in the World, present and manifest to the knower appropriately embodied.

Like all perceptual knowledge, a scientific observation is not apodictic in the natural attitude; even well-trained perceivers are not able to sample at will the perceptual profiles of the object. However, like all perceptual knowledge, to the extent that profiles and invariants can be clearly articulated, it too is capable of aspiring to apodicticity in the reflective attitude. It is this capacity to become apodictic that establishes the possibility of genuine perceptual scientific horizons in the World. This capability distinguishes the hermeneutics of literary texts from the hermeneutics of perception and scientific observation: while both are underdetermined, the former is about meanings as unexhib-

276 Toward a Philosophy of Science

ited, possibly unexhibitable, possibilities, and the latter is about exhibited possibilities, that is, states of the World.

Returning to the question of the ontological status of scientific entities: if 'reading' a thermometer is a perceptual process, then because of the ontological primacy of perception, thermodynamic temperature and other scientific entities like it enter the World as recognizable objects of definite kinds defined by scientific theories, and become part of the furniture of the earth.

A striking example of the use of instrumentation to generate a new field of perceptual knowledge is the Tactile Visual Substitution System (TVSS) of Collins and Bach-y-Rita. This is a device designed to provide a blind person with essentially visual information, but in tactile form.

Identity Theories

Perception implies systems of physical interaction between subject and object, controlled on the side of the subject by codings in appropriate neurophysiological networks. Such codings cannot explain perception without presupposing intentions in the subject or in the researcher. The somatic information channel, or the codings that control this channel, cannot be identified as systems (in some scientific image) without an appeal to the perceptual intention (in the appropriate manifest image), and as scientific systems they would have no significance for the psychologist or the epistemologist, unless they were known to code for particular elements in some appropriate manifest image. The lack of a single common reference for objects of the two complementary images makes reduction without remainder impossible.

Hermeneutics and the History of Science and Perception

Traditions of scientific inquiry have a continuity over time despite great changes in the theoretical models used to describe the subject matter. Continuity and development take place hermeneutically; that is, through a continuous reinterpretation of data that reveal, at first obscurely and later more clearly, with the aid of "readable technologies," the profiles of a common perceptual essence. Continuity then is in the identification and isolation of the same set of individual perceptual referents, for the purpose of inquiry into their perceptual essence.

Such continuity of reference is consistent with changes in the theoretical models used to describe these objects.

Traditions of scientific inquiry can die out. One reason is that the alleged objects of the inquiry turn out to have no real perceptual essence; another reason is that cultural interests change. In the latter case, unlike the former, the abandoned inquiry can be successfully revived at a later time by new communities of inquirers.

The hermeneutical quality of scientific observation explains the variety and tentativeness of scientific traditions, and is connected with the possibility of alternative scientific 'readings' that were the concern of Duhem and Poincaré and are currently topics of lively discussion in the Conventionalist debate.

World-Building

Euclidean Visual Space as a Cultural Artifact: A study of the invariance properties of Euclidean space, and of the physical embodiments of a Euclidean frame of reference, reveals how implausible it is to expect that localized perceivers like ourselves could come to experience visual Worlds as Euclidean; it is implausible that, unaided, we could become physical frames of reference for Euclidean Worlds. A more natural form for unaided vision, however, is hyperbolic. Setting a priorism aside, the fact that we can and do in fact see in a Euclidean way indicates that some technological assistance has been provided. This in fact we have provided for ourselves by making and dispersing widely in all inhabited space engineered forms that can serve as rigid measuring rods and paradigms of rigid Euclidean shapes; these, together with the sensory organs and neurophysiological system, linked functionally by a (virtually) instantaneous medium of communication, constitute the visual Euclidean frame of reference. From the way these artifacts relate to objects, we 'read' the Euclidean structure that is normative for physical space, and impose this 'reading' on what we see; visual space becomes in this way normatively Euclidean. Euclidean visual space then depends for its existence on the ''carpentering'' of our environment, and is, as a consequence, an artifact of our scientific (Cartesian) culture.

Plurality of Worlds or World Horizons: Euclidean vision and hyperbolic vision use different sets of visual cues in the surrounding optical array. Each set of cues is systematically associated with one of the two

Toward a Philosophy of Science

spatial geometries. The two modes of vision are incompatible—what is seen as a plane cannot be seen at the same time as a concave surface. They are, however, complementary, since we experience smooth transitions between the two.

That the cultural Worlds of antiquity presupposed a finite space with roughly hyperbolic properties can be argued from various sources: from the works of philosophers such as Parmenides, from the resistance shown to the atomist notion that whatever *is* is made up of atoms moving in an infinite Void, and from certain indications in the art that has survived. Cultural space no doubt contained (as evidenced by city planning) local Euclidean oases, but the shape of space as a cosmological whole was not Euclidean, but finite and spherical, as suggested by the experience of hyperbolic visual space. The Middle Ages borrowed its notions of finite space and a spherical cosmology from Aristotle. Dante's universe, like Aristotle's, also suggests the connection with hyperbolic visual space. The establishment of cultural space as infinite and Euclidean was the achievement of the European Renaissance. Hyperbolic visual space was abandoned as the source of norms for realistic description, and in its place, Euclidean norms, connected with the scientific image of classical science, became entrenched.

Classical science and the Cartesian image associated with it have since been shown to be partial and inadequate. In the light of this crisis of the scientific image, we should ask whether a privileged cultural position should continue to be given to Cartesian/Euclidean space in everyday life and language. Should we not instead choose to recognize, as something more than a source of visual illusions, the existence of alternative visual norms, such as the hyperbolic?

The inclusion of hyperbolic visual space within an enlarged realism would make a positive statement about the primacy of perception, and the centrality to perception of physical embodiment in the (material fabric of the) universe. It would state that persons as perceivers are genuinely part of reality, like wave lengths, entropy or (possibly) black holes. The mating of Euclidean and hyperbolic visual spaces is not as problematic as would appear at first sight, since they merge smoothly in ordinary vision; they are like the "stereopsis" of sensed and spectral color, of felt and thermodynamic heat, and of all pairs of manifest and scientific images. Such a "stereopsis" of realistic spaces would enlarge the scope of possible Worlds (or of the horizons of *one* great World?) by adding to our current Cartesian World, horizons deliberately related to the *in-der-Welt-Sein* character of human perceivers.

Such an enlargement of the notion of human realistic experience would also help us recover the cultural "myths" our ancestors lived by. To 'read' the artistic and architectural remains left by our ancestors, we need to be able to enter hermeneutically into their World. If our judgment about many ancient artifacts is correct, ancient Worlds were often dominated by the structure of hyperbolic visual space.

Such an enlargement of human experience would help us, moreover, to understand, to relate to, and even to construct "natural" environments to which we could respond, like our Greek and medieval ancestors, without the real or assumed need of a particular technological shell.

If this study has shown anything, however, it is that what is "natural" is not limited to unaided perception; it is as natural for us to structure our Worlds with the assistance of standard "readable" technological praxes among which are scientific instruments, as it is natural to structure our Worlds without any. But even in the latter case, we are not without a praxis and physical intermediaries that shape what we perceive. In this latter case, however, the intermediaries we use were fashioned not by human invention but by events in the biological, evolutionary past of our race. We are inescapably embodied. Neither science nor unaided perception give us anything other than a World-to-and-for-human-embodied-subjects. Science, however, offers the possibility of making manifest and present in perceptual experience objects that are beyond the ken of unaided experience, and that are among the antecedent and necessary conditions of possibility of all Worlds and of perception itself. With the assistance of these objects, new Worlds (or just new horizons of one great World?) can be created. Science can rid us of human perspectives only to the extent that it establishes the physical conditions and infrastructure necessary for all human perspectives. The reality of scientific entities is then only a part of a greater reality, that part that is necessary for Worlds to exist, and for their client perceivers to exist. The essence of a World is to-be-to-and-for-human-subjects, and the essence of being a human perceiver is to-be-in-a-World. The circularity is informative because it is hermeneutical.

Appendix
Hyperbolic Visual Map
of Physical Space

Introduction

A theory of non-Euclidean space was first proposed by Rudolf Luneburg (1947, 1948, 1950).[1] Subsequent experimental work by Blank (1953, 1958a, 1958b, 1958c, 1961), Hardy et al. (1953), Indow (1967, 1968, 1974, 1975), Indow, Inoue, and Matsushima (1962, 1963), Shipley and Williams (1968), Foley (1964, 1965, 1967a, 1967b, 1968, 1969, 1972, 1978, 1980) and others made it clear that if visual space is organized geometrically, and if the geometry belongs to a two-parameter family of Riemannian geometries, then the specific values of the parameters are not prewired independently of the stimulus configuration. Acting on a suggestion by Luneburg, work was done by Linksz (1952), Shipley and Williams (1968), Hardy et al. (1953), and Foley (1967a, 1967b) to determine the degree of dependency of the visual estimate of distance on the farthest point in the visual field. Some clear dependence was found in experiments involving points of light in a dark background. No systematic inquiry, however, was pursued into a hyperbolic visual model possibly dependent both on psychophysical features of the stimulus configuration and on cognitive factors in the field of vision. Foley (1972) summarized the status of research on the geometry of visual space by saying that the results "show that no one geometrical model can be appropriate to all stimulus situations, and suggest that the geometry may approach Euclidean with the introduction of cues to distance."[2] He also surmised that geometrical judgments in the environments in which the experiments were done (weak pointlike or rodlike light sources in an otherwise dark room), were not the product of a single coherent process, but the product of two independent processes. There has been little theoretical development of the hyperbolic space model since 1972. Foley (1978) summarized the scientific data considered to have been established. Suppes (1977) reviewed the Luneburg program from the point of view of its possible interest for the philosophy of perception.

I am about to propose a model of hyperbolic visual space—I shall call it the "hermeneutical Luneburg model"—in which the parameters of the space are assumed to be dependent on cognitive features of the visual field as well as on basic psychophysical variables.

Hermeneutical Luneburg Model: Configuration-Dependent Hyperbolic Space

All of the work referred to above is—in Indow's term[3]—"frameless": that is, the stimuli were pointlike (or rodlike) sources presented in a homogeneous field outside the context of the framework of the everyday experiential World. In contrast, the hermeneutical Luneburg model assumes that in any configuration of visual clues and stimuli sufficiently complex and stable to be perceived as a coherent geometrical structure, certain key configurations, "primary clues," exist in the field of vision. The relative geometrical relationships, such as size, distance, and orientation, as perceived enter into the constitution of the spatial background; in which and against which whatever happens, happens as a foreground event with geometry predetermined by its background. These primary clues play the role of boundary conditions. If a fixed geometry and a constant scale factor are to be maintained, then these primary clues must be kept constant. Only under such conditions will a perceiver see the things that surround him as having a "rational" form and observe events unfold in a "rational" way, that is, following the rules of a definite geometry.

In relation to the visual estimation of distance and depth, two psycho-physical processes are relevant: *stereopsis* and *accommodation*. The former uses binocular clues to depth and to distance, both functions, it is thought, of *binocular parallax*. The latter is a monocular process: it brings to focus on the retina the diverging pencil of light emanating from the object; this is done by an adjustment of the curvature of the crystalline lens. Both processes respond to some measure of the infinitesimal divergence of a pencil of rays, or of the infinitesimal parallax of a point relative to the visual observer. That both *monocular* and *binocular* spatial judgments seem to be related to the divergence of a pencil of rays (or some measure of parallax) suggests that there may be a single perceptual process fitted by a common model; however, it is clear from the work of Julesz (1971) and others that depth perception is markedly influenced by stereopsis. The appropriate psychophysical variable then for the visual estimation of depth is taken to be *some appropriate measure of the infinitesimal parallax of a point relative to some small base distance*. The magnitude of this base—whether it be the interpupillary distance or some other— does not in fact affect the geometrical transformation for small values of the parallax.

The model makes the following assumptions: that, the results of Foley (1972) notwithstanding, visual space is homogeneous and constitutes a constant curvature geometry; that the relevant psychophysical variable for visual distance is an infinitesimal parallax (not necessarily based on the interpupillary distance); and that the orientation (e.g., direction cosines) of any object relative to the observer is the same in the visual field as it is in physical space.

Following the work of Luneburg, formulae for the transformation of physical to visual space will be derived, in particular such as would be useful to compute the visual transforms of humanly engineered objects bounded by Euclidean lines and planes such as the walls of a room, a floor or ceiling, external architectural elements, the edges of a road or the lines of a railway, and so forth, which characterize the "carpentered" environments in which we live. As representative of a typical manmade object in physical space, a small reference tripod (P,QRS) is taken, composed, say, of rods, with three of its mutually orthogonal edges, PQ, PR, and PS, oriented in the directions of the x-, y-, and z-axes, respectively (see figure A.1). The reference tripod (P,QRS) can be taken as part of a small reference cube of which PQ, PR, and PS are three orthogonal edges. The following functions will then be derived for the transforms within hyperbolic visual space: the lengths of these edges, the orientation of the edges, and the tilt of the faces relative to the visual vertical and horizontal axes at the vertex of the tripod.

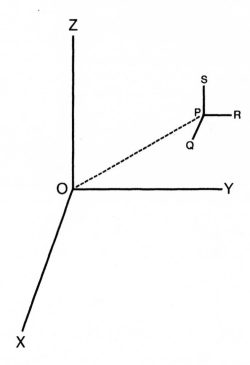

Figure A.1: Reference tripod in physical space.

Physical Space

It is assumed that visual space is topologically equivalent to physical space and that the coordinate labels used to identify and distinguish points (or objects) in physical space can also be used to identify and distinguish the same material points (or objects) in visual space. Points in physical space will be denoted by unprimed Roman letters, sometimes with subscripts. The addition of a prime to the designation of a point (or object) means that that point (or object) is being considered within the context of visual space. Furthermore, quantities designated by unprimed Roman letters refer to or are to be evaluated in physical space; quantities designated by Greek letters or by primed Roman letters refer to or are to be evaluated in visual space.

The origin of coordinates, O, in physical space is taken to be coincident with the midpoint between the observer's eyes. A right-handed tripod of Cartesian axes (x,y,z) has its origin at O, with z

vertical, y horizontal in the median plane, and x horizontal in the frontal plane. A point P can be located in this system by the use of bipolar coordinates (C,A,B) with foci at two points $L(-a,0,0)$ and $R(a,0,0)$ on the x-axis. In a binocular model, L and R would be the centers of rotation of the left and right eyes respectively, and $2a$ would be the interpupillary distance (see figures A.2 and A.3). The bipolar azimuth relative to L and R is A (positive to the observer's right); B is the angle of elevation of the plane containing the point P and the x-axis (positive for points above the horizon), and C is the angle of convergence (or parallax) of P relative to L and R.

Let $r_1 = PR$, $r_2 = PL$, and $2R = r_1 + r_2$

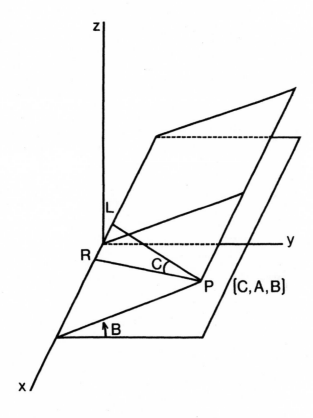

Figure A.2: Coordinates in physical space.

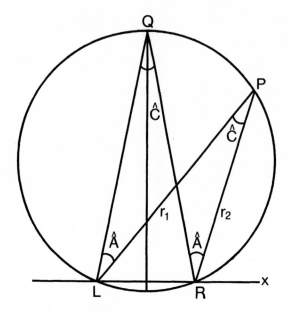

Figure A.3: Parallax (convergence) angle dC and bipolar azimuth A.

The relations between the Cartesian and bipolar coordinates are:

$x = a \sin 2A/\sin C$

$y = a \cos B (\cos 2A + \cos C)/\sin C$

$z = a \sin B (\cos 2A + \cos C)/\sin C$

$R = (r_1 + r_2)/2 = a \cos A/\sin C/2$

or for small values of C,

$C = (2a \cos A)/R$

Bipolar coordinates of the kind introduced here do not set up an orthogonal network in physical space. For small values of C, the metric expressed in terms of these coordinates is:

$$ds^2 = 4a^2 [\cos^2 A \cdot dC^2 + C^2 \cdot dA^2 + C^2 \cdot \cos^4 A \cdot dB^2$$
$$+ C \cdot \sin 2A \cdot dA \cdot dC]/C^4$$

The reference tripod (P,QRS) has its apex at $P(x,y,z)$ and has the following vertices: $P(x,y,z)$, $Q(x+dx,y,z)$ $R(x,y,+dy,z)$, $S(x,y,z+dz)$. In the (C,A,B) coordinate system, the coordinates of the apex and vertices are: $P(C,A,B)$, $Q(C+dC_x,\ A+dA_x,\ B+dB_x)$, $R(C+dC_y, A+dA_y,\ B+dB_y)$, $S(C+dC_z,\ A+dA_z,\ B+dB_z)$. The quantities dC_x, dA_x, etc. are given below:

Vertex Q, side $PQ = dx$,

$$dC_x = -C^2 \cdot \tan A \cdot dx/a \qquad (1)$$

$$dA_x = C \cdot dx/2a$$

$$dB_x = 0$$

Vertex R, side $PR = dy$,

$$dC_y = -C^2 \cdot \cos B \cdot (1-\tan^2 A) \cdot dy/2a \qquad (2)$$

$$dA_y = -C^2 \cdot \cos B \cdot \tan A \cdot dy/2a$$

$$dB_y = -C \cdot \sin B \cdot dy/2a \cos^2 A$$

Vertex S, side $PS = dz$,

$$dC_z = -C^2 \cdot \sin B \cdot (1-\tan^2 A) \cdot dz/2a \qquad (3)$$

$$dA_z = -C \cdot \sin B \cdot \tan A \cdot dz/2a$$

$$dB_z = C \cdot \cos B \cdot dz/2a \cos^2 A$$

Visual Space

Given the assumptions listed above about the structure of visual space, and assuming that visual space is hyperbolic rather than elliptic or flat, its metric has the form:

$$ds'^2 = \kappa^2(\sigma^2 d\gamma^2 + d\alpha^2 + \cos^2 \alpha d\beta^2)/\sinh^2 \sigma(\gamma + \tau)$$

where $\gamma = C/2a$, $\alpha = A$, $\beta = B$, κ is a scale factor, and σ and τ are parameters determining the curvatures (see figure A.4 for coordinates in visual space).[4] The scale factor κ is also the radius of curvature; the Gaussian curvature is κ^{-2}.

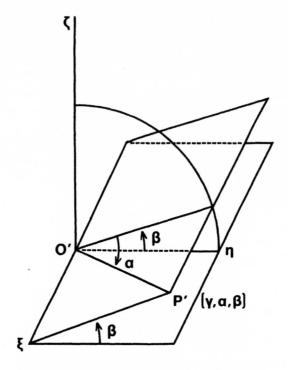

Figure A.4: Coordinates in visual space.

In the configuration-dependent model, κ, σ, and τ may all depend on certain features of the stimulus configuration and of the perceptual field. The precise character of the space is specified by filling at most two arbitrary geometrical conditions, and by giving the scale factor an appropriate value. The conditions that fix the values of these parameters are generally in this model relations between the space indicators mentioned above.[5] The indeterminateness of the scale factor κ adds another degree of freedom, though not a new structural one. The value of κ, however, would usually be chosen so that at the true point (see below), visual size and shape would be congruent with physical size and shape: under these conditions κ should be set equal to σ sinh $(1+\sigma\tau)$.

The reference tripod in visual space is (P',Q',R',S') and its vertices are: $P'(\gamma,\alpha,\beta)$, $Q'(\gamma + d\gamma_x, \alpha + d\alpha_x, \beta + d\beta_x)$, $R'(\gamma + d\gamma_y, \alpha + d\alpha_y, \beta + d\beta_y)$, $S'(\gamma + d\gamma_z, \alpha + d\alpha_z, \beta + d\beta_z)$. The quantities $d\gamma_x$, $d\alpha_x$, etc., are given by equations (3)−(5) with the substitutions $C \rightarrow 2a\gamma$, $A \rightarrow \alpha$, $B \rightarrow \beta$

Let

$$S' = \sinh \sigma(\gamma + \tau)$$

$$U' = (1 + 4\sigma^2 \gamma^2 \tan^2 \alpha)^{1/2}$$

$$V' = [1-\cos^2 \alpha\cos^2 \beta + \sigma^2\gamma^2 \cos^2 \alpha\cos^2 \beta (1-\tan^2 \alpha)^2]^{1/2}$$

$$W' = [1-\cos^2 \alpha\sin^2 \beta + \sigma^2\gamma^2 \cos^2 \alpha\sin^2 \beta (1-\tan^2 \alpha)^2]^{1/2}$$

Let

$P'Q' = dx'$, $P'R' = dy'$, and $P'S' = dz'$, and taking $PQ = dx$, $PR = dy$, and $PS = dz$, then

$$dx'/dx = P'Q'/PQ = \kappa\gamma U'/S' \tag{4}$$

$$dy'/dy = P'R'/PR = \kappa\gamma V'/S' \cos \alpha \tag{5}$$

$$dz'/dz = P'S'/PS = \kappa\gamma W'/S' \cos \alpha \tag{6}$$

and,

$$[dx'/dy'] \cdot [dy/dx] = [P'Q'/P'R'] \cdot [PR/PQ] =$$
$$[U' \cos \alpha]/V' \tag{7}$$

$$[dz'/dy'] \cdot [dy/dz] = [P'S'/P'R'] \cdot [PR/PS] = W'/V' \tag{8}$$

$$[dx'/dz'] \cdot [dz/dx] = [P'Q'/P'S'] \cdot [PS/PQ] =$$
$$[U' \cos \alpha]/W' \tag{9}$$

If the rods of the reference tripod are physically equal in length, that is, if $PQ = PR = PS$, then the visual ratios of the three rods to one another can be derived from the equations (7)–(9). Conversely, the same equations can be used to determine the physical ratios when the visual ratios are assigned, as for example, in size-depth-distance tests (see below).

The coordinates of Q', R', and S' can be expressed entirely in terms of visual quantities by substituting for dx, dy, and dz their values as functions of dx', dy' and dz' (equations [4]–[6]).

Vertex Q', side $P'Q' = dx'$:

$$d\gamma_x = -2 \gamma S' \tan \alpha dx'/\kappa U'$$

$$d\alpha_x = S' dx'/\kappa U'$$

$$d\beta_x = 0$$

Vertex R', side $P'R' = dy'$:

$$d\gamma_y = -\gamma S' \cos \beta \cos \alpha \ (1-\tan^2 \alpha) \ dy'/\kappa V'$$

$$d\alpha_y = -S' \cos \beta \sin \alpha dy'/\kappa V'$$

$$d\beta_y = -S' \sin \beta \ dy'/\kappa V' \cos \alpha$$

Vertex S', side $P'S' = dz'$:

$$d\gamma_z = -\gamma S' \sin \beta \cos \alpha \ (1-\tan^2 \alpha) \ dz'/\kappa W'$$

$$d\alpha_z = -S' \sin \beta \sin \alpha \ dz'/\kappa W'$$

$$d\beta_z = S' \cos \beta \ dz'/\kappa W' \cos \alpha$$

Radial Distance in Visual Space

The visual distance between two points P'_n (γ_n, α, β), the near point, and P'_f (γ_f, α, β), the far point, on the same radial geodesic (see figure A.8) is given by the equation:

$$P'_n P'_f = \int_{\gamma_n}^{\gamma_f} ds' = -\kappa \int_{\gamma_n}^{\gamma_f} \sigma d\gamma / \sinh \sigma(\gamma + \tau) \tag{10}$$

$$= \kappa \ln \tanh \ [\sigma \ (\gamma_n + \tau)/2] - \kappa \ln \tanh \ [\sigma \ (\gamma_f + \tau)/2]$$

The visual distance between the observer O' and P'_n is obtained by inserting for γ_n the value π, which is the value appropriate to O'. Putting tanh $[\sigma(\gamma_n + \tau)/2] = 1$, we get,

$$O'P'_f = -\kappa \ln \tanh \ [\sigma(\gamma_f + \tau)/2] \tag{11}$$

Letting P_f go to infinity in physical space, or γ_f tend to zero in visual space, we find that the radius of visual space is

$$-\kappa \ln \tanh \ \sigma\tau/2 \tag{12}$$

which remains finite as long as $\sigma\tau$ is not zero.

Hyperbolic visual spaces then are of two classes:

(1) *infinite spaces* with $\sigma\tau = 0$, or (since $\sigma > 0$) $\tau = 0$, and

(2) *finite spaces* with $\sigma\tau > 0$, or (since $\sigma > 0$) $\tau > 0$. In this case, the interior of the infinite Euclidean sphere is mapped on the interior of

a finite hyperbolic sphere, "the horizon sphere," with a radius given by equation (12).

Size-Depth Match

Equations (7)–(9) can be used to derive equations that relate the physical extent PR to a fixed vertical extent PS, when their visual transforms $P'S'$ and $P'R'$ appear to match in size (see figure A.5). Putting $P'S' = P'R'$ in equation (8), we have

$$PR = PS \cdot W'/V'$$

If the experiment matching size with depth is set up at the intersection of the median and xy-planes ($\alpha = \beta = 0$), then $V' = \sigma\gamma$, and $W' = 1$. Consequently, we have

$$PR = PS/\sigma\gamma = OP \cdot PS/\sigma \tag{13}$$

Equation (13) shows that it would be possible to determine σ from the slope of the curve PR vs. OP (with PS fixed). Size-depth matches (for small values of PS) are a particularly useful way of determining σ, since the experiments do not involve the variable parameter τ or the scale factor κ.

Equation (13) can be reexpressed in terms of the disparity between nearest and farthest points in the experiment. Let $P' = (\gamma_n, 0, 0)$ and $R' = (\gamma_f, 0, 0)$, then (13) is equivalent to

$$\gamma_f \cdot PS = \sigma \, (\gamma_n - \gamma_f) = \sigma \, \Gamma_{nf}$$

where Γ_{nf} is the disparity between P' and R'.

Contraction and Dilation of Space

It is of interest to consider the visual distortion of a physical cube.

Consider a small, flat, square physical object, such as a piece of stiff cardboard, located at $P(C, A, B)$ in physical space, and oriented in a plane perpendicular to the y-axis. Then, putting $PQ = PS$ in equation (9), we get:

$$dx'/dz' = P'Q'/P'S' = U' \cos \alpha/W'$$

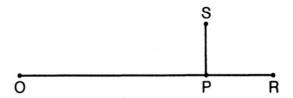

Figure A.5: Depth-size matches.

At the center of the field of vision ($\alpha = \beta = 0$), this ratio is unity, that is, $dx' = dz'$.

For very distant objects ($\gamma = 0$),

$$dx' > dz', \text{ if } |\sin \beta| > |\tan \alpha|$$

and

$$dx' < dz', \text{ if } |\sin \beta| < |\tan \alpha|$$

That is, there is a lateral contraction of objects in the visual field to right and left of the observer, and a vertical contraction above and below the observer.

For objects in the horizontal plane ($\beta = 0$),

$$dx' > dz', \text{ if } \sigma\gamma > 1/2,$$

and

$$dx' < dz', \text{ if } \sigma\gamma < 1/2$$

Consequently, there is a *near zone* defined by $\sigma\gamma > 1/2$, in which the field of perception in the plane of the horizontal is experienced as horizontally elongated, relative to the physical objects that are present in it. Outside of that zone, physical objects in the plane of the horizontal appear to be vertically elongated.

Consider a similar small square lying horizontally in the xy-plane. Then putting $PQ = PR$ in equation (7), we get

$$dy'/dx' = P'Q'/P'R' = V'/U' \cos \alpha$$

In the median plane ($\alpha = 0$),

$$dy'/dx' = [\sin^2 \beta + \sigma^2\gamma^2 \cos^2 \beta]^{1/2}$$

On the horizontal plane ($\beta = 0$), this becomes,

$$dy'/dx' = \sigma\gamma = \sigma/OP$$

Hence,

$$dy' > dx', \text{ if } \sigma > OP,$$

and

$$dy' < dx', \text{ if } \sigma < OP$$

Moreover, dy'/dx' is inversely proportional to the physical distance OP of the object from the observer. This last theoretically derived result agrees with a set of experiments performed by Gogel (1960) in which subjects were asked to adjust the depth component of a device with a fixed width but variable depth to make the depth appear equal to the width. In this case, putting $dy' = dx'$, in equation (7), and holding PQ constant, we get

$$PR = \sigma \cdot PQ/OP \tag{14}$$

Equations (13) and (14) imply that there is a definite point directly in front of every observer in the neighborhood of which small physical and visual shapes are similar. This point can be called the "true point" for the space. With an appropriate choice of scale factor κ, the physical and visual shapes can become congruent. Letting the distance, OP, from the observer to the true point be d, then the value of σ equals the value of d, that is

$$\sigma = d$$

The natural unit of visual length in this space is the radius of curvature of the space; this equals the scale factor κ. The assignment of a numerical value to κ depends on the type of coordination (or partial matching) that exists (or is brought about) between physical and visual space. For example, if the region of the true point is occupied by a small, visually significant physical object whose *shape and size* are preserved by the visual transformation, then the visual scale can be determined by its partial congruence with the physical scale at the true point. This gives the following relation among κ, σ, and τ:

$$\kappa = \sigma \sinh (1 + \sigma\tau)$$

The visual unit of length, say, the "visual centimeter," is now determined as that unit of length which is congruent with a physical centimeter located at the true point.

If, however, the location of the true point is not occupied, or is not occupied by a visually significant object, coordination with the physical scale may still be achieved in some other way, say, by matching frontal size at a distance, or average size in a domain. It is, of course, conceivable that no effective coordination relative to size occurs because of the peculiar character of the visual field. In all such cases, it may be more appropriate to approach the model as if κ, the scale factor, were one of the free parameters, the second free parameter being the dimensionless product $\sigma\tau = \mu$. The relations among κ, μ, σ and τ are as follows:

$$\kappa\tau = \mu\sinh (1 + \mu)$$

and   (15)

$$\sigma = \kappa/\sinh (1 + \mu)$$

Since size and shape coordination between physical and visual space is only partial, it is of interest to consider how certain extensions appear to be contracted or dilated in visual space relative to their physical dimensions.

Equations (4)–(6) give the relationships between visual and physical dimensions in the distant zone. Depth, as we have seen, vanishes. Whether the size of the facing side of a physical cube vanishes at the horizon sphere or tends to a finite limit depends on whether the space is finite ($\tau > 0$) or infinite ($\tau = 0$). For infinite spaces, at great distances ($\gamma \rightarrow 0$),

$$dx'/dx = \kappa/\sigma$$
$$dz'/dz = [\kappa/\sigma] \cdot J$$

where

$$J = [1 - \cos^2\alpha \; \sin^2\beta]/\cos \alpha$$

The shape of a small square of sides dx and dz, with $dx = dz$, in the xz-plane is a small rectangle of sides dx' and dz', where

$$dz'/dx' = J$$

This shape is vertically elongated, $dz' > dx'$, to the right and left of the observer, that is, in that part of the field where $|\tan \alpha| > |\sin \beta|$. It is elongated horizontally, i.e., $dx' > dz'$, above and below the observer, that is, in that part of the field where $|\tan \alpha| < |\sin \beta|$.

For finite spaces, at great distances ($\gamma \rightarrow 0$),

$$dx'/dx = \gamma \cdot [\kappa/\sigma\tau]$$
$$dz'/dz = \gamma \cdot [\kappa/\sigma\tau] \cdot J$$

and both these ratios tend to zero with γ. Thus, physical objects, as they move away from the observer, will appear to disappear down a funnel of space that contracts to a point on the horizon sphere.

Orientation of Sides of Reference Tripod Relative to Visual Horizontal and Vertical

In visual space, the side, $P'S'$, of the reference tripod will not in general be aligned parallel to the visual vertical axis (ζ-axis) at the location, P', of the tripod, nor will $P'Q'$, nor $P'R'$, be parallel to the two horizontal visual axes (ξ-axis in the visual frontal plane at P', and η-axis in the visual plane parallel to the median plane at P') at that point. To calculate the direction cosines in visual space of the sides of the tripod, we take the scalar product under the visual metric of unit vectors at P' in the directions of $P'Q'$, $P'R'$ and $P'S'$ and unit vectors along the three visual axes at P' (see figure A.6). The elementary triad of vectors along the three local visual axes at P' are given below:

$$\bar{\xi}d\xi:\ d\gamma_{\xi} = -S' \sin \alpha \cdot d\xi/\kappa\sigma \qquad (16)$$
$$d\alpha_{\xi} = S' \cos \alpha \cdot d\xi/\kappa$$
$$d\beta_{\xi} = 0$$
$$\bar{\eta}d\eta:\ d\gamma_{\eta} = -S' \cos \alpha \cos \beta \cdot d\eta/\kappa\sigma$$
$$d\alpha_{\eta} = -S' \sin \alpha \cos \beta \cdot d\eta/\kappa$$
$$d\beta_{\eta} = -S' \sin \alpha \cdot d\eta/\kappa \cos \alpha$$

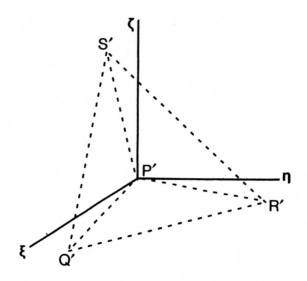

Figure A.6: Reference tripod in visual space.

$\bar{\zeta}d\zeta$: $d\gamma_\zeta = -S' \cos \alpha \sin \beta \cdot d\zeta/\kappa\sigma$

$d\alpha_\zeta = -S' \sin \alpha \sin \beta \cdot d\zeta/\kappa$

$d\beta_\zeta = S' \cos \alpha \cdot d\zeta/\kappa \cos \alpha$

Let $x'\hat{\xi}$, $x'\hat{\eta}$ and $x'\hat{\zeta}$ be the angles made by $P'Q'$ ($=dx'$) with the visual vertical and horizontal axes, $\bar{\xi}$, $\bar{\eta}$ and $\bar{\zeta}$ at P'; $y'\hat{\zeta}$, $y'\hat{\eta}$ and $y'\hat{\zeta}$, the angles made by $P'R'$ ($=dy'$) with $\bar{\xi}$, $\bar{\eta}$ and $\bar{\zeta}$ at P'; and $z'\hat{\xi}$, $z'\hat{\eta}$ and $z'\hat{\zeta}$, the angles made by $P'S'$ ($=dz'$) with $\bar{\xi}$, $\bar{\eta}$ and $\bar{\zeta}$ at P'. Then:

$$\cos x'\hat{\xi} = \cos \alpha \, (1 + 2 \, \alpha\gamma \tan^2 \alpha)/U' \qquad (17)$$

$$\cos x'\hat{\eta} = \sin \alpha \cos \beta \, (2 \, \sigma\gamma - 1)/U'$$

$$\cos x'\hat{\zeta} = \sin \alpha \sin \beta \, (2 \, \sigma\gamma - 1)/U'$$

$$\cos y'\hat{\xi} = -\sin \alpha \cos \alpha \cos \beta \, (1 - \sigma\gamma + \sigma\gamma \tan^2 \alpha)/V' \qquad (18)$$

$$\cos y'\hat{\eta} = [1 - \cos^2 \alpha \cos^2 \beta + \sigma\gamma \cos^2 \alpha \cos^2 \beta \, (1 - \tan^2 \alpha)]/V'$$

$$\cos y'\hat{\zeta} = -\cos^2 \alpha \sin \beta \cos \beta \, (1 - \sigma\gamma + \sigma\gamma \tan^2 \alpha)/V'$$

$$\cos z'\hat{\xi} = -\sin \alpha \cos \alpha \sin \beta \, (1 - \sigma\gamma + \sigma\gamma \tan^2 \alpha)/W' \qquad (19)$$

$$\cos z'\hat{\eta} = -\cos^2 \alpha \sin \beta \cos \beta \, (1 - \sigma\gamma + \sigma\gamma \tan^2 \alpha)/W'$$

$$\cos z'\hat{\zeta} = [1 - \cos^2 \alpha \sin^2 \beta + \sigma\gamma \cos^2 \alpha \sin^2 \beta \, (1 - \tan^2 \alpha)]/W'$$

A unit vector $\bar{\rho}$ in the direction of the radial geodesic has components $(S'/\kappa\sigma, 0, 0)$. The direction cosines of $P'Q'$, $P'R'$, and $P'S'$ relative to $\bar{\rho}$ are:

$$\cos x'\hat{\rho} = -2\ \sigma\gamma \tan \alpha/U'$$

$$\cos y'\hat{\rho} = -\sigma\gamma \cos \alpha \cos \cdot \beta\ (1 - \tan^2 \alpha)/V'$$

$$\cos z'\hat{\rho} = -\sigma\gamma \cos \alpha \sin \beta\ (1 - \tan^2 \alpha)/W'$$

The angles between the rods of the reference tripod in visual space are given by the following equations:

$$\cos x'\hat{y}' = -\sin \alpha \cos \beta\ [1 - 2\sigma^2\gamma^2\ (1 - \tan^2 \alpha)]/U'V'$$

$$\cos y'\hat{z}' = \cos^2 \alpha \sin \beta \cos \beta\ [1 - \sigma^2\gamma^2\ (1 - \tan^2 \alpha)^2]/V'W'$$

$$\cos z'\hat{x}' = -\sin \alpha \sin \beta\ [1 - 2\ \sigma^2\gamma^2\ (1 - \tan^2 \alpha)]/W'U'$$

When P is distant from the observer in physical space, that is, when $\gamma \to 0$, the three rods, $P'Q'$, $P'R'$, and $P'S'$ appear to be coplanar and orthogonal to the radial vector $\bar{\rho}$, and the angles between them are given by:

$$\tan x'\hat{y}' = -\tan \beta/\sin \alpha$$

$$\tan y'\hat{z}' = -\tan \alpha/\cos \alpha \sin \beta \cos \beta$$

$$\tan z'\hat{x}' = -1/\sin \alpha \tan \beta$$

while,

$$\cos z'\hat{\xi} = -\sin \alpha \cos \alpha \sin \beta/[1 - \cos^2 \alpha \sin^2 \beta]^{1/2}$$

$$\cos z'\hat{\eta} = -\cos^2 \alpha \sin \beta \cos \beta/[1 - \cos^2 \alpha \sin^2 \beta]^{1/2}$$

$$\cos z'\zeta = [1 - \cos^2 \alpha \sin^2 \beta]^{1/2}$$

Consider a small disk of physical diameter $2q$, orthogonal to the radius vector at P and in the median plane, $\alpha = 0$. Its vertical and horizontal axes will be defined by the infinitesimal vectors $(0, 0, q/R)$ and $(0, q/[R \cos \beta], O)$ respectively. The disk will appear in visual space as an oval with axes of the following magnitudes:

$$\text{length of vertical axis} = 2\kappa q \cos \alpha/R\ S'$$

$$\text{length of horizontal axis} = 2\kappa q/[R\ S' \cos \beta]$$

Hence, the ratio of the vertical axis to the horizontal axis is

$$\cos \alpha \, \cos \beta$$

which is always $<$ or $= 1$.

Consider the variation of the function $[2 \, \kappa q/R \, S']$, which is the diameter of the disk when it is viewed directly ($\alpha = \beta = 0$). Consider the following two cases.

(1) The disk subtends a fixed angle, θ, with the eye. $\theta \, (= 2q/R)$ is a constant. For objects at a very great distance, such as the moon, the following approximately hold: $\gamma = 0$ and $S' = \mu \, (= \sigma \tau)$. Let Δ be the visual diameter of the moon, then

$$\Delta = \kappa \theta / \mu \qquad\qquad\qquad (20a)$$

The moon, as we know, appears sometimes larger, sometimes smaller, depending on its position above the horizon: it is largest and appears closest when just above the horizon, and smallest and farthest away when it is at its zenith. This set of anomalous phenomena, inconsistent with a Euclidean visual space, can be reconciled with the family of hyperbolic visual space. Let the subscript "1" refer to the visual space of the horizon moon, and the subscript "2" to the visual space of the zenith moon. Then if ω is the "diameter" of visual space, the anomalous phenomena of the moon illusion comprise (a) $\Delta_1 > \Delta_2$, and (b) $\omega_1 < \omega_2$.

Equations 12 and 20a then lead to the following inequalities:

$$-\kappa_1 \ln (\mu_1/2) < -\kappa_2 \ln (\mu_2/2)$$
$$\kappa_1 \, \theta/\mu_1 > \kappa_2 \, \theta/\mu_2$$

and these are satisfied only if the following also hold:

$$\kappa_1/\mu_1 > \kappa_2/\mu_2$$
$$\kappa_2 > \kappa_1$$
$$\mu_2 > \mu_1$$
$$\sigma_2 > \sigma_1$$

See chapter 4 for further consideration of the moon illusion.

(2) The distant object has a fixed physical size: then the visual diameter of the disk $(2q')$ is,

$$2q' = [\kappa/\sigma] \, [1 - (e^{R'/\kappa}) \cdot \mu/2] \, 2q \qquad (20b)$$

This equation shows the character of the dependency of the visual diameter $(2q')$ on estimated visual distance (R'). If the space is an infinite space $(\mu=0)$, then the diameter of the disk tends to a fixed size,

$$[\kappa/\sigma] \, 2q.$$

If the space is finite, then it should be noted that the factor $[1 - (e^{R'/\kappa}) \cdot \mu/2]$ is a decreasing function of R', and approaches 0 as R' tends to its limit on the horizon sphere. This relation $(20b)$ is contrary to the size-distance constancy law operative in Euclidean space. We may take it to be the *size-distance* relation characteristic of hyperbolic visual space.

Discussion

It is of interest to study the way physical horizontals and verticals appear to bend in visual space. Equations $(17)-(19)$ describe these curvatures. Two qualitatively different zones can be distinguished:

(1) near zone: $\sigma\gamma > 1/2$, which has two regions, $\qquad\qquad$ (21)

$\quad$ (a) $1 < \sigma\gamma \, (1 - \tan^2 \alpha)$, the near near zone, and

$\quad$ (b) $1 > \sigma\gamma \, (1 - \tan^2 \alpha)$

(2) distant zone: $\sigma\gamma < 1/2$ $\qquad\qquad\qquad\qquad\qquad\qquad$ (22)

Each zone has mirror left-right symmetry relative to the median plane and mirror up-down symmetry relative to the horizontal plane of the observer. Hence, it is sufficient to describe the characteristics of the first quadrant (α >or=0, β >or=0): the characteristics of the other quadrants follow by symmetry from these. Let the term "x-line" stand for a physical line with an equation of the form, $y = $ constant, $z = $ constant: then y-lines and z-lines can be defined in similar ways.

In the near zone, region (a), as they approach the median plane, x-lines appear to bend inward at the center toward the observer by a combination of rotations in horizontal and vertical planes. z-lines will also appear to bend inward at the center toward the observer alone by a combination of rotations in vertical planes. This combination of curvatures of the x- and z-lines makes a vertical frontal surface (an xz-plane) appear to swell in convex fashion, "like a shield," to use Panofsky's phrase, as one approaches it. Moreover, y-lines will appear to diverge as they move away from the observer, suggesting that they be represented in a drawing by a system of reverse perspective.

In the near zone, region (*b*), x-lines behave as in region (*a*), but z-lines and y-lines change their orientations. An xz-surface, then, would appear with a saddlelike swelling facing the observer, falling away to left and right, but bending concavely from top to bottom; y-lines will appear to converge as they move away from the observer. Region (*b*) goes from a depth of σ to a depth of 2σ in the observer's median plane.

In the distant zone, all lines bend toward the plane normal to the radial vector: consequently, all become concave relative to the observer.

The contrast between the near near zone (region *a*) and the distant zone is in some respects like the contrast between the visual images seen in photographs taken with short- and long-focus lens. With a short-focus lens, objects seem to be pictured in a distended space; with a long-focus lens, objects seem to be pictured in a foreshortened and flattened space.

Tilting of Planes in Visual Space

The plane horizontal face *PQR* of the reference cube, when not at the center of vision, is generally seen as tilted in visual space about an axis in the visual $\xi\eta$-plane. Let ϕ_3 be the angle of tilt, and θ_3 the angle that the axis of rotation of the tilt makes with the positive direction of the visual ξ-axis (see figure A.5). These angles are given by the following equations:

$$\tan \theta_3 = \frac{\cos(x'\hat{\eta}) \cos(y'\hat{\zeta}) - \cos(y'\hat{\eta}) \cos(x'\hat{\zeta})}{\cos(x'\hat{\xi}) \cos(y'\hat{\zeta}) - \cos(y'\hat{\xi}) \cos(x'\hat{\zeta})} \qquad (23)$$

$$\tan \phi_3 = \frac{\cos(x'\hat{\zeta}) \sqrt{[\cos^2 (x'\hat{\zeta}) + \cos^2 (y'\hat{\zeta})]}}{\cos(x'\hat{\zeta}) \cos(x'\hat{\xi}) + \cos(y'\hat{\eta}) \cos(y'\hat{\zeta})}$$

To a first approximation,

$$\tan \theta_3 = -\cos(x'\hat{\zeta})/\cos(y'\hat{\zeta}) \tag{24}$$

$$\tan \phi_3 = \sqrt{[\cos^2(x'\hat{\zeta}) + \cos^2(y'\hat{\zeta})]}$$

The other plane faces, *PRS* and *PQS*, of the reference cube will also appear to be tilted in visual space. *PRS* will appear to have been rotated through an angle of ϕ_1 about an axis making an angle of θ_1 with the η-axis, and *PQS* will appear to have been rotated through an angle of ϕ_2 about an axis making an angle of θ_2 with the ζ-axis. The equations for θ_1, ϕ_1 and θ_2, ϕ_2 are obtained from equations (19) and (21) by permuting the variables (xyz) and $(\xi\eta\zeta)$ cyclically.

Discussion

It is of interest to consider qualitatively how physical planes will appear to bend in visual space. In the distant zone, horizontal xy-planes below the horizon curve upward and inward: an observer at the origin would appear to be at the center of a bowl looking down. Above the horizon, a horizontal plane will curve downward and inward: an observer at the origin would appear to be at the center of an inverted bowl looking up.

In the near near zone (region *a*), the opposite phenomenon is experienced: an observer at the origin would appear to be looking at the outside of an inverted bowl looking down, or at the outside of an upright bowl, looking up.

Vertical planes, whether frontal zx-planes or lateral zy-planes, behave analogously: they appear as concave in the distant zone and as convex in the near near zone.

Perspectival Projections in Visual Space

In Euclidean space, distant objects are projected in perspective on an intervening plane by considering that plane, the plane of projection, to be transparent to rays of light, and by marking on the surface of that plane the points where light rays from the points on an object intersect that plane on their way to the "sight point." Such a procedure, when converted into a mathematical technique, generates a system of perspective. A drawing in perspective then provides a set of two-dimensional cues on a plane surface which are generally sufficient to enable an observer to see the illusion of the three-dimensional object that is pictured by the drawing. The system so described is the one that pictures an object in three-dimensional Euclidean space. Is there an analogous system of perspective for visual space, that is, is there a different system of constructing two-dimensional cues on a visually flat surface so as to picture a scene construed in hyperbolic space?

Consider a small vertical rectangular area $d\xi_2 d\zeta_2$ located at P_2' and in a plane frontal to the observer, and let this area be projected on a visually vertical plane intermediate between the rectangle and the observer. Let P_1' be the projection of P_2'; that is, if $P_2' = (\gamma_2, \alpha_2, \beta_2)$ and $P_1' = (\gamma_1, \alpha_1, \beta_1)$, then $\alpha_1 = \alpha_2$ and $\beta_1 = \beta_2$. This rule of projection is the same as that used in Euclidean space. Whether or not it is also the appropriate one to follow in visual space may be argued, but it follows naturally from the initial assumption that the orientation of a point in visual space agrees with its orientation in physical space (see figure A.7).

If $d\xi_1$ is a projection of $d\xi_2$, then according to the rule of projection given above, $d\alpha_{\xi_1} = d\alpha_{\xi_2}$ and $d\beta_{\xi_1} = d\beta_{\xi_2}$. Using equation (16), we derive the following conditions:

$$d\zeta_1 = S_2' \, d\zeta_2/S_2'$$
$$d\zeta_1 = S_2' \, d\zeta_2/S_1'$$

Let $d\mu_1$, a small vector in the plane of projection, represent the projection of $d\eta_2$, a small element at P_2' orthogonal to the $\xi_2\zeta_2$-plane.

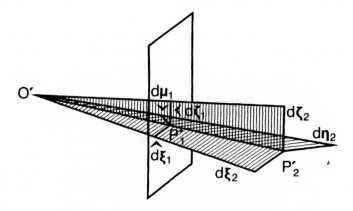

Figure A.7: Projection in visual space.

Then,

$$d\mu_1 = S'_2 [1 + \cos^2 \alpha \cos^2 \beta - \cos^4 \alpha \cos^4 \beta]^{1/2} d\eta_2/S'_1 \cos \alpha \cos \beta$$

and $d\mu_1$ makes an angle θ with the ζ-direction given by

$$\tan \theta = \sin \beta/\tan \alpha$$

Since, as expected, this rule of projection results in the tracing of identical projected segments on any plane of projection, then cues to a hyperbolic interpretation would have to be provided by the fine structure of the drawing, by shading and texture changes, perhaps by subtle refinements of the perspective lines, and by other indications that the viewer is to read the drawing in a different way.

To study how the size of the projected form varies, note that

$$d\xi_1 = \sinh \sigma(\gamma_2 + \dot{\tau}) \, d\xi_2/\sinh \sigma \, (\gamma_1 + \tau)$$

where γ_2 is the only variable. As P_2 moves farther from the observer in physical space, one would expect $d\xi_1$ to get smaller and approach zero. This does not happen unless $\tau = 0$. Consequently, in a wide open configuration with objects at all distances to the horizon, τ must be virtually zero. Luneburg has shown that the condition that the projection vanish for very distant objects cannot be satisfied by elliptic or flat Riemannian spaces. Visual space must then be hyperbolic for such open configurations. If the configuration, however, is closed and of finite dimensions, τ could conceivably be greater than zero, as it has been found to be in the analysis of some experimental data, but these latter possibilities are not entertained in this paper.

Depth-Distance Ratios in Visual Space

Depth-distance ratios in visual space are derivable from equations (10)−(11): they have the following form:

$$\frac{P'_n P'_f}{O' P'_n} = \frac{\text{visual depth}}{\text{visual distance}}$$

$$= \frac{\ln \tanh \sigma(\gamma_f + \tau)/2 - \ln \tanh \sigma(\gamma_n + \tau)/2}{\ln \tanh \sigma(\gamma_n + \tau)/2} \tag{25}$$

Case 1: P_n and P_f in the median plane ($\alpha = 0$), $OP_n \gg P_nP_f$ and $\tau = 0$. Then,

$$P'_nP'_f/O'P'_n = [-1/\ln(\sigma\gamma_n/2)] \cdot P_nP_f/OP_n \tag{26}$$

Equation (26) gives the relation between the ratio of the visual quantities and the ratio of the physical quantities, provided $OP_n \gg P_nP_f$. Where this latter condition is not fulfilled, as in depth-distance matches, equation (25) must be used.

Case 2: Depth-distance matches. P_n and P_f are adjusted until P_nP_f is judged to be a given multiple or fraction, say m, of $O'P'_n$. Typical values given to m are, ¼, ½, 1, 2, 3, etc. Let $P'_nP'_f/O'P'_n = m$. Then,

$$(m + 1) \ln \tanh \sigma(\gamma_n + \tau)/2 = \ln \tanh \sigma(\gamma_f + \tau)/2 \tag{27}$$

Case 3: $m = 1$. P'_n is adjusted so that it bisects the distance between O' and P'_f; then $m = 1$, and equation (27) reduces to

$$[\tanh \sigma(\gamma_n + \tau)/2]^2 = \tanh \sigma(\gamma_f + \tau)/2 \tag{28}$$

or what is equivalent,

$$\cosh \sigma(\gamma_n + \tau) = e^{\sigma(\gamma_f + \tau)}$$

For small values of γ_n and γ_f, (28) reduces to

$$\gamma_n - 2\,\gamma_f/[1 - e^{-2\sigma\tau}] - 1/\sigma = 0$$

and in addition for small values of $\sigma\tau$, we have,

$$\sigma\gamma_n - \gamma_f/\tau - 1 = 0$$

If a series of depth-distance matching experiments are done, and if γ_n is plotted against γ_f for small values of both, then the intercept on the γ_n–axis is $1/\sigma$, and the slope of the line is $1/\sigma\tau$.

$$O \qquad\qquad P_n \qquad\qquad P_f$$

Figure A.8: Depth-distance measurements.

For such a series of experiments, however, the conditions must be such that σ and τ are constant throughout the series. The visual environment, then, must not be dominated by the variable elements, P'_n and P'_f; conversely, visual space must have a definite structure independently of the positions of P'_n and P'_f. The depth-distance matches performed by Foley (1967a, 1967b) probably did not fulfill this condition. It was possible, however, to fit the data by taking τ to be a linear function of the form ($a\ \gamma_n + b\ \gamma_f + c$), with $a = 0$, although the significance of this is not clear.

Illustration of Visual Shapes

Cross-section profiles have been calculated for various physical configurations:

(1) two parallel lines orthogonal to the frontal plane, 100 centimeters to right and left of the observer. The visual coordinates have been calculated for the eye level plane, $z = 0$, and in a plane 150 centimeters below eye level: the latter would simulate the viewing of railway tracks. Tables 1, 2, and 3 list the appropriate visual coordinates in units of "visual centimeters"; a visual centimeter being a unit of length in visual space which is equal to a physical centimeter at the true point. Values have been calculated for three models: (i) $\sigma = 100$ cms, $\tau = .00154$ cm^{-1}, (ii) $\sigma = 100$ cms, $\tau = .000154$ cm^{-1}, and (iii) $\sigma = 100$ cms, $\tau = .00000154$ cm^{-1}. The values for the eye level plane are mapped in figure A.9: subscript "1" refers to model (i), subscript "2" to model (ii), and subscript "3" to model (iii).

(2) Tables 1, 2, and 3 also list the coordinate values in visual space for the profile of the floor or ground plane, 150 centimeters below eye level, in the median plane of the observer. Figure A.10 maps this profile for the same three space models.

(3) Tables 4, 5, and 6 list computed values for the points of intersection of a physical grid or checkerboard pattern of lines in a frontal plane directly in front of the observer. The perimeter of the grid pattern in each of the following cases constitutes a visual cone of half-angle 45°, that is, the diameter of the circular pattern subtends an angle of 90° with the sight point.

In table 4, the grid is at a distance of d/2 or 50 centimeters from the observer, and the lines of the grid are 10 centimeters apart. The hyperbolic space has $\sigma = 100$ cms and $\tau = 0$.

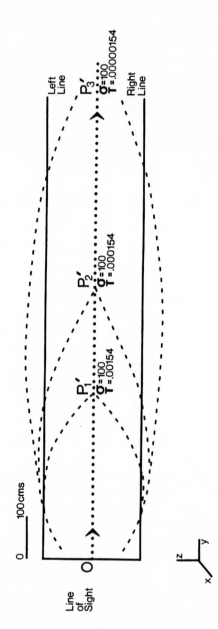

Figure A.9: Visual shapes of two parallel orthogonals to a frontal plane, 100 cms to left and right of the observer in the eye level plane, $z = 0$. Subscript "1" refers to model (i) with $\sigma = 100$ cms, $\tau = .00154$; subscript "2," to model (ii) with $\sigma = 100$ cms, $\tau = .000154$, and subscript "3," to model (iii) with $\sigma = 100$ cms, $\tau = .00000154$.

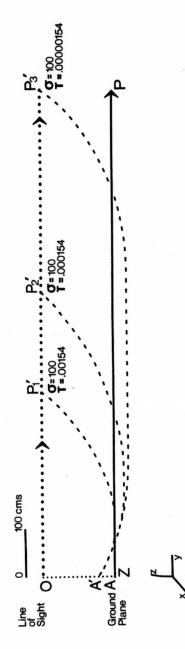

Figure A.10: Visual shape of the profile of the floor or ground plane, 150 cms below eye level, as calculated for the three models (i), (ii), and (iii) referred to in the text.

In table 5, the grid of lines is at a distance of $y=d=100$ centimeters from the observer, and the spacing of the lines is 20 centimeters. The plane of the grid passes through the true point, and $\tau=0$.

In table 6, the grid is at a distance of 10d or 1000 centimeters from the observer, and the spacing of the grid is 200 centimeters. The calculation is made for two hyperbolic spaces, $\tau=0$ and $\tau=.00154$. The very large difference between the two models in the distant zone should be noted: there is only a small difference between the models in the near zone.

The visual profiles of the checkerboard pattern have been mapped on a small scale in figure 4.7. *Caution*: Figure 4.7 is too small to enable a viewer to verify by inspection the visual effect of the hyperbolic transforms. Each pattern represents a field of vision which subtends 90° with the sight point. For diagrams of the size printed in this book, the correct sight point would be at a distance from the frontal plane of the paper of less than 5 centimeters—and the theory we are proposing probably does not hold at such a close range.

(4) Tables 7 to 11 show the points of intersection of a series of grids subtending at the periphery a half-angle of 11° with the sight point. The grids are placed at physical distances of 50, 100, 200, 300, 400, and 1000 centimeters from the observer; the true point is at a distance of 100 centimeters.

(5) Table 12 gives the visual linear dimensions of objects (in the xz-plane) that subtend a constant angle with the viewer (0.01 rads), but are situated at different distances ($y=50$, 100 cms, etc.). These values are represented in figure 4.1.

Illustration: Determination of σ, τ, and κ

Consider the problem of deriving the parameters for the visual space of a closed room. Premeasurement estimates can be made on the basis of various assumptions. The following assumptions suggest themselves as at least plausible: (1) the interior of the room as seen at any moment falls entirely within the near zone of the visual space; (2) objects of equal size and shape in different locations within a visual cone of half-angle 15° are also perceived as being roughly equal in size. In addition, the theory prescribes that at the true point, the measure of visual size is equal to the measure of physical size.

If the distance to the far wall (considered frontally) is d, then assumption (1) together with equations (21)−(22) lead to the conclusion that $d/2 <$ or $= \sigma <$ or $= d$. Taking $\sigma = d$, then a value for κ can be estimated from assumption (2) by sampling a variety of locations in the room within the visual cone, looking for a value for τ which gives the least spread of visual sizes relative to a fixed physical size. Equations (4)−(6) give the visual dimensions of an object in terms of its physical dimensions. From the values of σ and τ, the value of κ can be calculated.

TABLE 1*

$y=$	$x=0$ $z=0$	$x=100$ $z=0$	$x=0$ $z=-150$	$x=100$ $z=-150$
	$y'=$	$x'=$	$z'=$	$x'=$
50	33	67	−136	104
60		79		104
70		87		104
80		93		103
90		97		103
100	93	100	−143	102
200	165	101	−144	98
300	204	94	−136	90
400	230	86	−126	83
500	249	79	−117	77
600	263	73	−108	71
700	273	68	−101	67
800	282	63	−94	63
900	289	60	−88	58
1000	295	56	−83	55
2000	326	35	−52	
Inf.	366	0	0	

*$\sigma = 100$ cms, $\tau = 0.00154$ cm^{-1}, $\kappa = 142.7$ cms, d = 100 cms.
(x', y', z') are coordinates, analogous to Cartesian coordinates, in visual space

TABLE 2*

y=	x=0 z=0	x=100 z=0	x=0 z=−150	x=100 z=−150
	y'	x'	z'	x'
50	32	65	−135	108
100	91	100	−145	108
200	165	111	−159	112
300	211	112	−164	112
400	243	112	−165	112
500	268	110	−164	110
600	288	109	−162	109
700	304	108	−161	108
800	319	106	−159	106
900	331	105	−157	105
1000	342	104	−155	104
2000	410	92	−137	
3000	446	82	−123	
4000	468	74	−111	
Inf.	584	0	0	

$*\sigma = 100$ cms, $\tau = 0.000154$ cms^{-1}, $\kappa = 120$ cms.

TABLE 3*

y=	x=0 z=0	x=100 z=0	x=0 z=-150	x=100 z=-150
	y'	x'	z'	x'
50	32	64	-135	108
100	91	100	-145	112
200	165	112	-162	116
300	212	116	-169	116
400	245	116	-172	116
500	271	116	-173	116
600	292	116	-174	116
700	310	116	-175	116
800	326	116	-175	116
900	340	116	-175	116
1000	352	116	-175	116
2000	433	116	-175	
5000	540	116	-175	
10000	621	116	-174	
50000	803	108	-164	
100000	876	100	-153	
1000000	1054	46	-48	
Inf.	1113	0	0	

*$\sigma = 100$ cms, $\tau = 0.00000154$ cms^{-1}, $\kappa = 117.5$ cms.

TABLE 4*

In plane of y = d/2 = 50 cms

x	z	x'	z'		x	z	x'	z'
0	0	0	0		20	40	21	36
0	10	0	7		30	40	35	40
10	10	8.6	7.3		40	40	51	44
10	0	8.5	0		40	30	54	34
					40	20	56	23
0	20	0	15		40	10	58	12
10	20	9	16		40	0	59	0
20	20	21	17					
20	10	21	8		0	50	0	45
20	0	21	0		10	50	9	46
					20	50	21	47
0	30	0	24		30	50	34	50
10	30	9	25		40	50	49	54
20	30	21	26		50	50	61	61
30	30	35	29		50	40	70	51
30	20	37	20		50	30	73	39
30	10	38	10		50	20	77	27
30	0	38	0		50	10	80	14
					50	0	81	0
0	40	0	34					
10	40	9	35					

*True point at y = 100 cms: $\sigma = 100$ cms, $\tau = 0$ cms^{-1}, $\kappa = 117.5$ cms.

TABLE 5*

In plane of y = d = 100 cms

x	z	x'	z'	x	z	x'	z'
0	0	0	0	40	80	43	88
				0	80	67	96
0	20	0	20	80	80	93	107
20	20	21	20	80	60	93	83
20	0	20	0	80	40	94	56
				80	20	95	28
0	40	0	40	80	0	96	0
20	40	20	41				
40	40	42	45	0	100	0	100
40	20	42	22	20	100	22	102
40	0	42	0	40	100	44	108
				60	100	67	117
0	60	0	60	80	100	92	130
20	60	21	62	100	100	118	145
40	60	43	66	100	80	120	120
60	60	67	74	100	60	122	93
60	40	67	50	100	40	124	63
60	20	67	25	100	20	126	32
60	0	67	0	100	0	127	0
0	80	0	80				
20	80	21	82				

*True point at y = 100 cms: $\sigma = 100$ cms, $\tau = 0$ cm^{-1}, $\kappa = 117.5$ cms.

TABLE 6

$y = 10d = 1000$ cms

	(i)[1]				(ii)[2]	
x	z	x'	z'		x'	z'
0	0	0	0		0	0
0	200	0	235		0	111
200	200	235	239		110	111
200	0	235	0		111	0
0	400	0	465		0	219
200	400	235	474		106	218
400	400	470	501		211	216
400	200	470	253		217	109
400	0	469	0		220	0
0	600	0	683		0	318
200	600	235	696		101	317
400	600	469	736		201	314
600	600	705	797		295	307
600	400	705	542		309	211
600	200	705	274		318	107
600	0	705	0		322	0
0	800	0	884		0	405
200	800	235	902		95	404
400	800	470	953		189	401
600	800	705	1032		279	394
800	800	940	1133		362	383
800	600	940	875		382	298
800	400	940	596		398	204
800	200	940	301		410	103
800	0	940	0		414	0
0	1000	0	1068		0	479
200	1000	235	1089		89	479
400	1000	470	1150		177	476
600	1000	705	1246		262	470
800	1000	940	1368		341	459
1000	1000	1175	1511		415	445

TABLE 6 (Continued)

$$y = 10d = 1000 \text{ cms}$$

	(i)[1]			(ii)[2]	
x	z	x′	z′	x′	z′
1000	800	1175	1251	438	369
1000	600	1176	967	460	286
1000	400	1176	658	478	195
1000	200	1176	332	490	98
1000	0	1176	0	494	0
2000	0	2369	0	796	0
2000	2000	2360	4078	565	619
0	2000	0	1824	0	753

[1]True Point at $y = 100$ cms: $\sigma = 100$ cms, $\tau = 0$ cms^{-1}, $\kappa = 117.5$ cms.
[2]True Point at $y = 100$ cms: $\sigma = 100$ cms, $\tau = 0.00154$ cms^{-1}, $\kappa = 142.7$ cms.

TABLE 7*

$y = d/2 = 50$ cms

x	z	x'	z'	x	z	x'	z'
0	0	0	0	5.0	7.5	3.6	5.2
0	2.5	0	1.7	5.0	10.0	3.6	7.0
0	5.0	0	3.4				
0	7.5	0	5.1	7.5	0	5.5	0
0	10.0	0	7.0	7.5	2.5	5.5	1.7
				7.5	5.0	5.6	3.5
2.5	0	1.7	0	7.5	7.5	5.6	5.3
2.5	2.5	1.7	1.7	7.5	10.0	5.6	7.1
2.5	5.0	1.7	3.4				
2.5	7.5	1.7	5.2	10	0	7.7	0
2.5	10.0	1.7	7.0	10	2.5	7.7	1.8
				10	5.0	7.8	3.6
5.0	0	3.5	0	10	7.5	7.8	5.4
5.0	2.5	3.5	1.7	10	10	7.8	7.3
5.0	5.0	3.6	3.4				

*$\sigma = 100$, $\tau = 0.00154$, $\kappa = 142.7$

TABLE 8*

y = d = 100 cms

x	z	x'	z'	x	z	x'	z'
0	0	0	0	10	15	10	15
0	5	0	5	10	20	10	20
0	10	0	10				
0	15	0	15	15	0	15	0
0	20	0	20	15	5	15	5
				15	10	15	10
5	0	5	0	15	15	15	15
5	5	5	5	15	20	15	20
5	10	5	10				
5	15	5	15	20	0	21	0
5	20	5	20	20	5	21	5
				20	10	21	10
10	0	10	0	20	15	21	15
10	5	10	5	20	20	21	21
10	10	10	10				

*$\sigma = 100$, $\tau = 0.00154$, $\kappa = 142.7$

TABLE 9*

y = 2d = 200 cms

x	z	x'	z'
0	0	0	0
0	40	0	40
40	40		
40	0	41	0

*$\sigma = 100$, $\tau = 0.00154$, $\kappa = 142.7$

TABLE 10*

y = 3d = 300 cms

x	z	x'	z'
0	0	0	0
0	60	0	56
60	60		
60	0	56	0

*$\sigma = 100$, $\tau = 0.00154$, $\kappa = 142.7$

TABLE 11*

$y = 10d = 1000$ cms

x	z	x'	z'	x	z	x'	z'
0	0	0	0	100	150	55	83
0	50	0	28	100	200	55	109
0	100	0	55				
0	150	0	82	150	0	83	0
0	200	0	110	150	50	83	28
				150	100	83	55
50	0	28	0	150	150	82	82
50	50	28	28	150	200	82	109
50	100	28	55				
50	150	28	83	200	0	110	0
50	200	27	110	200	50	110	28
				200	100	110	55
100	0	55	0	200	150	109	82
100	50	55	28	200	200	109	109
100	100	55	55				

*$\sigma = 100$, $\tau = 0.00154$, $\kappa = 142.7$

TABLE 12*

Objects of Constant Angular Size

y	x (z=0) or z (x=0)	x' (z'=0) or z' (x'=0)	y'
50	.5	.67	33
100	1.0	1.0	93
200	2.0	1.02	164
300	3.0	.94	204
400	4.0	.86	230
500	5.0	.79	247
1000	10.0	.56	295

*$\sigma = 100$, $\tau = 0.00154$, $\kappa = 142.7$
All values are either in cms or in visual cms

Notes

Preface

1. I owe a special debt of gratitude to Professor Eugene Wigner of Princeton University and to Professor Jean Ladrière of the University of Louvain for much formative influence during this period.

2. Birkhoff and von Neumann (1936).

3. See Heelan (1965), (1967), (1970*a*), (1970*b*), (1971), and (1972*b*). During that period, my view of the role of perception changed in relation to the establishment of an ontology, and by the end of this period, I had accepted the primacy of perception. My interpretation of the *primacy of perception* goes beyond the limitations that M. Merleau-Ponty attached to this principle, in significant ways that will become clear as the thesis of this book is developed.

4. Ibid. (1972*b*), (1972*c*), and especially (1975).

5. Ibid. (1972*a*) and references in chapter 6 below. I was particularly helped in this study by the following scholars: John Foley (then of MIT, and now at the University of California at Santa Barbara), Richard Held (MIT), Irma B. Jaffe (Fordham University), Marx Wartofsky (Boston University), Professor and Mrs. Laurence Wylie (Harvard University), and Richard Zegers (then of Fordham University, and now deceased).

6. For a collection of recent papers critical of the paradigm of normal analytic epistemology and analytic theory of knowledge, see Morick, ed. (1980). See also the critique of Rorty (1979).

1. Phenomenology, Hermeneutics, and Philosophy of Science

1. It has generally been assumed that phenomenological and hermeneutical philosophy has little of interest to say about natural science, except to deplore the alienation from the natural World which it has introduced

into our culture. Significant contributions, however, have been made by such writers as S. Bachelard, O. Becker, J. Compton, T. Kisiel, J. Kockelmans, J. Ladrière, E. Ströker, H. Weyl, F. H. Zucker and others; see, for example, Kockelmans and Kisiel (1970), Compton (1969) and (1979), Ströker (1977) and (1979). The contribution of phenomenology and hermeneutics to the social and psychological sciences is, however, notable.

2. For a review and critique of the vast literature belonging to this tradition, see Heffner (1975), and Block (1981).

3. Rorty (1979). Morick (1980) brings together the significant papers of recent years within the empiricist tradition by the authors mentioned and by others who have challenged the central positions of empiricism. Praetorius (1981) presents an excellent critique of Mind-Body dualism, Behaviorism, and materialistic identity theories, from a point of view that bridges phenomenological and analytic interests. See also the work of Sigmund Koch, for a critique of contemporary empiricism in experimental psychology by this renowned scholar.

4. Block (1981) covers the range of contemporary views and problems. See also Puccetti and Dykes (1978), especially for the commentators' views.

5. Uttal (1978), p. 685.

6. See, for example, the works of P. Feyerabend, M. Hesse, J. Piaget, A. Shimony, S. Toulmin, M. Wartofsky, the later L. Wittgenstein, as well as various schools of pragmatists, Marxists, and evolutionary psychologists; see chapter 10 below. The works of M. Polanyi, S. Koch, M. Grene, C. Hooker, R. Neville, and N. Praetorius fall into a special category that has close resemblances in terms of basic positions to those articulated in phenomenological and hermeneutical terms in this book.

7. For a review of the literature and a defense of the causal theory of perception, see Heffner (1975) and (1980).

8. Husserl (1960), pp. 12−13; Husserl (1965), p. 96; also taken up by Heidegger (1962), p. 501.

9. Ibid. (1931), pp. 237−245; (1960); (1965); and (1970*a*). For an excellent account of phenomenological practice, see Ihde (1977), to which the outline of this chapter is much indebted; see especially pp. 43−51. The work of Aron Gurwitsch on the perceptual field is of the first order of importance, especially Gurwitsch (1964). See also Kohak (1978), Peursen (1972), and Zaner (1970) and (1981) for good introductions to phenomenology. For the relevance of phenomenology to the experimental psychology of perception, see, for example, Føllesdal (1974), Giorgi (1970) and (1977), and Gurwitsch (1964).

10. For an account of the *Erlanger Program*, see Klein (1939). For Husserl's relationship with his distinguished colleagues in mathematical physics, see Constance Reid's biographies of Hilbert and Courant: Reid (1976), p. 21, and (1970).

11. See the list of references for the works of Heidegger and Merleau-Ponty relevant to this study. For work in hermeneutical phenomenology, see, for example, Heidegger (1962), pp. 61−62; Ricoeur (1967) and (1980); and Gadamer (1975); for an excellent survey of hermeneutical philosophy, see Bleicher (1980).

12. For the complexity of the notion of *horizon*, see Husserl (1970*a*), pp. 158−164; Gurwitsch (1956−57), pp. 429−430; Gurwitsch (1964), pp. 237−245; Merleau-Ponty (1962), pp. 68−69; also Peursen (1977).

13. See Merleau-Ponty (1962), p. 4.

14. For an account of *apodicticity* in the context of phenomenological inquiry into perceptual phenomena, see Ihde (1977), p. 72.

15. See Husserl (1970*a*), pp. 145, 329; see also Merleau-Ponty (1962), Preface.

16. For the notion of *World*, *Lebenswelt*, or *lived World*, see Husserl (1970*a*), pp. 142−150; Merleau-Ponty (1962), part II, and Heidegger (1962), pp. 88−89; also Natanson (1964), Schutz (1973), pp. 207−259 and particularly, Schutz and Luckman (1973). For Husserl's Life World in relation to science, see Ströker, ed. (1979).

17. Sellars (1963), p. 6. The position I take is, of course, contrary to that of Sellars: I hold the primacy of manifest images over scientific images, but how I understand this claim, as compared with how Sellars would understand it (cf. p. 26 of the work referred to), awaits the clarification of chapter 10 below.

18. "[The everyday World] shall mean the intersubjective world which existed long before our birth, experienced and interpreted by others, our predecessors, as an organized world. Now it is given to our experience and interpretation. All interpretation of this world is based upon a stock of previous experiences of it, our own experiences and those handed down to us by our parents and teachers, which in the form of 'knowledge at hand' function as a scheme of reference. To this stock of experience belongs our knowledge that the world we live in is a world of well-circumscribed objects with definite qualities, objects among which we move, which resists us and upon which we act." Schutz (1973), p. 208.

19. The literature on *intentionality* is immense: see Merleau-Ponty (1962), p. 329; Heidegger (1962), pp. 24−28, 411; Husserl (1970*a*), sections 68 and 70; see also Peursen (1972); and for comparison with analytic usage, see Carr (1975), Quinton (1975).

20. Sellars (1963), pp. 127−196. See also the collection, *Challenges to Empiricism*, ed. Morick (1980), in which are to be found the major papers dealing with the breakdown of contemporary logical empiricism in the philosophy of science.

21. See Husserl (1960) and (1969), where these questions are raised.

22. For the nature of hermeneutics and the hermeneutical circle, see

Husserl (1970*a*), p. 145; Heidegger (1962), pp. 188−195; Gadamer (1975), pp. 135−140, 235−245; Bleicher (1980); and Dreyfus (1980). For a fuller treatment of the variety of hermeneutical activity, deliberate and indeliberate, in perception and in the interpretation of texts, see chapter 10 below.

23. Heidegger (1962), pp. 78−90; Merleau-Ponty (1962), p. 79, and part III.

24. Merleau-Ponty (1962), part I.

25. Ibid., pp. viii−ix; for other comments of Merleau-Ponty about science, see his essay "Eye and Mind," in Merleau-Ponty (1964), p. 159.

26. Habermas (1971), pp. 306−307.

27. Husserl (1970*a*), p. 15.

28. Habermas writes of Husserl's critique of modern science in the *Crisis*, "it is directed in the first place against the objectivism of the sciences, for which the world appears objectively as a universe of facts whose lawlike connection can be grasped descriptively. In truth, however, knowledge of the apparently objective world of fact has its transcendental basis in the prescientific world. The possible objects of scientific analysis are constituted a priori in the self-evidence of our primary life-world. In this layer, phenomenology discloses the products of a meaning-generative subjectivity." Habermas (1971), p. 304. See also Gadamer (1975), pp. 404−405, 411; Kockelmans (1963); and Grene (1966).

29. Heidegger (1962), pp. 413−414.

30. Husserl (1970*a*), p. 69.

31. Boehm (1964), p. 425.

32. Habermas (1971), p. 309.

33. See Heelan (1972*b*).

34. Heidegger (1962), p. 77. For the views of the early Heidegger on science, see especially *Introduction*, section I, and part I, section I, of *Being and Time*. (Pagination is from the English translation where the *loci* in the later German editions are given marginally.) See also W. Richardson's magnificent book, Richardson (1963), for a mainstream interpretation of Heidegger. For a defense of the uncommon view that *Being and Time* presents a philosophy of science, see Seigfried (1978).

35. Ibid., p. 31: italics in text.

36. Ibid., p. 76: italics in text.

37. Heidegger (1977), p. 157. For the later Heidegger's views about science, see, for example, "The question concerning technology," ibid., pp. 36−52, and "Science and Reflection," ibid., pp. 155−182. In this, Heidegger seeks the modern sense of *theory* in the Latin *contemplatio* and the German *Betrachtung*, while he places its deeper but forgotten ontological roots in the Greek *theoria*.

38. Ricoeur (1978), p. 225; see also pp. 227−229. He writes, "A philosophical critique of science, it seems to me, does not at all consist of

criticizing the results or the methods of science. There the scientist has nothing to learn uniquely from the philosopher. Scientific knowledge is a proper mode of knowledge and a proper mode of results, principles, laws, etc. But the task for which the philosopher is responsible is to understand how scientific understanding takes place within the comprehension of my existence in the world" (pp. 227 f.).

39. Heidegger would not admit that science has a transcultural essence (see Heidegger [1977], pp. 115—154, "The Age of World Picture"). Understanding why Heidegger takes this position enables us anew to ask this question critically, even within the general ontological perspective of his early interests, and to come to an answer different from the one Heidegger came to with respect to these concerns. How and why this is the case should be made clear in what follows.

40. For example, Habermas (1971).

41. Husserl (1970*a*).

42. Ibid., part II; also p. 127.

434. For the notion of *profile* (*Abschattung*), see Husserl (1960), pp. 57, 132; cf. also Gurwitsch (1964), pp. 173—181 and Sokolowski (1974), pp. 89—93; for the notion of *eidetic essence*, see Husserl (1969), pp. 70—71.

44. For this use of the term "model," see van Fraassen (1980) or any formalist philosopher of science, such as, for example, Suppes.

45. For the semantical approaches to scientific explanation, see Suppe (1974), Introduction.

46. Husserl (1970*b*), pp. 376—377.

47. The view that theoretical explanation is metaphoric redescription of the explanandum was first made by Black (1962); see also Hesse (1980), chapter 4, and Heelan (1979*a*).

48. For the underdetermination of theory relative to empirical data, see, for example, Duhem (1954), part II; Quine (1960), chapter 2; Hesse (1980), *passim*; and van Fraassen (1980), chapter 3.

49. For a masterly discussion of this point, see Lonergan (1957), pp. 25—32 on the "empirical residue."

2. Introduction to Visual Space

1. Loran (1963), pp. 46—47; cf. also pp. 49 f. and pp. 76—79. For further interesting and relevant comments on Cézanne's pictorial space, see Finkelstein (1979).

2. Arnheim (1954), p. 266.

3. Ballard (1970). The illusions associated with the Arc are discussed in Coren and Girgus (1978), pp. 43—44.

4. Ibid., p. 191.

5. Ibid., p. 192.
6. Ibid., p. 200.
7. Ibid., p. 198.

3. Visual Space: Search for a Model

1. See Newton (1686).

2. See the masterly work of Adolf Grünbaum, in which he untangles many of the confusions that underlie various treatments, particularly the conventionalist one, of the geometry of space: Grünbaum (1973).

3. Kant (1929): the quotations are from N. Kemp-Smith's translation.

4. Euclidean geometry is distinguished from all others by any one of the following characteristics: (1) that Pythagoras's Theorem holds in all domains large and small; (2) that the geometrical properties of figures are independent of their size, that is, figures of different sizes can be *similar*; (3) that through any point off a straight line, one and only one straight line can be drawn parallel to the first line (follows from Euclid's Fifth Postulate).

5. Vuillemin (1969).

6. Klein (1939).

7. Strawson (1966), pp. 284–285. "Kant's fundamental error, for which, at that stage of the history of science, he can scarcely be reproached, lay in not distinguishing between Euclidean geometry in its phenomenal interpretation and Euclidean geometry in its physical interpretation. . . . Because he did not make this distinction, he supposed that the necessity which truly belongs to Euclidean geometry in its phenomenal interpretation also belongs to it in its physical interpretation" (p. 285).

8. While the field of phenomena is not restricted to visual phenomena, most commentators—and all recent ones—assume that visual phenomena are what Kant has principally in mind when he speaks of space as being an a priori structuring of empirical intuition or the field of perceived objects. I think it is doubtful that Kant intended to place the intuition of space exclusively in visual phenomena.

9. Reichenbach (1956). Reichenbach uses the term "visualization" for the more usual Kantian term "intuition." He distinguishes among *physical* or *empirical visualization, pure visualization, perception* or *perceptual space* and *physical objects.*

10. Lucas (1969). The same view is also held by J. Hopkins (1973). Ralph C. S. Walker (1978) distinguishes between the geometry of two-dimensional visual space, which he believes to be Euclidean and—in some appropriate sense—a priori, and the addition of a third dimension that in his view need not be added in a Euclidean way. He argues from the kind of anomalous visual perceptions associated, for example, with Escher's drawings and with certain

kinds of optical illusions. About the latter, see Gregory and Gombrich (1973), and below, chapter 4.

11. Buchdahl (1969), pp. 612−613.

12. For a brilliant analysis of *pictorial vision*, see Snyder (1980). Snyder writes, "The choice of any particular pictorial model of vision depends on numerous factors, purpose of representation and expressive requirements being two of the most important" (p. 504).

13. Cf. the important critique of the notion of Mind as Mirror of Nature, Rorty (1979).

14. Panofsky (1924−1925). Max Wartofsky takes a similar position; see Wartofsky (1980), and chapter 6 below for comments on this position.

15. For the concept of *model*, see Harré (1970) and Hesse (1963), and chapter 13 below.

16. Arnheim (1974), pp. 289−290.

17. Ibid., p. 289.

18. Ibid., p. 290.

19. Bonola (1955), Eisenhart (1949), or Klein (1939); for a historical account, see Richards (1979).

20. Clifford (1946), p. 1; cited in Richards (1979), p. 155. See Grünbaum (1973) for the classic philosophical treatment of congruence.

21. For the philosophical question, see the classic work of Grünbaum (1973), and the recent and excellent work of Nerlich (1976).

22. See chapter 11 below.

23. Helmholtz (1876).

24. Hillebrand (1902).

25. Blumenfeld (1913).

26. Mach (1959); Merleau-Ponty (1962), pp. 203−205, 101−103, and chapter on Space.

27. Rudolf Luneburg was a mathematician of the Courant Institute of New York University. His works on the geometry of binocular visual perception are listed among the references. A review of the literature on the Luneburg theory is to be found in Suppes (1977), and of the present status of experimental work in Foley (1978) and (1980).

28. The parallax of a point is the angle subtended at that point by a common base line. Binocular parallax, sometimes called "convergence," is the angle at a point object subtended by a base line terminated by the left and right eyes of an observer. *Accommodation* for monocular viewing is sometimes called "monocular parallax"; see Foley (1978) and the references in the Appendix.

29. Metric spaces of constant curvature, called "Riemannian spaces," are of three kinds—elliptic, hyperbolic, and flat; see Eisenhart (1949). (Note: sometimes only elliptic spaces are called Riemannian; not so in this work.)

Luneburg argued—Luneburg (1950), p. 642—in favor of a hyperbolic space because only in a hyperbolic space would the visual size of a distant object (as measured by its projection on a nearby transparent screen) decrease monotonically with the object's distance from the observer. This argument is good for a perceptual field that extends to the limits of visual space. Experimental data suggest that visual spaces of elliptic curvature sometimes occur, perhaps when the perceptual field is limited. In this book, I shall ignore the elliptic and flat possibilities.

30. Luneburg (1948), pp. 228–230: "A metric space permits free movability of objects if to any two congruent configurations a motion of the space exists which moves the one configuration into the other" (p. 228).

31. See the Appendix for references.

32. See the Appendix for references.

33. See the Appendix.

34. For a careful study of the evidence in favor of the hyperbolic geometry of binocular vision, see Heffner (1975); he makes the claim that the evidence is "overwhelmingly favorable to the [Luneburg] theory" (p. v).

35. Foley (1978).

36. By "local," I mean in the immediate neighborhood of the observer. In many psychophysical measurements of relative magnitude, e.g., of loudness, subjects seem to perform more consistently if permitted to choose their own subjective unit, rather than an arbitrary unit assigned by the experimenter. This suggests that hyperbolic viewers would tend to prefer an intrinsic measure, like the radius of curvature of the space, to an arbitrary assigned measure.

37. Sometimes small rodlike light sources are used. When used, they could provide Euclidean-type clues to a visual observer; however, any one of them could provide a local standard for a different member of the family of hyperbolic spaces. The use of such light sources rather than point sources could result in nonsystematic ambiguities under the hermeneutical model.

38. The need for feature-specific information in the optical field relative to the neurophysiological structure of the visual cortex, has been shown experimentally for cats by the work of Lettvin, Hubel, and Wiesel. For a summary and evaluation of current neurophysiological work on perception, see Uttal (1978), pp. 442–482. For a very readable account, see Frisby (1980).

39. The case that clues appealing to our preunderstanding, or a process of "unconscious inference"—which I take to refer to the same thing—may be necessary to achieve stability for visual space, is argued, for example, by Rock (1977), pp. 357–369. Grünbaum has also criticized the design of Luneburg's experiments for failure to ensure the presence of contextual clues; Grünbaum (1973), p. 155. Grünbaum's criticism of Luneburg's program (ibid., pp. 154–157) is worth pondering, and I shall attempt to answer his

queries in this book, particularly in chapter 10, insofar as they also relate to the hermeneutical model to be presented below.

40. See Appendix for definitions of these variables.

41. For references, see chapter 1, n. 22, and chapters 11 and 13.

42. *Cue* is less cognitive and more stimulus-oriented than *clue*. I shall take "cue" to denote a physical structure of the stimulus field (while connoting a hermeneutical use), and "clue" to denote a hermeneutical element (while connoting a physical structure of the stimulus field).

43. The cognitive recognition of primary clues would go hand in hand with the stimulation by and through the optical field of some appropriate learned or innate neural structure.

44. The relationship between foreground and background is complex and involves the whole set of contextual factors, including near and far, inner and outer horizons, etc. To what extent "nearsightedness" and "farsightedness" are capable of affecting the geometry of visual space is uncertain: for this see Trevor-Roper (1970), pp. 14–37 on the effect of "blunted vision" on representations.

45. Foley (1980).

46. Uttal (1978).

47. Ibid., p. 432. For a discussion of theories of fusion and suppression, see Kaufman (1974), chapter 8. Important for this discussion is Julesz (1971): Julesz has shown, using random dot patterns, that stereopsis as a function of binocular perception occurs before object formation.

48. Merleau-Ponty (1962), p. 232. Part II, chapters 1 and 2 are especially relevant to this discussion; immediately relevant are pp. 232 f.

49. Foley has found that under experimental conditions such as I have described, visual judgments are not consistent with a homogeneous space. I am provisionally setting this aside for the reasons given in the text, cf. Foley (1972).

50. These assumptions are, for example, that except possibly for very close objects, any object has the same orientation, that is, the same azimuth and the same elevation relative to the observer in both visual and physical space; that the perceived radial distance of an object is a function only of its parallax (γ); that the resultant space is orthogonal under the coordinate network parametrized by azimuth (α), elevation (β), and binocular parallax (γ), quantities that are measurable in physical space.

51. Eiseikonic transformations are linear transformations that change the binocular parallax of all points by a fixed amount without changing their directional orientation from the observer. The spectacular demonstrations of A. Ames, Jr. in which he showed that, on the assumption that depth is estimated in vision by interocular parallax, "rooms" geometrically distorted in accordance with this principle can be made to appear rectangular and

regularly shaped when viewed from a certain binocular sight point suggests, as Luneburg pointed out, that visual space—under some conditions—possesses eiseikonic invariance, that is, invariance under eiseikonic transformations. If this is generally the case, the value of σ would have to be a personal constant.

52. See Heffner (1975) for an excellent summary of the scientific-explanatory potential of the Luneburg theory.

4. Hyperbolic Space: The Model

1. Epstein (1977), pp. 108—110.

2. Mark Peterson uses this paradoxical feature, exemplified more perfectly in a 3-sphere, to explain the structure of Dante's universe; see Peterson (1979), and chapter 15 below.

3. See Appendix, "Illustration: Determination of σ, τ, and κ."

4. See Appendix, equations (21) and (22), which define the near and distant zones.

5. See, for example, Gombrich (1961), figure 246, p. 300.

6. See the checkered diagrams in Coren and Girgus (1978), p. 84, fig. 4.6, which illustrate this phenomenon for monocular vision.

7. Panofsky (1924—1925).

8. See the artists' manual of the Middle Ages, Cennino Cennini (1933).

9. Ballard's study was discussed in chapter 2; see Ballard (1970).

10. See Pirenne (1970), pp. 102—103, on the telephoto effect in the work of Canaletto, and Bate (1974).

11. Robert Hansen, a painter, has graphically described the visual experience of being in motion among carpentered objects: "a street or corridor veritably ripples as I walk along it, its apparently largest section accompanying me precisely as I move. Standing on a railway platform watching a passing train presents the simplest experience of the swelling of the nearest section; but every vehicle, pedestrian, or animal moving past us on a straight street exhibits the same curved passage," pp. 148—149 of Hansen (1973). Hansen's views are discussed below in chapter 6.

12. The diagrams were modeled on the following values of the parameters: $\sigma = 100$, $\tau = 0.000154$. Different values (in the plausible range for visual spaces) would only alter the figures in degree, not in kind.

13. For application to the architectural refinements of Greek Doric temples, see below, chapter 6.

14. Panofsky (1924—1925).

15. Rothenstein and Butlin (1964), p. 8; also see particularly ibid., plate XII, *Petworth Park: Tillington Church in the Distance*, where the sense of curving surfaces is very pronounced.

16. Ihde (1977), p. 72.

17. Schutz (1973) has an excellent treatment of the Life World, see especially pp. 37—42.

18. Mead (1938), cited in Schutz (1973), p. 42.

19. These themes will be taken up again below in chapter 15.

20. Lynch (1960).

21. Ross (1974).

5. Evidence from Perceptual Illusions

1. "The confounded eye," by R. L. Gregory in Gregory and Gombrich (1973), pp. 49—96. See also Gregory (1970), and Gombrich (1961). The literature on visual illusions is immense; consult, for example, the excellent presentations in Held and Richards (1972) and Held (1974). For a recent study, see Coren and Girgus (1978).

2. Gregory (1970) and (1963). See also Gombrich (1961), p. 300.

3. See, for example, Coren and Girgus (1978), pp. 62—63.

4. Ihde (1977), pp. 82—85.

5. See, for example, Coren and Girgus (1978), pp. 30, 124, etc.

6. Gregory (1966) and (1970).

7. Segall et al. (1966).

8. See Coren and Girgus (1978), pp. 140—141. For a review of cross-cultural studies on visual illusions, see Robinson (1972) and Dawson (1971).

9. See, for example, Coren and Girgus (1978), pp. 52, 126, etc.

10. Ibid., pp. 33, 121, etc.

11. Ibid., pp. 50—51.

12. The retinal image may be part of a neurophysiological explanation underlying perception, but whether the retinal image is curved or not bears no obvious relation to whether or why straight lines appear in perception to be curved. The confusion between the phenomenon and a particular way of attempting to explain the phenomenon mars much literature by psychologists about perceptual illusions; to mention a few examples, Boring (1952), Kaufman (1979), and Coren and Girgus (1978).

13. Thouless (1931). See also Rock (1977), pp. 346—347, and Wartofsky (1972). Note the tilted oval shape of the polygonal table in the *Annunciation* scene of the Mérode Altarpiece by the Master of Flemalle (Cloisters Museum of the Metropolitan Museum of Art, New York): this was painted c. 1425 and before mathematical perspective became entrenched in the north of Europe as the correct way of representing objects.

14. Gregory, in Gregory and Gombrich (1973), p. 83—84. See also, A. Ames, Jr. (1952), and Ittelson and Kilpatrick (1952), reprinted in Beardslee and Wertheimer (1958).

15. See Ronchi (1957), pp. 124—204 and Ronchi (1963).

16. See Ronchi (1957), p. 204; also Feyerabend (1975), chapter 10 for an excellent study of Galileo's use (and misuse) of the telescope.

17. Helmholtz (1962), III, nos. 28 and 30, pp. 4–5, 25–41; Helmholtz's view is discussed very fairly by Heffner (1975), pp. 357–382.

18. See Kaufman (1979), pp. 322–332; also cf. Merleau-Ponty (1962), pp. 37, 259.

19. For example, Arnheim (1954), p. 254; Panofsky (1924–1925). Vitruvius—an ancient authority (30 b.c.)—explicitly says that the purpose of the refinements is to correct for visual distortions; see Vitruvius (1931).

20. See particularly, Dinsmoor (1975), pp. 165–175; also Hauck (1879). See also Gombrich (1961), p. 258; Janson (1962), p. 99

21. There are other possible explanations: (1) that the refinements were the outcome of trial and error experimentation with small models: what works visually with small models does not necessarily work when the dimensions are blown up; (2) that the refinements represent an attempt to model in stone the rounded oval curves of the visual cone diagrammed in figure 4.7, sector 4.

22. Escher (1971), plates 243 and 250. Other versions of the "impossible triangle" have been given by R. Gregory, L. S. and R. Penrose; see Frisby (1980), pp. 21–24, 65.

23. The physical form of the "impossible triangle," devised by R. Gregory, is discussed in Frisby (1980), p. 21.

6. Evidence from the History of Art

1. The following paragraphs draw from an important paper by Joel Snyder, particularly for the analysis of *pictorial vision*; Snyder (1980). See also Danto (1979), and the other useful papers in Nodine and Fisher (1979).

2. See, for example, Santillana and von Dechend (1977), McClain (1977). For a stimulating discussion by a philosopher of science as to how the World of cosmology of a people is mirrored in their language and art, see Feyerabend (1975), chapter 17.

3. Snyder (1980), and Panofsky (1924–1925).

4. Margaret E. Hagen enunciates some general principles that any adequate theory of visual perception must, she says, include. Her assumption that the variant aspects (or profiles) and the generating law (or invariant) of a perceptual object are given only within a common projective geometry, on account of (what is called) "natural perspective," seems to me to be too general. In addition to natural perspective, that is, the solid angular analysis of perceived objects, there is the metric estimation of distance: this is never absent and can be cued to a variety of natural or artificial cues in the visual field; cf. Hagen (1979), and for a contrary opinion, chapters 9 and 14 below.

5. The classic work about perspective was that of Alberti (1435). For the history of systems of perspective, see Panofsky (1924–1925), Francastel (1977), and White (1957), particularly chapters 14 and 17 in the last-mentioned work.

6. Gregory and Gombrich, eds. (1973), p. 208.

7. Wölfflin (1950).

8. For a fascinating study of the controversial role played by Aristotelian psychology in pre-Copernican science and in Galileo's attempt to overcome it, see Feyerabend (1977), pp. 145—161.

9. The gradual adoption beginning in the fifteenth century of an infinite Euclidean cosmological space has been studied, for example, by Koyré (1957). The first at that time to extend Euclidean geometry to cosmological space was probably Nicholas of Cusa (1401—1464).

10. Panofsky (1924—1925). The same view is held by Ernst Cassirer (1957), especially part II, chapters 3 and 6, and part III, chapter 5.

11. See Morgan (1977), p. 10 for citation from Descartes's *Dioptique*; p. 52 for citation from Denis Diderot's *Lettre sur les aveugles, à l'usage de ceux qui voient*, and p. 78 for reference to Etienne Bonnot, Abbé de Condillac.

12. For Cézanne, see Lindsay (1969), p. 252; for Turner, see chapter 4 above, and for van Gogh, see below in this chapter.

13. Cf. Mach (1959), in which Mach gives an account of "physiological space."

14. Panofsky (1924—1925), figure 5.

15. Ibid.

16. Cornford (1936).

17. Gombrich (1961).

18. Following the usage of ordinary speech, I shall often refer to the object-as-pictured and the pictorial space of a painting or photograph as the *illusion* of an object and the *illusion* of a space, although the account in chapter 8 below suggests that this phenomenon may be more akin to a *hallucination* than to an *illusion*.

19. See Gombrich's discussion of "gates" or "grills" in Gombrich (1961), pp. 250—251; also Arnheim (1954), chapter 5; Pirenne (1970), pp. 151—160, and Ittelson and Kilpatrick (1952).

20. This seems also to be the view of M. Hagen, see Hagen (1979). Goodman, however, takes the opposite view, as probably would Quine as an extension of his "underdetermination thesis," cf. Goodman (1968) and Quine (1960).

21. Gombrich (1961), p. 73; also see Goodman (1968), p. 12. For a pointed critique of Gombrich's underlying theory of perception, and one with which I would in general concur, see Hagen (1979), pp. 196—197.

22. Polanyi (1970) and Pirenne (1970).

23. See chapters 11 and 13 below for an elaboration of this notion in connection with the 'reading' of instruments and "readable technologies."

24. See n. 19 above.

25. In my paper, Heelan (1972a), I took the position that van Gogh was looking for a new system of linear perspective to portray hyperbolic visual

space. I was not myself then clear that this search would be in principle vain and that greater emphasis should have been placed on the interaction between linear and painterly clues in van Gogh's paintings.

26. Panofsky (1924–1925); White (1957); and Gioseffi (1957).

27. White (1957), pp. 207–215; see also the interesting paper, Haber (1979).

28. Reid (1804).

29. For a recent account of Reid's "visibles," see Angell (1974). According to Angell, "visibles" are the two-dimensional appearances of visual objects when distance along the line of sight is considered irrelevant: they are the shapes that the painter and draftsman try to reproduce by means of drawing on a flat surface—not the shapes on the flat surface, but the shapes made visible by their drawing technique, ibid. pp. 89–91. See also the insightful discussion in Nerlich (1976), pp. 79–82. Experimental evidence seems to be conclusive that from earliest childhood we naturally perceive in depth as well as in length and breadth: the contrary opinion (e.g., in Berkeley) stems from the view that space is not given visually, but is essentially *haptic* (or tactile).

30. White (1957), pp. 274–276. In White's view synthetic perspective is not "accurate," and "can never claim to represent exactly what is seen, no matter what conditions may be postulated" (p. 275). White's analysis, however, does not take into account many of the factors raised in this book.

31. Hansen (1973). Hansen uses the term "hyperbolic" to describe both the visual character of straight lines in three-dimensional space and the way appropriate to his curvilinear system of representing these lines on a flat surface. Hansen's use of this term then does not coincide with mine, but the evidence he adduces about the visual appearances of objects suggests that these refer to phenomena that are also *hyperbolic* in my sense of the term. White reports that the painter Ivon Hutchins used a system of synthetic (curved) perspective developed by him to represent his visual experience of the curvature of objectively straight lines: White (1957), p. 276.

32. Hansen (1973), p. 154.

33. Fleck (1979) states the extreme position, that "there is no visual perception except by ideovision and there is no other kind of illustration than ideograms" (p. 141). Fleck fails to make the distinction between perception in the strict sense and "ideovision," in which the pictorial function is subordinated to the exhibition of conceptual (scientific) relationships.

34. If Hansen's aim is to construct an image of a visually curved World, then he has fallen into what J. J. Gibson called the "El Greco fallacy." For an account of the El Greco Fallacy, see Wartofsky (1980), pp. 27–33.

35. Wartofsky (1980), p. 23.

36. Ibid., pp. 30–31.

37. See chapter 11 below.

38. See the fascinating studies of Vasco Ronchi on the visual illusions associated with the use of optical equipment; Ronchi (1957) and (1963). See also Feyerabend (1975), chapter 10 for a study of Galileo's use of telescopic evidence in the absence of a theory about the visual characteristics of optical images.

39. Wheelock (1977).

40. Gombrich (1961), p. 256.

41. Ibid.

42. Ibid., p. 257.

43. For example, Schapiro (1978), pp. 87–100; Heelan (1972a); Ward (1976); Gaffron, Mercedes, in Koch (1962), pp. 607–608.

44. Schapiro (1978), p. 88.

45. Cf. Heelan (1972a), p. 478, and Schapiro (1978), pp. 87–88.

46. Van Gogh (1958), vol. III, p. 518; letter No. B19 to Emile Bernard from Arles, October 1888.

47. Ibid., vol. I, pp. 431, 433, letters Nos. 222 and 223, and vol. II, p. 539, No. 473.

48. Ibid., vol. III, p. 533, No. 469.

49. I must thank Lawrence Wylie, Douglas Dillon Professor of the Civilization of France at Harvard University, and Mrs. Anne Styles Wylie for having expedited the collecting of this information. For others who assisted me in this work, see Heelan (1972a), pp. 478, 480, and 484.

50. Ward (1976). John Ward disagrees with several items of my reconstruction, particularly the width of the bed (I take it to be a *double* bed, he argues for a *single* bed), the width of the window in relation to the end wall, and the position of the artist. These are all matters that can be disputed and the differences are small. The major difference relates to the visual interpretation of the pictorial elements.

51. I must thank Ward for pointing out an error in figure 9 of my original paper (this corresponds with figure 6.5 above): it should have shown 13.6 tiles across instead of 11.6. However, Ward is incorrect in stating that my projection of the foreground chair would place the height of its seat at only ten inches from the floor: the height is roughly forty-four centimeters or seventeen inches (one-third of the height of the bedpost).

52. Heelan (1972a), p. 485 and Ward (1976), p. 597.

53. Heelan (1972a), pp. 484–486.

54. Vincent's room looked out on the public gardens of Arles, with the walls, towers and rooftops of Arles in the background. It is an interesting fact that in the paintings of his house, the *Yellow House*, the shutters of his bedroom window—second to the left from the corner—are always painted closed.

55. Ward (1976), pp. 594–600.

56. Ibid., p. 597.

57. Ward often seems to equate "curvature" in the object-as-depicted with "curvature" in the lines of the drawing (p. 599, n. 24). This results in a systematic ambiguity in his writing. This ambiguity is minimized if it is understood that the curved lines of the drawing are always to be interpreted within an overall Euclidean space (for example, as representing a projection on an infinite spherical field); however, this assumption makes it difficult for him and for the reader to confront the focal points of difference between his interpretation of van Gogh's painting and mine.

58. Ward (1976), p. 599.

59. However, it is not possible *without changing the geometry of pictorial space* to assert, as Ward (1976) does, that "the fact that infinitely long parallel lines project as great circles on a spherical visual field and that they converge at points at opposite sides of this sphere does not mean that they should *appear* curved, however" (p. 598, n. 19; reference should be to *Hansen* and *not* to *Hanson*). The sentence appears plausible only because of the confusion between the object-as-depicted and the lines of the drawing.

60. Van Gogh (1958), vol. III, p. 86, No. 554 to Theo.

61. Ward (1976), p. 600.

62. Many have noted the strong sense of reality one experiences when the original painting is viewed from the proper distance: it is not perceived as a distorted view; cf. Gaffron, in Koch (1962), pp. 607–608, and Schapiro (1978).

63. *Bulb Fields*, numbered F186 (H20) in the de la Faille catalogue (1970 edition).

64. The photographs are reproduced in Rewald (1956): *The Iron Bridge of Trinquetaille* (F481 in the de la Faille catalogue) on p. 219, and *Railway Bridge over Avenue Montmajour* (F480 in the de la Faille catalogue) also on p. 219. Photographs of other paintings by van Gogh will be found there and in Rewald (1942), as well as in *Art News Annual*, xix (1950), *Art News*, November, 1949, and *L'Amour de l'Art*, xvii (1936), p. 294.

65. See Novotny (1953), p. 37; Pach (1936), p. 38.

66. I tried to use such a frame once to catch the special quality of projections in visual space. I sat in front of it for hours looking at the horizontal and vertical lines made by the floor, walls, and bookshelves, but only found the experience very frustrating.

7. Nature of Perception

1. See, for example, Armstrong (1961) and (1973); Hamlyn (1961) and (1968); Sellars (1963); and Rorty (1979) for third-person accounts of perception; and see, for example, Husserl (1931) and (1970a), and Merleau-Ponty (1962) for first-person approaches.

2. Under certain circumstances, often but not exclusively pathological, a

person can come to behave towards objects exactly as if he or she had perceptual knowledge of objects without, however, being able to name those objects. In such cases, the content of that act can come to linguistic expression through a third-party interpreter. I want to bypass the question as to whether such (for the subject) nonlinguistic cognitive acts are to be properly accounted as perception.

3. If *sensations* are to be taken as belonging to individual senses (touch, sight, hearing, and so on), then a *percept* is not a sensation, but a way of combining possibly multiple sensations (touch *and* sight, for instance) in an "internal representation" or *percept* of an individual perceptual object. The status of percepts will be discussed below.

4. The same distinction expressed in a naturalistic theory of perception is made by Neville (1981), chapters 7 and 8.

5. For an excellent study of the philosophical positions, called "empiricism" and "nativism," and how they have shaped debates about perception in experimental psychology, see Morgan (1977). He concludes that the terms of this debate do not provide a satisfying theoretical background for current work in psychology. He writes, "The major conclusion from studies of sensory substitution [for example, touch for retinal stimulation] is that perception is not necessarily to be equated with the input from particular sense organs, still less with inputs along particular nerve fibres, as the obsolete law of specific energies proclaimed. Perception is the recognition of certain properties in the input, properties which are important to the organism in its behavioral interactions with the environment" (p. 207).

6. "Third, an argument which would suffice by itself to destroy Locke's thesis [that we infer visual depth from touch], is that it is impossible for us to bring these judgments into consciousness," Abbé de Condillac, *L'essai sur l'origine des connaissances humaines*, Part I, section vi, cited in Morgan (1977), p. 65. For a relevant distinction between "explicit inference" and "tacit inference," see Polanyi's essay, "The logic of tacit inference," pp. 138–158 in Grene (1969).

7. See Denis Diderot's *Lettre sur les aveugles*, as cited in Morgan (1977), p. 51, and Condillac's *L'essai sur l'origine*, "We know now that the eyes of the statue [once endowed with the sense of sight] must learn to see," cited in Morgan (1977), p. 76.

8. Gibson (1966), p. 5 (emphasis added). Gibson was for many decades the leader of those psychologists who repudiated the view that the perceived object was in some way constructed out of sense data, etc., that alone were immediately and directly given. While this position is by now generally accepted among psychologists, Gibson has recently come under fierce criticism for ambiguities latent in his views. For an account of the range of the controversy, see Ullman (1979) and the commentaries that follow the main paper: with respect to differing uses of the terms "immediate" and "direct," see the comment of Johansson et al., ibid., p. 388.

9. On "raw feels," see Meehl and Sellars (1956), pp. 249–250; Feigl (1958), pp. 378, 427, 437, 475 and passim; Sellars (1963), pp. 127–196; Rorty (1979), pp. 99–102; Uttal (1978), pp. 27 and 200.

10. Schrödinger (1967), p. 104.

11. This view is shared, for example, by Armstrong (1961), Sellars (1963), and Rorty (1979).

12. Van Fraassen (1980), p. 12.

13. By "physical causality," I mean causality in the generative sense; that is, Aristotle's *efficient causality. Causality*—except when qualified, as in *physical causality*—will be taken in the minimal sense in which it is used in the philosophy of science, as implying no more than a regular spatio-temporal correlation between events. See Heffner (1980) for a review of the literature on causal theories of perception, and also for an interesting proposal of a limited causal theory.

14. Gibson (1966). Gibson stresses the importance of textural clues and gradiants—higher order invariants—in the optical field as bearers of perceptual information. For criticism of the view that the retinal stimulation is sufficient, apart, say, from the operation of some "internalized geometric logic," see Epstein (1977), Rock (1977), and Ullman (1980).

15. Casey (1976) distinguishes memory, hallucination, and fantasy from imagination and perception, not principally by their relatedness to the everyday World, but by other internal characteristics. See also Merleau-Ponty (1962), pp. 1–13, and passim.

16. Cf. Casey (1977) where imagination and memory are contrasted.

17. Cf. Casey (1977) and (1979), where perception, imagination, and memory are compared.

18. The *pathological* character of hallucinations is stressed by Merleau-Ponty (1962), p. 335; Casey (1976), on the contrary, points out that hallucinatory experiences are not always pathological, but some are (like dreams) paranormal. Others define a hallucination by the absence of a real object in a quasi-perceptual experience; for example, Armstrong (1961), pp. 126–127.

19. Cf. Gregory and Gombrich (1973); Gregory writes "illusions should be regarded as *systematic* deviations from facts," p. 51.

20. Cf. Gombrich (1961), pp. 5–7, Armstrong (1961), pp. 80–100.

21. *Direct realism* is the epistemological position that claims that there is a congruence between physical objects (and some or all of their measured properties) and these same objects as perceived: generally, realism is asserted only of primary (i.e., spatial) qualities. *Scientific realism* states that reality is that (and only that) which has structures, properties, and processes independently of human history and culture, and that such structures, properties, and processes are exclusively those given by the explanatory content of scientific theories. Most direct realists are also scientific realists.

22. For the meaning of "Euclidean," see chapter 3, n. 4.

23. See references.

24. See Gregory and Gombrich (1973), pp. 49–96; see also chapter 4 above.

25. Quine (1960), chapter 2, and (1969), chapter 2.

26. For conventionalist positions, see Quine (1960), Duhem (1954), Poincaré (1946) and (1952), and Reichenbach (1956). Reichenbach's conventionalism is not as radical as Quine's or Poincaré's; against Reichenbach's view, see chapter 9 below.

8. Causal Physiological Model of Perception

1. I shall use the term "coding" for mappings between physical events or structures (e.g., brain states) and the content of psychological acts. I shall use the term "representation" for "cognitive representation"; this follows the decoding of brain states by and through intentionality-structures, and is the content of a cognitive act.

2. On views about neurophysiological-psychological equivalence, from the point of view of a neurophysiologist, see, for example, Uttal (1978), chapters 1 and 2; also chapters 6 and 7, pp. 491, 498–500, and 513.

3. See chapter 7 above.

4. See, for example, Pylyshyn (1980) where such models are discussed.

5. Certain brain states—alpha-waves and the like—are associated with certain perceivable "states of consciousness." Note that a "state of consciousness" is a special kind of psychological activity, not the content of a cognitive act. Moreover, the physical description of the associated brain state is indirect and a function of scientific apparatus such as EKG machines. However, with the use of a suitably designed "readable technology," we can come to perceive our own brain state simultaneously with the experience of the appropriate state of consciousness: this is the situation exploited by biofeedback. The ontological and epistemological conditions of possibility of this experience are explored in chapter 11.

6. Helmholtz (1963). For a contemporary defense of "unconscious inference," as the operation of an internalized logic isomorphic with reality, see Rock (1977) and Gregory (1979).

7. I use it because of an analogy with the observation question in quantum mechanics: see Heelan (1975).

8. Cf. Uttal (1978), chapter 9 and pp. 470–472.

9. Cf. Merleau-Ponty (1962), p. 143, and Polanyi (1964), p. 61 on the "blind man's cane." Polanyi's distinction between "focal" and "subsidiary awareness" is very relevant to this discussion.

10. See Rock (1977), p. 324, and Gyr et al. (1979), in which numerous cases are discussed where despite constant visual input changes in perception nevertheless occur.

11. For examples of multistable illusions, see Attneave (1971), and their analysis by Ihde (1977), pp. 67–80.

12. Kaufman (1974), pp. 409–460.

13. Kuhn (1970), p. 193.

14. Ibid., p. 192.

15. See Uttal (1978), chapters 8 and 9 on learning.

9. Perception as Mirroring: Realism

1. Cf. Rorty (1979).

2. See chapter 3 for the meaning of the terms "pictorial," "pictorial vision," "pictorial reality," and "pictorial space." The usage of the term "pictorial" is *not* related to Wittgenstein's use of it.

3. See chapter 10, n. 10, and text.

4. Merleau-Ponty (1962), part II, chapter 2, on Space.

5. Cf. for example, Merton (1973), p. 270.

6. Reichenbach (1956).

7. Ibid., p. 91.

8. Poincaré (1952).

9. Reichenbach (1956), pp. 24–37; see also Grünbaum's trenchant criticisms of Reichenbach and Poincaré in Grünbaum (1973), pp. 81–105.

10. Grünbaum (1973), p. 85. The disk illustration is used by Reichenbach (1956), chapter 1.

11. See above chapter 3, nn. 23–25.

12. See below, especially chap. 11.

13. This is also the position of, for example, Armstrong (1961), Sellars (1963), and Rorty (1979). They would also assert that sensations, sense impressions, and percepts are not real moments, but just moments in the theoretical reconstruction of the formation of perceptual judgments.

14. The arguments for knowledge being merely the possession of a mental representation or image are thoroughly and effectively refuted by Rorty (1979). So-called computer simulation models of perceptual knowing, for example Kosslyn et al. (1979), appropriate the traditional vocabulary of mental representations, e.g., "interpretation," "mind's eye," "looking at," for purely neurophysiological network functions, thus leaving them and their critics literally speechless (as well as confused and irritated!) when the subject matter really requires a distinction between the (neurophysiological) code and the (cognitive) object, representation, or message. Far from "demystifying" the nature of imagery, as Kosslyn et al. claim they are doing, they are exacerbating problems of understanding and communication, and putting unnecessary obstacles in the way of research into an important and difficult area.

15. For an excellent account of the process of "bracketing," also called

"epoché" or "phenomenological reduction," see Schutz (1973), I, pp. 104—109.

16. Allusion is to the phenomenon of regression to the real; see chapter 4, n. 13.

17. Merleau-Ponty (1962), pp. 334—336.

18. A geometry can be defined as the study of those shapes or functions that remain invariant under motions of the object or of the reference system (observer's frame): see Klein's *Erlanger Programm*, Klein (1939).

19. See Grünbaum (1973) for comprehensive treatment.

10. Horizonal Realism

1. For example, Armstrong (1961); Ellis (1979); Sellars (1963); Smart (1963); Feigl (1958); Pitcher (1971); and Uttal (1978).

2. Ellis (1979), p. 28.

3. Van Fraassen (1980), e.g., pp. 16, 18, 64.

4. Ibid., e.g., p. 12.

5. Ibid., e.g., p. 18.

6. Ibid., e.g., pp. 41, 61, 64.

7. See Merleau-Ponty (1962), pp. 67—72, Husserl (1970*b*), and Schutz (1973).

8. Worlds can be multiple; each, in Schutz's words, a different "finite province of meaning": Schutz (1973), I, pp. 229—234; Schutz (1973), pp. 22—25.

9. For example, James (1890), Mead (1938), Neville (1974) and (1981); also the "language game" of Wittgenstein (1958), the "epistemological behaviorism" of Rorty (1979), the "historical epistemology" of Wartofsky (1979), the "evolutionary naturalistic realism" of Hooker (1976), (1975), and (1978), the "constructive empiricism" of van Fraassen (1980), as well as for the most part those, such as Hesse and Bloor, who hold the strong thesis of sociology of science: see Hesse (1980), chapter 2. Feyerabend's "epistemological anarchy" ("*Anything goes*") also belongs here, because of its emphasis both on the praxical dimension of knowledge and on the necessity for freedom to search for and develop new contexts of inquiry, both within and outside of science. For general overview, see Radnitzky (1973).

10. See Sellars's magnificent essay, "Philosophy and the Scientific Image of Man," in Sellars (1963), pp. 1—40. Although I find myself in disagreement with Sellars's basic positions, his analysis nevertheless is extremely insightful.

11. Ibid., p. 6. In keeping with the view that there is a plurality of Worlds and scientific images, I shall speak in the plural of manifest and scientific images.

12. Ibid., p. 9; also, pp. 7, 18—25.

13. Ibid., p. 6.

14. Ibid., p. 7.

15. The term "incommensurable" is sometimes used of these circumstances, for example, by Feyerabend: "let us call a discovery, or a statement, or an attitude *incommensurable* with the cosmos (the theory, the framework) if it suspends some of its universal principles . . . (phenomena corresponding to [the old] facts may persist for a considerable time as not all conceptual changes lead to changes in perception and as there exist conceptual changes that never leave a trace in the appearances; however, such phenomena can no longer be *described* in the customary way and cannot therefore count as observations of the customary 'objective facts')," Feyerabend (1975), p. 259. For Feyerabend, however, incommensurability connotes a covert resistance to the possibility of an alternative incompatible context.

16. Heelan (1971); see also chapters 11 and 13 below.

17. For example, Laudan (1977), p. 144, who holds that problems can always be defined in some theoretical language independent of the language(s) in which they have solutions. I distinguish the *motivation* or *origination* of a problem from its *definition*. A problem may be motivated by or have its origins in a constant condition affecting a community of inquirers, but it may get different definitions within the different frameworks within which there is a possible solution, and within which, consequently, the motivating or originating condition is removed.

18. Heidegger (1962), p. 191.

19. The binary functor " $-$ " joins two sets A and B in such a way that x is in $(A-B)$ if and only if x is in A but not in B. The binary functor " U " forms the union of two sets: thus, x is in $(A U B)$ if and only if x is in either A or B.

20. A *quantum logic* is the name given to the partial ordering among subspaces of the Hilbert space of quantum mechanical state-functions: this is an orthocomplemented nondistributive lattice. In the usual form of quantum logic, first proposed by Birkhoff and von Neumann, the empirical sentences of quantum mechanics are said to follow such a logic under a natural extension of the meanings of "logical sum" and "logical product." I have argued that the quantum logic applies, not to the empirical sentences of quantum mechanics, but to the context-dependent descriptive languages and horizons of quantum mechanics: cf. Heelan (1970*a*), (1970*b*), and (1971), and Hooker (1979). For information on lattices, see, for example, Birkhoff and MacLane (1967), and Finkelstein, D. (1979).

21. Sellars (1963), p. 14.

22. Locke makes the distinction between primary or original qualities that resemble their objects and are somehow directly transmitted from objects to the mind, and secondary qualities that have no resemblance to the qualities owned by the objects themselves but are merely annexed by the mind to the motions in the senses produced by the object. See Locke's *Essay Concerning*

Human Understanding, II, viii. See also the discussion in Morgan (1977), p. 13.

23. That primary and secondary qualities should have the same epistemological status is argued, for example, in both Rorty (1979), and Merleau-Ponty (1962).

24. Schutz (1973), I, p. 61.

25. See Sellars (1963), p. 7.

26. Gregory (1973), p. 125. Emphasis added.

27. Gregory (1973), pp. 117–130.

28. Heffner (1980).

29. Sellars (1963), p. 26.

30. Ibid., pp. 35–36.

31. Gadamer says very appositely of the archaic phrase "the setting sun," that "to our vision, the setting of the sun is a reality (it is 'relative to being there')"; Gadamer (1975), p. 407.

32. Sellars (1963), p. 31.

33. Ibid., p. 26. Sellars expresses the options slightly differently; his refutation of the last possibility stands within the perspective of his own thought.

34. Ibid., p. 20.

35. Heffner (1980).

11. Perception of Scientific Entities

1. Heelan (1975*a*). This hermeneutical account of theory-laden observation in science provides a "deeper" rationale for the *learning model* of science as proposed, e.g., by Hesse: see Hesse (1980), chapters 5 and 6.

2. Heidegger (1962), pp. 188–195; see also his analysis of the use of equipment, pp. 98–114. For an excellent study of hermeneutical philosophy, see Bleicher (1980); see also the list of references on hermeneutics, in chapter 1, n. 22.

3. Ibid., p. 194.

4. Ibid., p. 191.

5. Gadamer (1975), pp. 235–240 and passim.

6. "This is what we mean when we say that we *read* a text and why we do not say that we *observe* it," Polanyi (1964), p. 92. Polanyi (ibid., chapter 4) has an analysis of the use of instruments and of the reading of texts which has been very influential and has certainly influenced me; see also his essay, "Sense-giving and Sense-reading," in Grene (1969). In addition, it is worth consulting Innis (1977) and (1981) on Polanyi's views. By "text," I mean to designate a string of syllables or other linguistic signs connoting the possibility of a reading.

7. Schrödinger (1967), p. 104.

8. In Aaron Sloman's commentary on Ullman's paper, "Against direct perception," Ullman (1980), p. 403.

9. See Ihde (1976), pp. 154–155, and Ricoeur (1980), chapters 4 and 5.

10. Ihde (1979), p. 13. Ihde makes a distinction between the "hermeneutic" of the instrument-world complex, where nonperceptual, indirect, or inferential knowledge is obtained by the knower from the instrument-world relationship, and "embodiment in an instrument," which is the process that makes a new form of perception possible. The success of the former is a necessary condition for the latter. A new embodiment, then—and with it, a new form of perception—is a possible outcome of the hermeneutical enterprise and a kind of crowning achievement of it. I shall call all such transformations "hermeneutical."

11. See the many examples of human-instrument or man-machine relations analyzed by Ihde (1979).

12. For a classic study, but from the positivist point of view, of the differences between sensed-heat and thermodynamic heat, see Mach (1886).

13. Cf. Morgan (1977), pp. 197–208 where there is an account of the work of Bach-y-Rita, Collins et al. on sensory substitution.

14. Morgan (1977), p. 203.

15. By G. Hartmann and G. Vattimo during a conference on hermeneutics and the humanities at the Center for Twentieth Century Studies, Milwaukee, in 1980.

16. Fleck (1979), p. 83 (italics in original). Fleck's work was first published in 1936.

17. Ibid., p. 97 (italics in original).

18. Cf. Santillana and von Dechend (1977), p. 68.

19. By the "received view" (also called the "standard view" by some), I mean the deductive-nomological account of explanation, as proposed and developed, for example, by Rudolf Carnap, Carl Hempel, and others. See Suppe's excellent study of the "received view" and the problems it faces, pp. 1–254 of Suppe (1976).

20. Cf. van Fraassen (1980) for an excellent study of these problems.

21. For example, Maxwell (1962), Toulmin (1960) and (1972), Hanson (1958), Feyerabend (1965), (1975), and (1978).

22. For the notion of *entrenchment*, see Hesse (1980), pp. 73–83.

23. Einstein (1949), p. 13.

24. Eddington (1928), pp. xi–xiii.

25. Sellars (1962), p. 97n.

26. The view of science I am opposing is one held by the later Heidegger (1977), Gadamer (1975), Habermas (1971), and many others. See chapter 1 above.

27. Heelan (1975).

28. A piece of information can be sent in relays only if its negentropy

(within the appropriate information channel at every stage of the relay) remains virtually unchanged. *Negentropy*, however, when applied to the kind of relays referred to in the text, is obscure; I suspect that under such circumstances it is generally used circularly—in the vicious sense.

29. Cf. Heelan (1965), (1970*a*), (1970*b*) and (1975*b*) where some earlier views of the author are expressed, foreshadowing the views expressed here.

30. See Heelan (1965), Petersen (1968), and Teller (1981) for references.

31. The TVSS system (see above) provides, for example, a Cartesian array of stimulators.

32. The "problem of measurement" is created by regarding the measuring instrument and interaction as such, as subject to quantum mechanics, implying a prima facie infinite regress in the observer-observed (subject-object) "cut," from which the only exit is to postulate an influence of human consciousness on the outcome of a measurement; cf. Heelan (1965), pp. 71—80 for an account of the problem with references to the basic literature.

33. For the plasticity of neuronal network codings vis-à-vis stable features of the environment presented to the organism, see Uttal (1978), pp. 469—470.

34. For many ways in which technologies influence our perception, see Ihde (1978).

35. Heelan (1977).

12. Identity Theories and Psychobiology

1. Feigl (1958). See Borst (1970) for a collection of basic papers on the mind-brain identity problem written in the 1960s; contributors include D. M. Armstrong, K. Baier, R. Coburn, J. Cornman, H. Feigl, P. Feyerabend, J. M. Hinton, N. Malcolm, T. Nagel, U. T. Place, R. Rorty, J. Shaffer, J. J. C. Smart, E. Sosa, J. T. Stevenson, and C. Taylor.

2. Uttal (1978), pp. 79—85, 681—695.

3. Ibid., p. 683.

4. Ibid., p. 685.

5. Ibid., p. 687.

6. Ibid., p. 688.

7. Ibid., pp. 684—685.

8. These arguments are to be found, for example, in the referenced works of Feigl, Sellars, Smart, Armstrong, and Rorty to mention just a few; see Borst (1970) for many of the earlier classic papers.

9. Uttal (1978): Psychobiology is thus "monistic in assuming that there is only one form of ultimate reality—that of matter and its arrangements in time and space. Furthermore, it is reductionistic in assuming that terms and objects of psychological discourse, as well as all other biological processes, are

346 *Notes to Pages 217—224*

reducible in theory, if not in practice, to the terms and components of the physical world . . . such a view is antiemergenistic in that it assumes that, if it were possible to have total knowledge of the parts, including their interactions, we would know all that is to be known about the whole." p. 82. See also p. 686.

10. See Merleau-Ponty (1962), pp. viii—ix, and my discussion of this passage in chapter 1 above.

13. Hermeneutics and the History of Science

1. See Heidegger (1962), p. 194.

2. Gadamer's view of "bias and prejudice" is that they serve a necessary role as containing the authority of a tradition; they are then inevitable, but are in need of continual criticism. See Gadamer (1975), pp. 241—253. See n. 5 below.

3. Gadamer (1975), p. 236.

4. Heidegger (1962), p. 194.

5. Ibid., section 32. On this point, see the excellent paper by Dreyfus (1980) where the author rightly points out that the central importance *Vorhabe* has for Heidegger is muted in Gadamer, and that this muting has led to misunderstandings, e.g., by Rorty and others, as to the conditions for a hermeneutic.

6. Gadamer (1975), p. 236.

7. Heidegger (1962), p. 194.

8. Mannheim (1936), pp. 116, 168; Merton (1973), p. 270. This mainline tradition of the sociology of science is being strongly challenged today, both by sociologists, see Bloor (1976) and Mulkay (1979), and by philosophers and historians of science, such as Hesse, Kuhn, Laudan, Rorty, C. Taylor, and Dreyfus to name a few.

9. For example, this is the case with Locke. "Our modern Western world has been made by scientists, merchants, statesmen, industrialists; Locke was the first philosopher to expound their view of life, to articulate their aspirations and justify their deeds," Cranston (1961), p. 3. See also, Mannheim (1936), pp. 165—169 for the role of the bourgeoisie in the development of science.

10. See Habermas (1971).

11. See Gouldner (1970); Hesse (1980), chapter 10.

12. For current discussion about the difference between the natural and social sciences, cf. for example, Mulkay (1979), Bloor (1976), as well as the excellent discussion on this topic among Taylor (1980), Rorty (1980), and Dreyfus (1980).

13. The core of this problem is the critique of conventionalism in chapter 9, the notion of the context of perception as one of embodiment in

chapter 10, and the study in chapter 11 of the epistemological influence of scientific instruments and readable technologies.

14. A reminder from chapter 11 that the empirical responses of scientific instruments or readable technologies are the 'texts' that are to be interpreted, that is, 'read' in a perceptual way.

15. Fleck (1979).

16. All textual interpretation is *to some extent* indeterminate (or underdetermined); the same is true of the perceptual interpretation of (the combination of) texts and 'texts.' The former corresponds to the Quine-Duhem indeterminacy thesis; see Quine (1960), chapters 1 and 2, and Duhem (1954), part II, chapters 6 and 7. The latter is the appropriate form of this thesis for a Heideggerian framework where *Vorhabe* is a significant factor in the hermeneutic.

17. Kuhn (1977), Preface, pp. ix–xxiii.

18. Fleck (1979).

19. Feyerabend (1975) and (1978).

20. Hanson (1958).

21. Polanyi (1964).

22. Fleck (1979). For Fleck, a *thought style* is a "given stock of knowledge and level of culture" used by a "thought collective"; this latter is "a community of persons mutually exchanging ideas or maintaining intellectual interaction," it is the "special 'carrier' for the historical development of any field of thought" (p. 39).

23. See "Reflections on my critics" by T. S. Kuhn in Lakatos and Musgrave (1970), pp. 231–278. Kuhn writes, "for ['paradigm'] I should now like some other phrase, perhaps 'disciplinary matrix': 'disciplinary,' because it is common to the practitioners of a specified discipline; 'matrix,' because it consists of ordered elements which require individual specification"; among the latter are "shared symbolic generalizations," "shared models, whether metaphysical, like atomism, or heuristic, like the hydrodynamic model of the electric circuit," "shared values, like the emphasis on accuracy of prediction," and particularly, "concrete problem solutions," ("problem-solution paradigms" or "exemplars"), pp. 271–272.

24. Kuhn, cited from Lakatos and Musgrave (1970), p. 271.

25. Fleck (1979), p. 35.

26. Kuhn (1962) and (1977); see Gutting (1980).

27. Kuhn (1970), p. 182. Kuhn was much criticized for the vagueness of his concept of *paradigm*, and made the clarifications mentioned in the text in the "Postscript" of the second edition of *The Structure of Scientific Revolutions*; see also n. 23 above.

28. Kuhn (1970), p. 206.

29. See, for instance, the references to Quine, Duhem, and Hesse; Putnam (1976) deals with the strategy of reinterpreting the logical connectives.

30. Quine (1960) and (1969), Duhem (1954), and Rorty (1979). All subscribe to the view that theory is underdetermined by empirical data: this is the Quine-Duhem (conventionalist) indeterminacy thesis. Dreyfus (1980) calls such conventionalism "theoretical holism"—it neglects the importance of *Vorhabe*—as opposed to the "practical holism" of Heidegger. See n. 16 above. Polanyi—like the present writer—holds an indeterminacy thesis, but not a conventionalist one; see Polanyi (1964), and Innis (1977).

31. C. S. Peirce, W. Sellars, J. J. C. Smart, S. Kripke to name a few; see Margolis (1980) for a review of the current idealist-realist dispute, and van Fraassen (1980) for a dissenting voice.

32. Hesse (1980).

33. See Margolis (1980).

34. This process is illustrated by the present study of the Euclidean and hyperbolic aspects of visual forms, both of which, however, I would describe as authentic.

35. Cf. Popper (1959), Feyerabend (1965), Lakatos and Musgrave (1970), Lakatos (1978), Laudan (1977), to name a few characteristically Popperian works.

36. Laudan (1977). For Laudan's dependency on other Popperians, see Feyerabend (1981).

37. Laudan (1977), p. 11.

38. Ibid., pp. 14, 45.

39. Ibid., pp. 18, 19, 21.

40. Ibid., pp. 143–144.

41. Ibid., p. 81.

42. Ibid., p. 109.

43. Ibid., p. 68.

44. Ibid., p. 61.

45. See Heelan (1975a), (1979b) and (1971) where evidence from the history of science is introduced.

46. A set is *partially ordered* whenever there exists a transitive ordering relation between some pairs of members of the set.

47. The logic of growth in knowledge within a developing tradition has been addressed in Sellars (1967), Lakatos and Musgrave (1970), Heelan (1971) and (1979b), Radnitzky (1973), Feyerabend (1975), Laudan (1977), Fleck (1979), and elsewhere. Lakatos's T_1, T_2, etc. (Lakatos and Musgrave [1970], p. 118) correspond roughly—but within Lakatos's theoretical holism—to my L_{A1}, L_{A2}, etc. within the practical holism of my approach.

48. See the work of Kuhn, Lakatos, Feyerabend, and Laudan for various accounts of research traditions. Laudan, for example, gives the following working definition: "a research tradition is a set of general assumptions about the entities and processes in a domain of study, and about the appropriate methods to be used for investigating the problems and constructing the theories in that domain" (Laudan [1977], p. 81). This account lacks, as most

do, reference to the *implicit* elements—the *Vorhabe*—of a research program (see below).

49. Husserl (1969), p. 319, for the "unconscious."

50. In part, because they do not belong to the internal history of science, and in part, because they are not objectivized in language.

51. See Sokolowski (1964), pp. 137—138.

52. See chapters 9—11 above.

53. See Holton (1973) for a study of the influences on Einstein's thought during the crucial period of transition to special relativity, and Heelan (1975*b*) for a study of Heisenberg's thought during the transition to quantum mechanics.

54. What I call "the current and contemporary version of an old theory" is, of course, quite different from Lakatos's "rational reconstruction." This latter aims at providing a model of how scientists *ought in the past to have behaved* (had they followed Lakatos's principles of rationality) in order to compare and contrast this with *how they actually misbehaved*. It is inevitable, of course, that the latter account is affected by the former; together they constitute a hermeneutic of sorts.

55. The history of science began strongly under the influence of the positivist tradition. However, there is scarcely a historian of science today who is not aware of the hermeneutical problems in that field; Kuhn, notably, has exercised great influence. Among the philosophers of science, all the Popperians admit some form of historical reconstruction. As Lakatos writes, "history without some theoretical 'bias' is impossible," Lakatos (1978), p. 120 (original in italics).

56. Fleck (1979), chapter 3.

57. Kuhn (1962), p. 202; see also Kuhn's remarks in Lakatos and Musgrave (1970), pp. 267—279.

58. See n. 53 above for references.

59. My view agrees roughly with Bloor's "strong thesis of sociology of science" as cited by Hesse (1980): "true belief and rationality are just as much explananda of the sociology of knowledge as error and non-rationality, and hence that science and logic are to be included in the total programme" (p. 31). See Bloor (1976).

60. Hesse (1980), pp. viii—xi, and chapters 4 and 5. For a discussion of the use of models and metaphors in science, see Black (1962) and the work of Hesse (1980), especially chapters 4 and 5. For a hermeneutical approach, see chapter 10 of Ricoeur (1978).

61. See the work of P. Suppes, W. Salmon, F. Suppe, B. van Fraassen and others.

62. See Heelan (1979*a*).

63. This, for example, would resemble in some ways the extreme position adopted by Feyerabend, for which he coined the motto, "Anything goes!"; see Feyerabend (1975) and (1978). For science as a work of the imagination,

see Holton (1978).

64. Foucault (1970), Monod (1971), and Skinner (1971), to mention just a few, have argued that science has banished objective grounds for all personal and religious values. Cf. Dreyfus (1980) for the "nihilism" of identity theories.

65. Just as medieval and much early modern science was pursued with a religious and theological interest, so in the work of P. Teilhard de Chardin, F. Capra, G. Zukav, to mention a few, we see the revival of a religious orientation toward scientific cosmology: See Jaki (1979).

66. See Gadamer (1975), pp. 98−99, where it is argued that human play involves players and spectators in the closed world of the game, and that the perfection of play is its transformation into the structure of a work of art; similarities with scientific model building would not be hard to express. See also Holton (1978).

14. Euclidean Space as a Scientific Artifact

1. Heidegger (1962), pp. 98−114 for ready-to-handedness of equipment and signs. Polanyi's views are worth consulting on this, cf. Innis (1980).

2. Cassirer (1957).

3. Goodman (1978).

4. The relation between visual distance and visual size is given by equation 20 of the Appendix.

5. Ballard (1970).

6. Husserl (1970b).

7. Ibid., p. 178.

8. Fisher (1968), p. 133; Segall et al. (1963) and (1966).

9. I am touching here on the experience of alienation from nature and reality which is a component of the antitechnological movement of the 1960s and 1970s, as illustrated, say, in the writings of J. Ellul, B. Commoner, T. Roszak, L. Mumford, C. Reich, R. Dubos, and others. See Ihde (1979) and Florman (1976) for interesting responses to this critique.

10. Florman (1976), p. 101.

11. For the relevance of the aesthetic component to my argument, see Gadamer (1975), pp. 91−150. Cf. also the relevant passages, Merleau-Ponty (1962), pp. viii−ix, Merleau- Ponty (1964), p. 154, and Heidegger (1977), pp. 36−52, 155−182.

15. World Possibilities

1. For the problem of multiple Worlds, see James (1980), II, chapter xxi, pp. 283−322; Mead (1934); Schutz (1973), I, pp. 207−259. For a particularly insightful contemporary treatment, see Neville (1981), chapters 5 and 6.

2. For a systematic development of the notion of a *cosmology* within a naturalistic philosophy, see Neville (1974), chapters 1—3.

3. See Eliade (1959), Berger (1969), Santillana and von Dechend (1977).

4. Cornford (1936).

5. Cf. Frey (1929) and Mumford (1963), pp. 18—28 for evidence relevant to the thesis that the World of the Middle Ages was based on hyperbolic visual space.

6. White, trans. (1948), p. 178, from Canto 28 of the *Paradiso*.

7. Ibid., p. 177, from Canto 27 of the *Paradiso*.

8. Ibid., p. 178, from Canto 28 of the *Paradiso*.

9. See chapter 5 where the phenomenon of the radiant sun is discussed.

10. I was led to consider the structure of Dante's universe by reading a fine little essay by Peterson who compares it to a 3-sphere, a non-Euclidean three-space that has some of the formal properties I have described: see Peterson (1979).

11. See Koyré (1957), and chapter 5 above.

12. I sympathize in general with Feyerabend's attack on the imperialism of science in our present-day culture: scientific knowledge is just one possible context within which things and persons can be understood. Whether the imperialism of science would ever justify the kind of strong political action recommended by Feyerabend and others to regain freedom of inquiry, is a matter of judgment; cf. Feyerabend (1977), pp. 295—309.

13. See chapter 11 above.

14. Butterfield (1951), p. 1; Butterfield's account of the scientific revolution is most insightful.

15. For a magisterial, though Whiggish, account of the transformation from ancient and medieval physics to the World-Picture of the machine, see Dijksterhuis (1961).

16. See Ihde (1979), pp. 23—26 for a study of the peculiar and reduced proximity of persons to one another which the telephone affords.

16. Retrospective

1. See chapter 11 above.

2. Heidegger (1962), p. 194; see also Bleicher (1980).

3. Schrödinger (1967), p. 104.

4. Single quotation marks signify that the art of 'reading' the instrumentally 'written' 'text' is similar to but not identical with the art of reading texts written in natural languages: see chapter 11 above.

Appendix: Hyperbolic Visual Map of Physical Space

1. For titles of works, see the list of references.

2. Foley (1972), p. 328.

3. Indow (1974), p. 50.

4. In Luneburg, Blank, Indow, Foley, and others, σ is a dimensionless constant. I have chosen to use and call σ what they would define as $2a\sigma$—having then the dimension of a length. In fact, my σ is the distance to the true point, and is directly measurable. Likewise, I use and call τ, what is commonly defined as $(\tau/2a)$: my τ then has the dimension of a reciprocal of length. The product $\sigma\tau$, as a consequence, remains dimensionless. Indow (1974) uses the value $e^{\sigma\tau}$ for the scale factor κ: I use the value $\sigma \sinh (1 + \sigma\tau)$, since this is the only value that guarantees shape and size congruence at the true point.

5. The phenomenon of eiseikonic invariance illustrated, for example, by Ames's rooms, has suggested to Luneburg (1947) and Hardy et al. (1953) that σ is a constant whose value is characteristic of the individual observer. However, the constancy of σ may be only for eiseikonically related visual configurations that permit interpretations in terms of familiar space indicators, such as doors, windows, and chairs. I am not therefore assuming that in general σ is a constant for each observer. If, however, σ were to be assumed constant for each observer, then this would be equivalent to the fulfillment of one of the arbitrary geometrical conditions, leaving only one unfulfilled.

References

Alberti, L. B. 1435. *De Pictura*. See also Grayson (1972).

Ames, A., Jr. 1952. "The rotating trapezoid." In Kilpatrick (1952).

———. 1953. "Reconsideration of the origin and nature of perception." In Ratner (1953).

———. 1955. *An Interpretative Manual: The Nature of Our Perception, Prehensions and Behavior*. Princeton: Princeton University Press.

Angell, R. B. 1974. "The geometry of visibles." *Nous* 8: 87–117.

Armstrong, D. H. 1961. *Perception and the Physical World*. London: Routledge and Kegan Paul.

———. 1973. *Belief, Truth, and Knowledge*. London: Cambridge University Press.

Arnheim, R. 1954. *Art and Visual Perception*. Berkeley and Los Angeles: University of California Press. New version (1974).

Arnold, W. J., ed. 1975. *Nebraska Symposium on Motivation*. Lincoln: University of Nebraska Press.

Asquith, P. D., and Giere, R., eds. 1981. *PSA 1980*. 2 vols. East Lansing, Mich.: Philosophy of Science Association.

Attneave, F. 1971. "Multistability in perception." *Scientific American* 225: 62–71.

Ballard, E. 1970. "The visual perception of distance." In Smith, F. J., ed. (1970).

Barnes, B., and Shapin, S., eds. 1979. *Natural Order: Historical Studies of Scientific Culture*. London: Sage.

Bate, M. 1974. "The phenomenologist as art critic: Merleau-Ponty and Cézanne." *Brit. Jour. Aesthetics* 14: 344–350.

Battro, A. M. 1977. "Visual Riemannian space versus cognitive Euclidean space: A revision of Grünbaum's empiricism and Luneburg's geometry of visual space." *Synthese* 35: 423–430.

Beardslee, D. C., and Wertheimer, M., eds. 1958. *Readings in Perception*. Princeton: Van Nostrand.

Beck, L. W. 1969. *Kant Studies Today*. LaSalle, Ill.: Open Court.

Berger, P. 1969. *The Sacred Canopy*. New York: Doubleday Anchor.

Birkhoff, G., and von Neumann, J. 1936. "The logic of quantum mechanics." *Ann. Math.* 37: 823–843.

Birkhoff, G., and MacLane, S. 1967. *Algebra*. New York: Macmillan.

Black, M. 1962. *Models and Metaphors*. Ithaca: Cornell University Press.

Blank, A. A. 1953. "Luneburg theory of binocular visual space." *J. Optic. Soc. Amer.* 43: 717–727.

———. 1958*a*. "The axiomatics of binocular vision: the foundations of metric geometry in relation to space perception." *J. Optic. Soc. Amer.* 48: 328–334.

———. 1958*b*. "Analysis of experiments in binocular space perception." *J. Optic. Soc. Amer.* 48: 911–925.

———. 1958*c*. "The Luneburg theory of binocular space perception." In *Psychology: Study of a Science*, ed. S. Koch, Study 1, I: 395–426. New York: McGraw-Hill.

———. 1961. "Curvature of binocular visual space: an experiment." *J. Optic. Soc. Amer.* 51: 335–339.

Bleicher, J. 1980. *Contemporary Hermeneutics*. London and Boston: Routledge and Kegan Paul.

Block, N. 1981. *Readings in Philosophy and Psychology*. 2 vols. Cambridge, Mass.: Harvard University Press.

Bloor, D. 1976. *Knowledge and Social Imagery*. London: Routledge and Kegan Paul.

Blumenfeld, W. 1913. "Untersuchungen über die scheinbare Grösse in Sehraume." *Zeit. f. Psychologie u. Physiologie der Sinnesorgane* 65: 241–404.

Boehm, G. 1968. *Studien zur Perspektivität*. Heidelberg: Winter.

Boehm, R. 1964. "Les sciences exactes et l'idéal husserlien d'un savoir rigoureux." *Arch. de Philo.* 27: 424–438.

Bonola, R. 1967. *Non-Euclidean Geometries*. New York: Dover.

Boring, E. G. 1952. "Visual perception as invariance." *Psych. Rev.* 59: 141–148. See also reply by Gibson, ibid., pp. 149–151.

Borst, C. V., ed. 1970. *The Mind-Body Identity Theory*. London: Macmillan-St. Martin's Press.

Buchdahl, G. 1969. *Metaphysics and the Philosophy of Science: The Classical Origins, Descartes to Kant*. Oxford: Blackwell.

Butterfield, H. 1951. *The Origins of Modern Science: 1300–1800*. London: Bell.

Capek, M., ed. 1976. *The Concepts of Space and Time: Their Structure and Their Development*. Dordrecht and Boston: Reidel.

Carnap, R., and Gardner, M. 1966. *The Philosophy of Physical Science*. New York: Basic Books.

Carr, D. 1975. "Intentionality." In Pivcevic, ed. (1975), pp. 18–36.

Carterette, E. C., and Friedman, M. P., eds. 1974–1979. *Handbook of Perception*. Vols. I-X. New York: Academic Press.

Casey, E. 1976. "Comparative Phenomenology of Mental Activity: Memory, Hallucination, and Fantasy Contrasted with Imagination." *Research in Phenomenology* 6: 1–25.

———. 1977. "Imagining and remembering." *Review of Metaphysics* 31: 187–209.

———. 1979. "Perceiving and remembering." *Review of Metaphysics* 32: 407–436.

Cassirer, E. 1957. *Philosophy of Symbolic Form, III: Phenomenology of Knowledge*. New Haven: Yale University Press.

Cennini, Cennino. 1933. *The Book of the Art of Cennino Cennini*. Trans. C. Herringham. London: Allen and Unwin.

Clifford, W. K. 1946. *The Common Sense of the Exact Sciences*. New York: Knopf. First publ. in 1885.

Cohen, R. S., and Wartofsky, M., eds. 1974. *Logical and Epistemological Studies in Contemporary Physics*. Boston Studies in the Philosophy of Science. Vol. 13. Dordrecht and Boston: Reidel.

Colodny, R., ed. 1965. *Beyond the Edge of Certainty*. New York: Prentice-Hall.

Compton, J. J. 1969. "Natural science and the experience of nature." In Edie, ed. (1969), pp. 80–95.

———. 1979. "Reinventing the Philosophy of Nature." *Review of Metaphysics* 23: 3–28.

Coren, S., and Girgus, J. 1978. *Seeing is Deceiving: The Psychology of Visual Illusions*. Hillsdale, N.J.: Lawrence Erlbaum Assoc.

Corner, M. A., and Swaab, D. F., eds. 1976. *Perspectives in Brain Research*. Progress in Brain Research. Vol. 45. Amsterdam: Elsevier.

Cornford, F. 1936. "The invention of space." Reprinted in Capek (1976), pp. 3–15.

Cranston, M. 1961. *John Locke*. Harlow, Essex: Longmans Green.

Crombie, A., ed. 1963. *Scientific Change*. New York: Basic Books.

Danto, A. 1979. "Pictorial representation and works of art." In Nodine and Fisher (1979), pp. 14–16.

Davidson, D. 1970. "Mental events." In Foster and Swanson, eds. (1970).

Dawson, J. L. M. 1971. "Theory and research on cross-cultural psychology." *Bull. Brit. Psychol. Soc.* 24: 291–306.

Dijksterhuis, E. J. 1961. *The Mechanization of the World Picture*. London: Oxford University Press.

Dinsmoor, W. B. 1975. *The Architecture of Ancient Greece*. 3d rev. ed. New York: Norton.

Drake, S., ed. and trans. 1957. *Discoveries and Opinions of Galileo*. Garden City, N. Y.: Doubleday Anchor.

Dretske, F. I. 1969. *Seeing and Knowing*. London: Routledge and Kegan Paul.

Dreyfus, H. 1980. "Holism and hermeneutics." *Review of Metaphysics* 34: 3–24.

———. 1979. *What Computers Can't Do*. 2d ed. New York: Harper Row.

Duhem, P. 1954. *Aim and Structure of Physical Theory*. Eng. trans. Princeton: Princeton University Press. First published in 1906.

Eddington, A. 1928. *The Nature of the Physical World*. Cambridge: Cambridge University Press.

Edie, J., ed. 1969. *Phenomenology in America*. Chicago: Quadrangle Books.

Einstein, A. 1949. "Autobiographical notes." In Schilpp, ed. (1949).

Eisenhart, L. P. 1949. *Riemannian Geometry*. Princeton: Princeton University Press.

Eliade, M. 1959. *The Sacred And The Profane*. New York: Harcourt, Brace and World.

Ellis, B. 1979. *Rational Belief Systems*. Oxford: Blackwell.

Elliston, F., and McCormick, P., eds. 1977. *Husserl Expositions and Appraisals*. South Bend, Ind.: University of Notre Dame Press.

Epstein, W., ed. 1977. *Stability and Constancy in Visual Perception*. New York: John Wiley.

Escher, M. 1971. *The World of M. C. Escher*. New York: Abrams.

Feigl, H. 1958. "The 'mental' and the 'physical'." In Feigl et al. (1958), pp. 370–497.

Feigl, H., and Scriven, M., eds. 1956. *Minnesota Studies in the Philosophy of Science*. Vol I. Minneapolis: University of Minnesota Press.

Feigl, H., Scriven, M., and Maxwell, G., eds. 1958. *Minnesota Studies in the Philosophy of Science*. Vol II. Minneapolis: University of Minnesota Press.

Feyerabend, P. 1965. "Problems of empiricism." In Colodny (1965), pp. 145–260.

———. 1975. *Against Method*. London: New Left Press.

———. 1978. *Science in a Free Society*. London: New Left Press.

———. 1981. "More clothes from the emperor's bargain basement." *Brit. Jour. Phil. Sci.* 32: 57–71.

Finkelstein, D. 1979. "The physics of logic." In Hooker, ed. (1979), pp. 141–160.

Finkelstein, L. 1979. "Unpicturelikeness of seeing." In Nodine and Fisher (1979), pp. 61–83.

Fischer, G. 1968. *Frameworks for Perceptual Localization*. University of Newcastle-upon-Tyne: Department of Psychology.

Fisher, J., ed. 1980. *Perceiving Artworks*. Philadelphia: Temple University Press.

Fleck, L. 1979. *The Genesis and Development of a Scientific Fact.* Trans. F. Bradley and T. T. Trenn. Chicago: University of Chicago Press. First published in 1935.

Florman, S. 1976. *The Existential Pleasures of Engineering.* London: Barrie and Jenkins.

Foley, J. M. 1964. "Desarguesian property of visual space." *J. Optic. Soc. Amer.* 54: 684–692.

———. 1965. "Visual space: a scale or perceived relative direction." *Proc. Amer. Psych. Assoc.* 1: 49–50.

———. 1967a. "Binocular disparity and perceived relative distance: an examination of two hypotheses." *Vision Research* 7: 655–670.

———. 1967b. "Disparity increase with convergence for constant perceptual criteria." *Perception and Psychophysics* 2: 605–608.

———. 1968. "Depth, size and distance in stereoscopic vision." *Perception and Psychophysics* 3: 265–274.

———. 1969. "Distance in stereoscopic vision: the three-point problem." *Vision Research* 9: 1505–1521.

———. 1972. "Size-distance relation and intrinsic geometry of visual space: implications for processing." *Vision Research* 12: 323–332.

———. 1978. "Primary distance perception." In *Handbook of Sensory Physiology.* Vol. VII: *Perception,* ed. R. Held, H. W. Leibowitz, and H.-L. Tauber, pp. 181–213. Berlin and Heidelberg: Springer Verlag.

———. 1980. "Binocular distance perception." *Psychological Review* 87: 411–434.

Føllesdal, D. 1974. "Phenomenology." In Carterette and Friedman, eds. (1974–1979), Vol. I, pp. 377–387.

Foss, B., ed. 1966. *New Horizons in Psychology.* Baltimore: Penguin.

Foster, L., and Swanson, J. W., eds. 1970. *Experience and Theory.* Amherst, Mass.: University of Massachusetts Press.

Foucault, M. 1970. *The Order of Things.* New York: Random House.

Fraassen, B. van. 1980. *The Scientific Image.* Oxford: Clarendon Press.

Francastel, P. 1977. *Peinture et Société.* Paris: DeNoel/Gonthier.

French, P., Uehling, T., Jr., and Wettstein, M., eds. 1980. *Midwest Studies in Philosophy V: Studies in Epistemology.* Minneapolis: University of Minnesota Press.

Frey, D. 1929. *Gotik und Renaissance als Grundlagen der modernen Weltanschauung.* Augsburg: Filser.

Frisby, J. P. 1980. *Seeing.* New York: Oxford University Press.

Gadamer, H.-G. 1975. *Truth and Method.* Trans. of *Wahrheit und Methode.* New York: Seabury Press.

Galilei, G. 1623. *The Assayer.* Trans. Stillman Drake. In Drake (1957), pp. 217–280.

Gibson, J. J. 1950. *The Perception of the Visual World.* Boston: Houghton Mifflin.

————. 1966. *The Senses Considered as Perceptual Systems.* Boston: Houghton Mifflin.

————. 1979. *The Ecological Approach to Visual Perception.* Boston: Houghton Mifflin.

Giorgi, A. 1970. *Psychology as a Human Science: A Phenomenologically Based Approach.* New York: Harper & Row.

————. 1977. "The implications of Merleau-Ponty's thesis of 'The Primacy of Perception' for perceptual research in psychology." *Jour. Phenomenological Psychology* 8: 81–102.

Gioseffi, D. 1957. *Perspectiva Artificialis.* Trieste: Istituto di Storia dell'arte antica e moderna, No. 7.

Gogel, W. C. 1960. "Perceived frontal size as a determiner of perceived binocular depth." *J. Psych.* 50: 119–131.

————. 1961*a.* "Convergence as a cue to absolute distance." *J. Psych.* 52: 287–301.

————. 1961*b.* "Convergence as a cue to the perceived distance of objects in a binocular configuration." *J. Psych.* 52: 302–315.

————. 1963. "The visual perception of size and distance." *Vision Research* 3: 101:120.

————. 1964. "Perception of depth from binocular disparity." *J. Exper. Psych.* 67: 379–386.

————. 1977. "The metric of visual space." In Epstein, ed. (1977), pp. 129–181.

Gogh, V. van. 1958. *Complete Letters of Vincent van Gogh.* 3 vols. Greenwich, Conn.: New York Graphic Society.

Gombrich, E. 1961. *Art and Illusion.* 2d. ed. Princeton: Princeton University Press. Bollingen Series.

————. 1973. "Illusion and art." In Gregory and Gombrich, eds. (1973), pp. 193–245.

Goodman, N. 1968. *Languages of Art.* New York: Bobbs-Merrill.

————. 1978. *Ways of Worldmaking.* Indianapolis: Hackett.

Gouldner, A. W. 1970. *The Coming Crisis of Western Sociology.* New York: Basic Books.

Grayson, C., ed. and trans. 1972. *On Painting and Sculpture: The Latin Texts of "De Pictura" and "De Statua."* London: Phaidon.

Gregory, R. L. 1963. "Distortion of visual space as inappropriate scaling constant." *Nature* 117: 678.

————. 1966. "Visual illusions." In Foss (1966).

————. 1970. *The Intelligent Eye.* New York: McGraw-Hill.

————. 1973. *Eye and Brain.* 2d ed. New York: McGraw-Hill.

————. 1974. *Concepts and Mechanisms of Perception.* New York: Scribner's.

————. 1979. "Space of pictures." In Nodine and Fisher (1979), pp. 228–245.

Gregory, R. L., and Gombrich, E., eds. 1973. *Illusion in Nature and Art*. New York: Scribner's.

Grene, M. 1966. *The Knower and the Known*. New York: Basic Books.

―――. 1976. "To have a mind. . ." *Jour. of Medicine and Philosophy* 1: 177–199.

―――. ed. 1969. *Knowing and Being: Essays by M. Polanyi*. Chicago: University of Chicago Press.

Grünbaum, A. 1973. *Philosophical Problems of Space and Time*. 2d enlarged ed. Boston Studies in the Philosophy of Science. Vol. 12. Dordrecht and Boston: Reidel.

Gurwitsch, A. 1956–1957. "The last work of Husserl." *Philosophy and Phenomenological Research*. Vols. XVI and XVII. Reprinted in Gurwitsch (1966), pp. 397–447.

―――. 1959. "Contribution to the phenomenological theory of perception." *Zeit. f. philos. Forschung*. Vol. XIII. Reprinted in Gurwitsch (1966), pp. 332–349.

―――. 1964. *Field of Consciousness*. Pittsburgh: Duquesne University Press.

―――. 1966. *Studies in Phenomenology and Psychology*. Evanston: Northwestern University Press.

Gutting, G., ed. 1980. *Paradigms and Revolutions*. South Bend, Ind.: University of Notre Dame Press.

Gyr, J., Willey, R., and Henry, A. 1979. "Motor-sensory feedback and geometry of visual space: a replication." *Behavioral and Brain Sciences* 2: 59–64.

Haber, R. 1979. "Perceiving the layout of space in pictures: a perspective theory based upon Leonardo da Vinci." In Nodine and Fisher (1979), pp. 84–99.

Habermas, J. 1971. *Knowledge and Human Interests*. Trans. J. J. Shapiro. Boston: Beacon Press.

Hagen, M. 1979. "A new theory of the psychology of representational art." In Nodine and Fisher (1979), pp. 196–212.

Hagino, G., and Yoshioka, J. 1976. "A new method for determining the personal constants in the Luneburg theory of binocular visual space." *Percept. Psychophys.* 19: 499–509. See correction, vol. 21, 1977, p. 14.

Hamlyn, D. W. 1961. *Sensation and Perception*. London: Routledge and Kegan Paul.

―――. 1978. *Experience and the Growth of Understanding*. London and Boston: Routledge and Kegan Paul.

Hansen, R. 1973. "This curving world: hyperbolic linear perspective." *J. Aesth. Art Criticism* 32: 147–61.

Hanson, N. R. 1958. *Patterns of Discovery*. London: Cambridge University Press.

Hardy, L., Rand, G., Rittler, C., Blank, A., and Broedner, P. 1953. *The*

Geometry of Binocular Space Perception. New York: Knapp Memorial Laboratory, Institute of Opthalmology, Columbia University, College of Physicians and Surgeons.

Harré, R. 1970. *Principles of Scientific Thinking.* Chicago: University of Chicago Press.

Harris, Errol E. 1970. *Hypothesis and Perception.* New York: Humanities Press.

Hauck, G. 1879. *Die subjektive Perspektive und die horizontalen Curvaturen des dorischen Styls.* Stuttgart.

Heelan, P. 1965. *Quantum Mechanics and Objectivity: A Study of the Physical Philosophy of Werner Heisenberg.* The Hague: Nijhoff.

————. 1967. "Horizon, objectivity and reality in the physical sciences." *Internat. Philos. Qrtly.* 7: 375–412.

————. 1970*a*. "Complementarity, context-dependence and quantum logic." *Foundations of Physics* 1: 95–110. Also in Hooker (1979), pp. 161–180.

————. 1970*b*. "Quantum logic and classical logic: their respective roles." *Synthese* 22: 3–33; also in Cohen and Wartofsky (1974), pp. 318–349.

————. 1971. "The Logic of framework transpositions." *Internat. Philos. Qrtly.* 11: 314–334.

————. 1972*a*. "Toward a new analysis of the pictorial space of Vincent van Gogh." *Art Bull.* 54: 478–492.

————. 1972*b*. "Nature and its transformations." *Theological Studies* 33: 486–502.

————. 1972*c*. "Towards a hermeneutic of natural science," and "Towards a hermeneutic of natural science: a reply to Wolfe Mays." *Jour. Brit. Soc. for Phenomenology* 3: 252–260 and 277–283.

————. 1975*a*. "Hermeneutics of experimental science in the context of the Life-World." In Ihde and Zaner, eds. (1975), pp. 7–50.

————. 1975*b*. "Heisenberg and radical theoretic change." *Zeit. f. allgem. Wissenschaftstheorie* 6: 113–138.

————. 1977. "Quantum relativity and the cosmic observer." In Yourgrau and Breck (1977), pp. 29–38.

————. 1979*a*. "Music as basic metaphor and deep structure in Plato and in ancient cultures." *Jour. Social and Biological Structures* 2: 279–291.

————. 1979*b*. "The lattice of growth in knowledge." In Radnitzky and Andersson (1979), pp. 205–211.

Heffner, J. 1975. "The epistemology of binocular vision." Ph.D. dissertation, Boston University. Ann Arbor: University Microfilms International.

————. 1980. "Causal relations in visual perception." Manuscript prepared for publication in *Internat. Philos. Qrtly.*

Heidegger, M. 1962. *Being and Time.* Trans. J. Macquarrie and E. Robinson. New York: Harper & Row.

————. 1966. *Discourse on Thinking*. Trans. from *Gelassenheit* by J. Anderson and E. M. Freund. New York: Harper & Row.

————. 1977. *The Question Concerning Technology and Other Essays*. New York: Harper Colophon.

Heil, J. 1982. "Seeing is Believing." *American Philosophical Quarterly* (forthcoming).

Held, R., ed. 1974. *Image, Object, Illusion*. San Francisco: Freeman.

Held, R., and Richards, W., eds. 1972. *Perception, Mechanisms and Models*. San Francisco: Freeman.

Helmholtz, H. von. 1876. "The origin and meaning of geometrical axioms." *Mind* 1: 301–321.

————. 1963. *Handbook of Physiological Optics*. Trans. J. P. Southal. New York: Dover.

Hempel, C. 1965. *Aspects of Scientific Explanation and Other Essays on the Philosophy of Science*. New York: Free Press.

Hesse, M. 1963. *Models and Analogies in Science*. London: Sheed and Ward.

————. 1974. *The Structure of Scientific Inference*. Berkeley and Los Angeles: University of California Press.

————. 1980. *Revolutions and Reconstructions in the Philosophy of Science*. Bloomington: University of Indiana Press.

Hillebrand, F. 1902. *"Theorie der scheinbaren Grösse bei binokularem Sehen," Wiener Akademieberichte: math-naturwiss. Klasse*.

Hochberg, J. 1979. "Things that paintings are." In Nodine and Fisher (1979), pp. 17–41.

Holton, G. 1973. *Thematic Origins of Scientific Thought: Kepler to Einstein*. Cambridge, Mass.: Harvard University Press.

————. 1978. *The Scientific Imagination: Case Studies*. Cambridge and New York: Cambridge University Press.

Hooker, C. 1974. "Systematic realism." *Synthese* 26: 409–497.

————. 1975. "Systematic philosophy and meta-philosophy of science." *Synthese* 32: 177–231.

————. 1978. "An evolutionary naturalist realist doctrine of perception and secondary qualities." In Savage, ed. (1978), pp. 405–440.

————., ed. 1979. *The Logico-Algebraic Approach to Quantum Mechanics. Contemporary Consolidation*. Vol. II. Dordrecht and Boston: Reidel.

Hopkins, J. 1973. "Visual geometry." *Philos. Review* 82: 3–34.

Husserl, E. 1931. *Ideas: A General Introduction to Pure Phenomenology*. Trans. W. R. Boyce Gibson. London: Allen and Unwin.

————. 1960. *Cartesian Meditations*. Trans. D. Cairns. The Hague: Nijhoff.

————. 1965. *Phenomenology and the Crisis of Philosophy*. Trans. Q. Lauer. New York: Harper & Row.

————. 1969. *Formal and Transcendental Logic*. Trans. D. Cairns. The Hague: Nijhoff.

————. 1970*a*. *Crisis of European Sciences and Transcendental Phenomenology*. Trans. D. Carr. Evanston: Northwestern University Press.

————. 1970*b*. "The origin of geometry." In Husserl (1970*a*), pp. 353–378.

Ihde, D. 1976. *Listening and Voice: A Phenomenology of Sound*. Athens: Ohio University Press.

————. 1977. *Experimental Phenomenology: An Introduction*. New York: Putnam, Capricorn Books.

————. 1979. *Technics and Praxis*. Dordrecht and Boston: Reidel.

Ihde, D., and Zaner, R., eds. (1975). *Interdisciplinary Phenomenology*. The Hague: Nijhoff.

Indow, T. 1967. "Multidimensional mapping of visual space with real and simulated stars." *Percept. Psychophys*. 3: 45–53.

————. 1974. "On geometry of frameless binocular perceptual space." *Psychologia—An International Journal of Psychology in the Orient* 17: 50–63.

————. 1975. "An application of MDS to study of binocular visual space." *U.S.–Japan Seminar*, August 20–24, 1975, University of California, San Diego.

Indow, T., Inoue, E., and Matsushima, K. 1962. "An experimental study of Luneburg's theory of binocular space perception: I and II." *Jap. Psych. Res*. 4: 6–24.

————. 1963. "An experimental study of Luneburg's theory of binocular space perception: III. The experiments in a spacious field." *Jap. Psych. Res*. 5: 10–27.

Innis, R. 1977. "In Memoriam Michael Polanyi." *Zeit. f. allgem. Wissenschaftstheorie* 8: 22–29.

————. 1980. "Notes on the semiotical model of perception." *Philosophical Inquiry* 2: 496–507.

————. 1981. "Heidegger's model of subjectivity." In Sheehan (1981), pp. 117–130.

Ittelson, W. H., and Kilpatrick, F. P. 1952. "Experiments in perception." *Scientific American* 185: 50–55.

Jaki, S. 1979. *The Road of Science and the Ways of God*. Chicago: University of Chicago Press.

James, W. 1890. *Principles of Psychology*. 2 vols. New York: Holt.

Janson, H. W. 1962. *History of Art*. New York: Prentice-Hall and Abrams.

Julesz, B. 1971. *Foundations of Cyclopean Perception*. Chicago: University of Chicago Press.

Kant, I. 1929. *Critique of Pure Reason*. Trans. N. Kemp-Smith. London: Macmillan.

Kaufman, L. 1974. *Sight and Mind: An Introduction to Visual Perception*. New York: Oxford University Press.

—————. 1979. *Perception: The World Transformed*. New York: Oxford University Press.

Kilpatrick, F. P., ed. 1952. *Human Behavior from a Transactional Point of View*. Hanover, N.H.

Kisiel, T. 1974. "Discussion: Hermeneutics of experimental science in the context of the Life-World." *Zeit. f. allgem. Wissenschaftstheorie* 5: 124−135.

Klein, F. 1939. *Elementary Mathematics from an Advanced Standpoint*. Vol. 2: Geometry. New York: Dover.

Koch, S., ed. 1962−1964. *Psychology: A Study of a Science*. Vols. 4−6. New York: McGraw-Hill.

—————. 1964. "Psychology and emerging conceptions of knowledge as unitary." In Wann, ed. (1964), pp. 1−45.

—————. 1975. "Language communities, search cells, and the psychological studies." In Arnold, ed. (1975), pp. 497−559.

Kockelmans, J. J. 1963. "L'objectivité des sciences positives d'aprés le point de vue de phénoménologie." *Arch. de Philo.* 27: 339−355.

Kockelmans, J. J., and Kisiel, T., eds. 1970. *Phenomenology and the Natural Sciences: Essays and Translations*. Evanston: Northwestern University Press.

Kohak, E. 1978. *Idea and Experience*. Chicago: University of Chicago Press.

Kosslyn, S., Pinker, S., Smith, G. E., and Schwartz, S. 1979. "On the demystification of mental imagery." *Behavioral and Brain Sciences* 2: 535−581.

Koyré, A. 1957. *From the Closed World to the Infinite Universe*. Baltimore: The Johns Hopkins University Press.

Kuhn, T. S. 1962. *The Structure of Scientific Revolutions*. 2d enlarged ed. 1970. Chicago: University of Chicago Press.

—————. 1977. *The Essential Tension*. Chicago: University of Chicago Press.

Lakatos, I. 1978. *The Methodology of Scientific Research Programs: Philosophical Papers*. Vol. I. London: Cambridge University Press.

Lakatos, I., and Musgrave, A., eds. 1970. *Criticism and the Growth of Knowledge*. London: Cambridge University Press.

Laudan, L. 1977. *Progress and its Problems*. Berkeley, Los Angeles, London: University of California Press.

Lettvin, J. Y., Maturana, H. R., McCulloch, W. S., and Pitts, W. H. 1959. "What the frog's eye tells the frog's brain." *Inst. of Radio Engineers* 47: 1940−1951.

Lindsay, J. 1969. *Cézanne: His Life and Art*. Greenwich, Conn.: New York Graphic Society.

Linksz, A. 1952. *Physiology of the Eye*. Vol. II: Vision. New York: Grune and Stratton.

Locke, J. 1689. *Essay Concerning Human Understanding*. Ed. Peter Niddich. Oxford: Clarendon Press, 1975.

Lonergan, B. J. F. 1957. *Insight: A Study of Human Understanding.* London: Longmans. Rev. ed. 1978, New York: Harper & Row.

———. 1972. *Method in Theology.* London: Darton, Longman and Todd.

Loran, E. 1963. *Cézanne's Composition.* Berkeley and Los Angeles: University of California Press.

Lucas, J. R. 1969. "Euclides ab omni naevo vindicatus." *Brit. Jour. Phil. Sci.* 20: 1–11.

Luneburg, R. 1947. *Mathematical Analysis of Binocular Vision.* Princeton: Princeton University Press.

———. 1948. "Metric methods in binocular visual perception." In *Studies and Essays: Presented to R. Courant on his 60th Birthday, 8 January 1948,* pp. 215–240. New York: Interscience Publ.

———. 1950. "The metric of visual space." *J. Optic. Soc. Amer.* 40: 627–642.

Lynch, K. 1960. *The Image of the City.* Cambridge, Mass.: MIT Press.

Mach, E. 1959. *Analysis of Sensations.* Trans. C. M. Williams. New York: Dover.

———. 1886. *Die Prinzipien der Waermelehre.* Leipzig.

Machamer, P. K., and Turnbull, R. G., eds. 1978. *Studies in Perception: Interrelations in the History and Philosophy of Science.* Columbus: Ohio State University Press.

McClain, E. 1977. *The Myth of Invariance.* Stony Brook: Nicolas Hayes.

Mannheim, K. 1936. *Ideology and Utopia.* New York: Harcourt, Brace and World.

Margolis, J. 1980. "Cognitive issues in the realist-idealist dispute." In French et al. (1980), pp. 373–390.

Mays, W. 1972. "Towards a hermeneutic of natural science: a reply to Patrick A. Heelan." *Jour. Brit. Soc. for Phenomenology* 3: 261–273.

Mead, G. H. 1938. *Philosophy of the Act.* Chicago: Open Court.

Meehl, H., and Sellars, W. 1956. "The concept of emergence." In Feigl and Scriven, eds. (1956), pp. 239–256.

Merleau-Ponty, M. 1962. *The Phenomenology of Perception.* Trans. Colin Smith. London: Routledge and Kegan Paul.

———. 1964. *The Primacy of Perception.* Ed. James Edie. Evanston: Northwestern University Press.

Merton, R. 1973. *Sociology of Science.* Chicago: University of Chicago Press.

Monod, J. 1971. *Chance and Necessity.* New York: Knopf.

Morgan, M. 1977. *Molyneux's Question: Vision, Touch and the Philosophy of Perception.* Cambridge: Cambridge University Press.

Morick, H., ed. 1980. *Challenges to Empiricism.* Indianapolis: Hackett.

Mulkay, M. 1979. *Science and the Sociology of Knowledge.* London and

Boston: Allen and Unwin.

Mumford, L. 1963. *Technics and Civilization*. New York: Harcourt, Brace and World.

Musatov, V. I. 1976. "An experimental study of geometric properties of visual space." *Vision Research*. 16: 1061–1069.

Nagel, E. 1961. *Structure of Science*. London: Routledge and Kegan Paul.

Nagel, E., Suppes, P., and Tarski, A., eds. 1962. *Logic, Methodology, and Philosophy of Science: Proceedings of 1960 International Conference*. Stanford: Stanford University Press.

Nagel, T. 1974. "What is it like to be a bat?" *Philos. Review* 83: 435–450.

Natanson, M. 1964. "The Lebenswelt." In Straus, ed. (1964).

Neisser, U. 1978. "Perceiving, anticipating and imagining." In Savage, ed. (1978), pp. 89–106.

Nerlich, G. 1976. *The Shape of Space*. London: Cambridge University Press.

Neville, R. 1974. *The Cosmology of Freedom*. New Haven: Yale University Press.

———. 1981. *Reconstruction of Thinking*. Albany: State University of New York Press.

Newton, I. 1686. *Philosophiae Naturalis Principia mathematica*. Trans. Motte, *The Mathematical Principles of Natural Philosophy and the System of the World*. Rev. F. Cajori. Berkeley: University of California Press, 1934.

Nodine, C. F., and Fisher, D. F., eds. 1979. *Perception and Pictorial Representation*. New York: Praeger.

Novotny, F. 1953. "Reflections on a drawing by van Gogh." *Art Bull*. 35: 34–43.

Pach, W. 1936. *Vincent van Gogh 1853–1890: A Study of the Artist and His Work in Relation to His Times*. New York: Artbook Museum.

Panofsky, E. 1924–1925. "*Perspektive als 'symbolische Form'*." In Saxl, ed. (1927), pp. 258–330.

———. 1960a. *Renaissance and Renascences in Western Art*. Stockholm: Almqvist and Wiksell.

———. 1960b. "*I primi lumi*: Italian trecento painting and its impact on the rest of Europe." In Panofsky (1960a).

Petersen, A. 1968. *Quantum Physics and the Philosophical Tradition*. Cambridge, Mass.: MIT Press.

Peterson, M. A. 1979. "Dante and the 3-sphere." *Amer. Jour. of Physics* 47: 1031–1035.

Peursen, C. van 1972. *Phenomenology and Reality*. Pittsburgh: Duquesne University Press.

———. 1977. "The horizon." In Elliston and McCormick, eds. (1977).

Pirenne, M. H. 1952. "The scientific basis for Leonardo da Vinci's theory of perspective." *Brit. Jour. Phil. Sci.* 3: 169–185.

———. 1970. *Optics, Painting and Photography.* London: Cambridge University Press.

Pitcher, G. 1971. *A Theory of Perception.* Princeton: Princeton University Press.

Pivcevic, E., ed. 1975. *Phenomenology and Philosophical Understanding.* New York: Cambridge University Press.

Poincaré, H. 1946. *The Foundations of Science.* Lancaster: The Science Press.

———. 1952. *Science and Hypothesis.* Part II: *The Value of Science.* New York: Dover.

Polanyi, M. 1964. *Personal Knowledge.* New York: Harper & Row.

———. 1970. "What is a painting?" *American Scholar* 39: 665–669.

Popper, K. 1959. *The Logic of Scientific Discovery.* London: Hutcheson.

Praetorius, N. 1981. *Fundamental Principles for a Theory of Consciousness.* Unpublished manuscript; University of Copenhagen, Copenhagen, Denmark.

Puccetti, R., and Dykes, R. W. 1978. "Sensory cortex and the mind-brain problem." *Behavioral and Brain Sciences* 3: 337–375.

Putnam, H. 1976. "What is 'realism'?" *Proc. Aristotelian Society,* 1975–76, pp. 177–194.

Pylyshyn, Z. 1980. "Computation and cognition: issues in the foundations of cognitive science." *Behavioral and Brain Sciences* 3: 111–169.

Quine, W. van O. 1960. *Word and Object.* Cambridge, Mass.: MIT Press.

———. 1969. *Ontological Relativity and Other Essays.* New York and London: Cambridge University Press.

Quinton, A. 1975. "The concept of phenomenon." In Pivcevic, ed. (1975).

Radnitzky, G. 1973. *Contemporary Schools of Metascience.* 3d enlarged ed. Chicago: Regnery.

Radnitzky, G., and Andersson, G., eds. 1979. *The Structure and Development of Science.* Dordrecht and Boston: Reidel.

Ratner, S., ed. 1953. *Vision and Action.* New Brunswick: Rutgers University Press.

Reichenbach, H. 1956. *The Philosophy of Space and Time.* Trans. M. Reichenbach. New York: Dover.

Reid, C. 1970. *David Hilbert.* New York: Springer-Verlag.

———. 1976. *Courant in Göttingen and New York: Story of an Improbable Mathematician.* New York: Springer-Verlag.

Reid, T. 1804. *An Inquiry into the Human Mind on the Principles of Common Sense.* 6th ed. Glasgow: Gray, Maver and Co.

Rewald, J. 1942. "Van Gogh vs nature: did van Gogh of the camera lie?" *Art News* 41: 8–11.

———. 1956. *Post-Impressionism: From van Gogh to Gaugin.* New York:

Museum of Modern Art.

Richards, J. L. 1979. "The reception of a mathematical theory: non-Euclidean geometry in England, 1868–1883." In Barnes and Shapin, eds. (1979), pp. 143–163.

Richardson, W. J. 1963. *Heidegger: Through Phenomenology to Thought.* The Hague: Nijhoff.

———. 1968. "Heidegger's critique of science." *New Scholasticism* 42: 511–536.

Ricoeur, P. 1967. *The Symbolism of Evil.* Trans. E. Buchanan. New York: Harper & Row.

———. 1978. *The Philosophy of Paul Ricoeur. An Anthology of His Work.* Boston: Beacon.

———. 1980. *Hermeneutics and the Human Sciences.* Cambridge: Cambridge University Press.

Roberts, F. S., and Suppes, P. 1967. "Some problems in the geometry of visual perception." *Synthese* 17: 173–201.

Robinson, J. O. 1972. *The Psychology of Visual Illusions.* London: Hutchinson.

Rock, I. 1977. "In defense of unconscious inference." In Epstein (1977), pp. 321–373.

Ronchi, V. 1957. *Optics: Science of Vision.* New York: New York University Press.

———. 1963. "Complexities, advances and misconceptions in the development of the science of vision." In Crombie, ed. (1963), pp. 542–561.

Rorty, R. 1979. *Philosophy and the Mirror of Nature.* Princeton: Princeton University Press.

———. 1980. "A reply to Dreyfus and Taylor." *Rev. of Metaphysics* 34: 39–46.

Rosenthal, S. B. 1977. "Activity and the structure of perceptual experience: Mead and Peirce revisited." *Southern Jour. of Philos.* 15: 207–213.

Ross, H. E. 1975. *Perception in Strange Environments.* New York: Basic Books.

Rothenstein, J., and Butlin, M. 1964. *Turner.* New York: Braziller.

Rudner, R., and Scheffler, I. 1972. *Logic and Art: Essays in Honor of Nelson Goodman.* Indianapolis: University of Indiana Press.

Santillana, G. de, and von Dechend, H. 1977. *Hamlet's Mill.* Boston: Godine.

Savage, C. W., ed. 1978. *Perception and Cognition; Issues in the Foundations of Psychology.* Minnesota Studies in the Philosophy of Science. Vol. IX. Minneapolis: University of Minnesota Press.

Saxl, F., ed. 1927. *Vorträge der Bibliothek Warburg: Vorträge 1924–1925.* Leipzig and Berlin: Teubner.

Schapiro, M. 1978. *Modern Art II.* New York: Braziller.

Schilpp, P., ed. 1949. *Albert Einstein: Philosopher-Scientist*. Evanston: Library of Living Philosophers.

Schrödinger, E. 1967. *What is Life? and Mind and Matter*. Cambridge: Cambridge University Press.

Schutz, A. 1973. *Collected Papers*. 3 vols. Vol. I: *The Problem of Social Reality*. The Hague: Nijhoff.

Schutz, A., and Luckman, T. 1973. *The Structures of the Life-World*. Trans. R. Zaner and T. Englehardt. Evanston: Northwestern University Press.

Segall, M. H., Campbell, D. T., and Herskovits, M. J. 1963. "Cultural differences in the perception of geometric illusions." *Science* 139: 769–771.

———. 1966. *The Influence of Culture on Visual Perception*. Indianapolis: Bobbs-Merrill.

Seigfried, H. 1978. "Heidegger's longest day: *Being and Time* and the sciences." *Philos. Today* 22: 319–331.

Sellars, W. 1963. *Science, Perception and Reality*. London: Routledge and Kegan Paul.

———. 1967. *Science and Metaphysics*. London: Routledge and Kegan Paul.

Shapere, D. 1974. "Scientific theories and their domains." In Suppe, ed. (1974), pp. 518–565.

Sheehan, T., ed. 1981. *Heidegger, The Man and The Thinker*. Chicago: Precedent Press.

Shipley, T., and Williams, D. 1968. "The relationship between retinal disparity and relative visual distance." *Vision Research* 8: 325–332.

Skinner, B. F. 1971. *Beyond Freedom and Dignity*. New York: Macmillan.

Smart, J. J. C. 1963. *Philosophy and Scientific Realism*. London: Routledge and Kegan Paul.

Smith, F. J., ed. 1970. *Phenomenology in Perspective*. The Hague: Nijhoff.

Snyder, J. 1980. "Picturing vision." *Critical Inquiry* 6: 499–526.

Sokolowski, R. 1964. *The Foundation of Husserl's Concept of Constitution*. The Hague: Nijhoff.

———. 1978. *Presence and Absence*. Bloomington: Indiana University Press.

Sperry, R. 1976. "A unifying approach to mind and brain: ten-year perspective." In Corner and Swaab (1976).

Strawson, P. F. 1966. *The Bounds of Sense*. New York: Methuen.

Ströker, E. 1977. *Philosophische Untersuchungen zum Raum*. Frankfurt am Main: Klostermann.

———., ed. 1979. *Lebenswelt und Wissenschaft in der Philosophie Edmund Husserls*. Frankfurt am Main: Klostermann.

Suppe, F., ed. 1974. *The Structure of Scientific Theories*. Urbana: University of Illinois Press.

Suppes, P. 1962. "Models of Data." In Nagel et al. (1962), pp. 252–261.
———. 1977. "Is visual space Euclidean?" *Synthese* 35: 397–422.
Synge, J. 1960. *Relativity: The General Theory*. Amsterdam: North-Holland.
Taylor, C. 1980. "Understanding in human sciences." *Rev. of Metaphysics* 34: 25–38.
Teller, P. 1981. "The projection postulate and Bohr's interpretation of quantum mechanics." In Asquith and Giere, eds. (1981).
Thouless, R. H. 1931. "Phenomenal regression to the real object, I and II." *British Jour. Psych.* 21: 339–359; 22: 1–30.
Toulmin, S. 1960. *Philosophy of Science*. New York: Harper Torchbook.
———. 1972. *Human Understanding*. Princeton: Princeton University Press.
Trevor-Roper, P. D. 1970. *The World Through Blunted Vision: An Inquiry into the Influence of Defective Vision on Art and Character*. Indianapolis: Bobbs-Merrill.
Tuller, A. 1967. *A Modern Introduction to Geometries*. New York: Van Nostrand.
Ullman, S. 1980. "Against direct perception." *Behavioral and Brain Sciences* 3: 373–416. Includes commentary by many authors.
Uttal, W. 1978. *Psychobiology of Mind*. Hillsdale, N.J.: Lawrence Erlbaum Associates.
Vitruvius, P. 1931. *De Architectura*. Trans. F. Granger. New York: Putnam.
Vuillemin, J. 1969. "The Kantian theory of space in the light of groups of transformations." In Beck (1969), pp. 141–159.
Walker, R. C. S. 1978. *Kant: The Arguments of the Philosophers*. London: Routledge and Kegan Paul.
Wann, T. W., ed. 1964. *Behaviorism and Phenomenology: Contrasting Bases for Modern Psychology*. Chicago: University of Chicago Press.
Ward, J. 1976. "A reexamination of van Gogh's pictorial space." *Art Bull.* 58: 593–604.
Wartofsky, M. 1972. "Pictures, representation and understanding." In Wartofsky (1979), pp. 175–210.
———. 1978. "Rules and representation: the virtues of constancy and fidelity put in perspective." In Wartofsky (1979), pp. 211–230.
———. 1979. *Models: Representation and Scientific Understanding*. Boston Studies in the Philosophy of Science. Vol. 48. Dordrecht and Boston: Reidel.
———. 1980. "Art history and perception." In Fisher (1980), pp. 23–41.
Weyl, H. 1922. *Space-Time-Matter*. Trans. H. L. Brose. New York: Dover.
Wheatstone, C. 1838. "Contributions to the physiology of vision—Part of the first: On some remarkable and hitherto unobserved phenomena of binocular vision." *Philosophical Transactions*. Royal Society (London), pp. 371–394.
———. 1852. "Contributions to the physiology of vision—Part of the

second: On some remarkable and hitherto unobserved phenomena of binocular vision." *Philosophical Transactions*. Royal Society (London), pp. 1–17.

Wheelock, A. K., Jr. 1977. *Perspective, Optics and Delft Artists Around 1650*. New York: Garland.

White, J. 1967. *Birth and Rebirth of Pictorial Space*. 2d. ed. Boston: Boston Book and Art Shop.

White, L. G., trans. 1948. *Dante Alighieri: The Divine Comedy*. New York: Pantheon.

Wittgenstein, L. 1958. *Philosophical Investigations*. Trans. G. E. M. Anscombe. 2d ed. New York: Macmillan.

Wölfflin, H. 1950. *Principles of Art History: The Problem of the Development of Style in Later Art*. New York: Dover.

Yourgrau, W., and Breck, A., eds. 1977. *Cosmology, History and Theology*. New York: Plenum.

Zajackzkowska, A. 1956a. "Experimental determination of Luneburg's σ and κ. *Qrtly. J. Exper. Psych.* 8: 66–78.

———. 1956b. "Experimental test of Luneburg's theory: horopter and alley experiments." *J. Optic. Soc. Amer.* 46: 514–527.

Zaner, R. 1970. *The Way of Phenomenology*. New York: Pegasus.

———. 1981. *The Context of Self*. Athens: Ohio University Press.

Index

Printed in the United States
23384LVS00007B/73-84

9 780520 057395